The Absolute Relevancy of Singularity In Terms of Life

In Mono Colour

ISBN-13: 978-1537733456

ISBN-10: 1537733451

AS

AS PART OF

MATTER'S TIME IN SPACE
– THE THESIS –

I. S. B. N.   0-9584410-8-1

TRANSLATED BY   PEET SCHUTTE
FROM THE ORIGINAL AFRIKAANS:
"MATERIE SE TYD IN RUIMTE"
WRITTEN BY PEET (P. S. J.) SCHUTTE
I. S. B. N.   0–620–27041–1

©KOSMOLOGIESE EN ASTRONOMIESE TEGNIKA

All rights are reserved.

No part, parts or the entirety of this book may be reproduced by publishing, electronically copied, duplicated by whatever means that form reproduction or duplication, without the prior written consent of the copy rite owner.

WRITTEN BY PEET (P. S. J.) SCHUTTE

WHOM IT MAY CONCERN,

I do find much pride in my status as being Afrikaner and would like to have my names used by pronouncing it in the manner Afrikaans dictates...therefore I would sincerely appreciate the courtesy when readers will take note that my name and last name are pronounced in Afrikaans, which is originally from Dutch and must be pronounced that way. Peet one would **pronounce** "here" which is the closest English to the pronouncing of the "ee". The "Sch" in Schutte is pronounced exactly as school is where both actually are pronounced Skutte or "skool". By pronouncing my name in Afrikaans you do me the utmost courtesy any one can. Being an Afrikaner is what I am most proud of. Another point I wish to highlight is that I feel compiled to produce this work in a comic-like format. I have found that the more intellectual and the more educated Academics are, the less they understand the most primitive or classical mistakes in science as well as physics.

As I said my mother tongue is Afrikaans and my second language is English. I have per suiting this theory that I partly present in this book, of which the investigating research was done the past thirty years. Then I compiled my presentation thereof for the past nine years on full time basis whereby I was tying to introduce my findings to many academics without much joy. This past nine years saw me go without any income as I tried to get my theorem recognised. Going without a steady income left me almost destitute and in order to find a manner to get my theory across to the attention of influential readers, I decided to publish these books electronically as to try and get around the stranglehold of Newtonian bias controlling science at present worldwide. I decided to publish these articles through LULU.com which I saw as way the only manner whereby I could generate funding by which I would be able to have the twenty seven books I already wrote linguistically edited and then to have the books published on a Print-On-Demand basis. With my first language not being English and the books not linguistically checked by an expert there are bound to be language errors that readers will notice. In the past I tried to check my work myself but after checking say one hundred and fifty pages for language corrections, instead of having corrected work I ended instead having four hundred pages of new written information which is still not language corrected but holds a lot more information. This is because my priorities lie elsewhere. I aim to spend money on correcting the work as far as language goes, as I receive money and in the hope that I will receive money. I will have all my work including the one you are reading edited professionally and corrected as I find money to do so. . .

I am P.S.J Schutte, nicknamed Peet. Being a white South African my mother tongue is Afrikaans and my second language is English. Because of a shortage of funding my books were never edited in any way. Where you do find language issues please rather think of this mistake in terms of my poverty than think of my stupidity. I do find much pride in my status as being Afrikaner and would like to have my names used by pronouncing it in the manner Afrikaans dictates...therefore I would sincerely appreciate the courtesy when readers will take note that my name and last name are pronounced in Afrikaans, which is originally from Dutch and must be pronounced that way. Peet one would pronounce "here" which is the closest English to the pronouncing of the "ee". The "Sch" in Schutte is pronounced exactly as school is where both actually are pronounced Skutte or "skool".

By pronouncing my name in Afrikaans you do me the utmost courtesy any one can. Being an Afrikaner is what I am most proud of. I submit article to well known physics magazines but my articles are rejected on the most unappeasable grounds and for the most outrageously ridiculous reasons the Newtonians can think of. I explain how gravity forms but I am rejected because they are of the opinion that my work does not meet. One such an article I may use because I said I was going to use the material as an open letter I gladly show.

Email as follows: orders@sirnewtonsfruad.com

IS THERE DEATH AFTER LIFE?

Yes, you have read the expression correct. I ask is there death after life because I say there can be no death at all. I hold the opinion that once something is part of the Universe it has to remain within the Universe and can after forming part of the Universe never again leave that which we think of as forming the Universe. The most adhering principle about cosmology is that only nothing can leave the Universe since nothing never was part of the Universe in any form to begin with. With this fact never disproved we have to accept that all things within the Universe remains within the Universe. Then the question that springs to mind is how can life leave the Universe after it once formed a part of the Universe. When we die we have to leave the Universe but if we were part of the Universe, we must find ourselves unable to leave or to vacate the Universe. Once something is in the Universe, that something is unable to leave the Universe because there is no other place to go but to be in the Universe, on the condition that the something was at any point part of the Universe. Looking at a cadaver it is plain to see the cadaver has no life left within because the cadaver has lost all ability to move and life is what drove the body in as much as the part that supplied the energy fuel. The body is still present but life that drove the body is absent. The presence that formed movement by free will has gone and with that being the case death entered the cadaver that was once filled with life. Since we can see life has the ability to leave the body and therefore the Universe by using death, the only reason for life to become relieved of the position it had in the Universe must then be because life was never part of the Universe.

To all Newtonians this must be the biggest load of rubbish any person could ever utter, but it is true. It has all to do with what is in the Universe, what is part of the Universe, what was never part of the Universe because it could never be in the Universe and what is the Universe we think of as being the Universe. Lets find what is in the Universe and what was never part of the Universe. The Universe we see as the Universe is an illusion and has no permanency and therefore it will end as it began; in singularity, but I am not going down that road at this point. If the Universe started in singularity and as it will end in singularity we better look at singularity. Singularity forming one point is never part of the Universe but only holds a relevancy to all other points that is also not being part of the Universe and only this relevancy of relative movement grants the singular point to become part of the Universe by movement or become the Universe. Everything we think of in terms of being the Universe is 1^0 or 4^0 or 1^{1000} or whatever could form a singular value. Singularity is the point where the Universe starts and is the point that forms the centre of the Universe.

There is no true Universe that we see. What we think we see is all mirages and images left by time as expanding markers of what was. We create a Universe in our minds by perception of what we think we see and not what is there to see. If we see the Moon, we don't see the Moon because we see light that left the Moon some number of seconds ago. When we look at the Sun but we don't see the Sun because we see light that left the Sun. Now using Newtonian standards that would be eight and a half minutes of light that travelled at the speed of light to the Earth and therefore to me. That is the perception also void from reality because the light is an image, But even that perception too is as Newtonian as there can be. We know that strong gravity bends light and that is something Newtonians accept but never got as far as intergrading the concept into Newtonian science because they have no idea what to calculate with such a perception in hand.

Newton science still caress the idea that the speed of light is a standard constant found anywhere to be the same while they know that strong gravity bends light, so strong gravity slows down the speed of light up to where light moves backwards. When the light leaves the Sun it will travel at the point of leaving the Sun at a rate of say a millimetre a day and then as it progresses from the Sun it will pick up speed. This has to be true because the space around the Sun is so compact and condensed that it makes the outer space cosmic gas turn into a flame filled cosmic liquid. With space that much denser it must take light much, much longer to travel through that compact space. I have a wild guess that on the rim of the Sun it will take light years if not centuries to progress one billionth of a millimetre because at that point light just cant travel straight as the Sun's gravity will take the light in a circle when the light departs. Therefore the Sun we see is not eight minutes away as Newtonians would say it is but is trillions of centuries away fighting gravity and millennia of ages all the way to reach the Earth. Why is this important? I mention this to show how much incorrectness there is about Newtonian concepts concerning time and just like theologians do, Newtonians cheat facts to reduce the cosmos or in the case of theologians to reduce the concept of God to get some understanding what they are dealing with. I mention this to try to show that Newtonians have no grasp of true time and the 13×10^9 years they put on the Universe as being the correct age of the cosmos could be as long as it takes the light coming from the closest galactica to reach us, and I stress the closest galactica making that the Milky Way and not the furthers such as they now wish to translate time as being. Also with me saying this then also take in mind that my Sun and my Solar system is one part of my Milky Way and therefore my light would take longer to reach me, I say me because I am serving the Milky Way in the capacity where I am forming that part as the controlling singularity being controlled by the Milky Way's part that forms the governing singularity and therefore the light within the Milky Way is still directly part of me. Time works this way and only concerning life can time be measured in that the Earth circles around the Sun. Calculating time in cosmic standards surpasses the ability of man completely.

The Newtonian concept of time has to be chucked and buried along with most other Newtonian concepts because to put it as mild as I possible can, it is outdated, rubbish and backwards. So too are the way science look at life. Every one since time began excluding one person, who is the one the Roman Catholics call Jesus Christ, but I believe has another name, thought of life as the body we hold. People that lived since man became intellectually civilised had the opinion that they or I could walk to a mirror and see myself in the mirror. They think the reflection that comes from the mirror carry the image of me. The body holding me is what I am and I don't control the body because I am the body. They think my body represents me as a life entity totally. I do intend to return to this line of thought in a debate later on but if the body was I, why did I then had to make use of electricity to control me which is the body. If the body was I, then my thoughts did not have to generate electricity to control the body but the thought alone would be sufficient to do what ever is required in the instant of time to initiate immediate reaction from the body because that would be how all ability of control over my person should be. If I had to use electricity to convey messages to my body via any of my body parts to get a result of finding control by movement, then I need not sacrifice time to get a message across space running through my body in order to get work done through commanding my body parts. The control of my body would then be in thought formed in the instant of time and not carried by generating electricity that has to cross the space to get a message across to get a job done by my body. Again Newtonians don't take into account that time leaves space as the history of time and if I had to use more time to cross space that is already time in the past, then the crossing of space alone tells me that the body I use is a vehicle I built and I borrowed to comfort me. A snake has the ability to build a body that has to serve the snake. If the snake had the ability to build the body human uses, the snake would not be satisfied with a legless spine covered with muscle and skin and fitted with a small brain on one end and a tail on the other. If given a choice, I am sure the snake would opt for a human body and brain.

Lets men to in end begin where all living things begin. I shall begin at the ego of man because in all (including women) that is where they see their Universe starts. I am going to start to show what man is and why man is what man is... or at least as far as the physics of man goes. Every human look at his or her body and refers to "me". Every person sees the body and say: that is I. It is the body that we see in the mirror, which is that we call I and that which we think of as I. If the body were I or "Me" then I or "Me" would begin life as a skeleton and not finally end as a skeleton because the skeleton holds the body and forms the most enduring part of a body. Then if the body was in control of life the body must begin by at first form the skeleton. Then the body would start to furnish the skeleton by providing flesh to cover the skeleton and as it furnished the flesh the skeleton would provide the gain in flesh with filling the flesh accumulated with suitable life. But the process starts where life begins as a cell uniting with another cell already filled with life. Life starts way before the male and female cells unite. However, when life disappears from the body, the body is unable to replenish the life that went missing and starts to detach in composition and structure. The process whereby life leaves the body we call death and the body ends as a skeleton a very long time after life left the body and the body then eventually ends as a skeleton. It is fine to look at DNA but it is life that controls DNA and when life abandons DNA it is not possible for DNA to replace life. That makes DNA the footprints life leaves behind and DNA does not fill the cells with life. If DNA were life, then the skeleton would take DNA and reconstruct life back into the skeleton.

Life starts at a point where a cell starts to grow. If life at any point leaves the cell, then the cell will decompose and lose form and structure. Therefore the first condition for life to be in control of the body is that a cell has to hold life and not just have the cell forming a presence as part of a body that captured life. It is life that provides the cell its structural composition and not the cell that provides life with anything, but security. The cell is merely a vehicle in which life will ride for a time. When life departs from the cell, the cell returns to an atomic structure or an atomic order. It is life gathering the body in the totally unique fashion that bodies are in and the body does not render life with any unique qualities. Life gathered the body atom by atom and arranged the formation as life intended the body to form. I can't touch any person or any thing other than to make contact to a few molecules set on the edge of a small point of the body structure that I am touching.

That what I would think of as I am not permanent but is formed by a hallucination giving an interpretation of what could be a reality only as far as the interpretation goes. It runs along the same lines as we think of what a mirage the Universe is. Should I think of my body as me then where am I? The factor called life gathers in a place by using a process we have no idea about, but life gathers a collection of cells to hold life. It is life that gains by duplicating and replacing the cell and not the cell that gains duplicity by giving more cells to accommodate life. If the cell was the factor representing the ability of life then as soon as life left the cell the cell should find the ability to again to fill the cell with new life. The "I" form the governing singularity while all the cells form the controlling singularity making the body to form the mutual singularity surrounded by the Earth forming the principle singularity. Life can only flourish in the principle singularity that favours life.

If the cell had control over life that would put the cell holding life in control of time and we would be able to live eternally. But that is not the case because as soon as life leaves the cell, the cell starts to decay and rot until it is in a structure composed by unconnected atoms. Again, if it were the body that held life and the body was I, and then as soon as life left the body then the body would replace what was is lost with life again. While it is true that a cell filled with life procreates and assembles more cells as need arrives and as time moves on, that proves that life is in charge of the task in performing the collecting of more cells and not the having cells performing the collection by forming more life by forming more cells to hold life. Life is spawning off by collecting new material, which will hold life, and that process is purely based on cosmology and moreover it is using time that we think of in terms of what is named as the Hubble constant. Some Hoggenheimers have the saying that time is money but they are immensely more stupid than Newtonians are and Newtonians are already immensely stupid to think mass brings about gravity! A Hoggenheimers is someone such as a banker or a money broker that lives for money and collects money, as money becomes the banker's god. A Hoggenheimer will sell off his sister or his wife or daughter for prostitution as long as there is money to be made, such as the mentality is of all bankers and moneylenders. You will find any stock exchange filled with the mentally impaired Hoggenheimers. I can and do explain the mentality of the demonian Hoggenheimer in much more clarity later on because in the Hoggenheimer is vested the lowest form of human thinking that could be possible and if we wish to venture into the human mindset, we have to reach the lowest form of human thinking as well. If you take away money from me you take away something some degenerate Hoggenheimer established as a yardstick to see how far the Hoggenheimer clan could defraud and corrupt my senses. Some Hoggenheimer print a paper, which all the Hoggenheimers then collectively say carries a worth of this or that. Just like the Universe, they create something worthless because it is just paper and ink without any intellectual information on it and then brainwashes me into a state of stupidity to believe it has more worth than merely being ink and paper and with me being that dysfunctional I will exchange my farm that has a lot more substance for a few bundles of these pieces of paper. I mention this to establish that there is criminals hiding in banks that take our preconditioned concept of worth and worthlessness and by doing that those Hoggenheimers enslave the rest of the human race. Think what would happen if one morning the world would wake up and refuses to accept the paper they call money as all of man suddenly realise the money they say has worth has no worth at all. Think how far would the politicians run and hide who was bought by the criminal Hoggenheimers with worthless money to help the Hoggenheimer defraud by law all people into believing the paper they offer as a currency has any worth. How would the Hoggenheimer force us to put a worth to the paper when we turn around and tell them to prove that paper they so freely print has any worth at all. Tell them to prove money has worth by giving it to a cow and have the cow go and buy fodder. If the cow takes it then the paper has worth to life but if not, they have to intoxicate my mind with criminal toxic bullshit to get me to believe the money is worth something. Tell those Hoggenheimers to buy me just one more hour of life when my last second of life has arrived. If I could by one more day while living my very last second to enjoy life one day longer I would agree money is worth something. Time is what has worth and money is fiction the Hoggenheimers create to mesmerise all the small-minded masses such as me. I rave on about this matter because to all of man money is the most precious commodity invented while money is the greatest fraud that came to mind. I challenge anyone to eat gold and not starve or to drink fossil oil and not die and show me that life can thrive on money. Food should be free and so must accommodation be free because being able to eat everyday and being to sleep secured every night is a human rite and to hell with voting being a human rite. To be able to live and not starve to death if not having money is a human rites issue. I wish to show that the normal thinking of man is so screwed and stitched upside down and back-to-front it will take many generation s just to unstitch and untangle the corruptness man believes is life! If man is brainwashed in believing money has worth, then think how far does man have to develop not to think of the body as being him or her. Money is the last thing that is equal to time because Hoggenheimers will enslave you further with the saying that time is money and that is hogwash at its best. It is another way the Hoggenheimers thought up to brainwash the senseless masses into believing bullshit is food. All this I bring to your attention to try to make everyone realise that our way of thinking about worth and value is screwed incorrectly and as my body has no life other than the life I placed there, other things too is worthless notwithstanding whatever worth I am told is there. Time is equal to life. Time is life. If you steal someone's time you steal that person's life.

You can't steal money because the Hoggenheimers are legally stealing money because that will eventually and unavoidably take place with giving money a value. The way that Hoggenheimers steal money is what they call taking profits and interest and duty charges and taxes but in the end it is all just theft unleashed on the mindless many out there. Money will always find a way to depart from the poor and land in the pocket of the rich. This process has been happening to me all my life and all my life I have been penniless because of the fact that I am not prepared to steal from the poor to become one of the rich. Also I will not steal from the poor because of principle proprieties. I am part of the poor so stealing from the poor would then start me off doing the unthinkable as I am going about stealing from me in person and that is madness. I believe money has no worth but life has all the worth there can be. I am positive that if Howard Hughes would have been prepared to pay a million dollars to stay on Earth one more living-day and when given that choice to do so in order to stay alive then he would have done so gladly. Life and time is equal. It is the factor called time that life uses to collect cells to last life a lifetime. Money can't buy time. Time is the essence of life because time is life and I prove that time is within singularity and then afterwards time becomes space as the history of time. I am going to show this statement in the following but brief argument. Money is replaceable any day of the year but time is as irreplaceable as life is.

We have this conception that every person we see is the person we touch because we think we see what we think we touch. Putting my hand on another person's body doesn't mean I am touching what I see. By making this obvious mistake we wish to translate that to the cosmos and there is where we truly fall short. No person can touch another person. That is a misconception mesmerising all that can think. I can't see another person nor can I touch another person. I can at best touch a selected few atoms that forms a smaller part of the larger construction that represents that person's body. Firstly I think I see what I touch but I see photons that are rejected by the atoms forming the body of the person and then by creating an illusion I form the perception that I am touching what I am seeing. At first between humans this sort of error does not translate to anything significant but when transforming this error to galactica that we see in space the error becomes widely exaggerated. I can't touch the Earth but at best I make contact with a few atoms that is part of the overall structure my intellect think of in terms of as the Earth. I can't ever see another person and this gets important. I can see light reflecting from a position that the body of another person was in when the light was reflected by the body of the other person. Now in the context where I see my wife in the daylight the error of seeing her or seeing light reflected from her body does not translate too much but then you get this Newtonian discovering a galactica on the brink of the Universe about 12 billion light years away and then that misreading of the truth gets to be a serious misjudgement of the truth. It is a vision of what could amount to a galactica that possibly filled some space that according to all the incompetence of the judgement of Newtonians was 12 billion light years away and that vision as sure as hell just does not even resemble what would now be there in the present time we now are experiencing. The time of 12 billion (and lets presume we could accept this as accurate, which I in person could never accept because it is about one trillion times more than the billion light years the Newtonian mind attaches to the reading of the age of the Universe) has so much time gone to form space in which the cosmos did develop from that which we see and what we see is not even comparable to an illusion. What the Newtonian think he or she sees compared to what currently is in place at that place they are seeing light pouring from is so much different that not even a Newtonian has the imagination to compare what they see to what is presently in place and when the truth be told, what ever was in place has developed many light-years to the left and the right and to the top and to the bottom that what they see could not be there any longer. It has developed to the extent that the material in the present must be totally unrecognisable that also it must be said that the space they now see is completely vacated. Lets return to my body to get some sort of an assessment of the truth because as I showed it is in seeing me where the cosmic misinterpretation between what is time and what is space starts. In order to fathom the Universe we have to get to the truth about what is time and what is space and what time brings about to become space.

We have to start at what I think of in terms of what I am. Lets get back to the "I in me" because where the "I in me" am I would find the time in which I am. Life uses time to gather a body and when life leaves the body then the body disintegrates to mere atoms. The "I in me" starts off by filling a few atoms forming a very few cells to start "me" of as a being "me" and that us all. Then the life uses time in progress to gather more cells to form more of the structure that eventually will form the body that will become the "I in me". It is life that performs the task of gathering material and not material accumulating cells filled with life. Therefore, placing the emphasis on the body when thinking in terms of life is totally Newtonian which means it is wholly backwards and Neanderthal. Its is just the same as filling the Universe with "nothing" till the Universe runs over with nothing by using more nothing in the expanding of the Universe and nothing is used for expanding by having been filled with "nothing" and this line of thought is brilliantly representing Newtonian thinking. It is on the same trend as thinking mass is creating gravity with having no proof to substantiate the claim in the very least. Without proof of contact between the two particles performing gravity by the measure of their mass, then still by magic applying between the two particles there is gravity and not only that, but the entire philosophy of physics holds this thinking as the basis of all physics. Can you see what I mean by being backwards in thinking? They think that the cell structure composing of material that disintegrates as a combined structure when life departs, holds the control it has on life. The "I in me" can never be the body that holds me because as soon as I holding life remove from the body the time component that gathered the structure we think of as forming the body no longer bonds the body. With my life leaving the body that which forms my body will desecrate and the structure will decompose and disintegrate into atoms. Where what form my life leaving the body this vacating of the body I thought of as me will allow the body too disintegrate into eventually becoming an unrecognisable atom debacle that could only portray the total fiasco that was my body because of the random scattering the process ends in.

The body I use as a vehicle to get me around during the time life is in some way connected to that body does not really exist. (If not for the life in me the body would not be because as soon as I leave my body riggers mortem steps in to destroy what took life one lifetime to maintain. The body I think is I am not I because that body will disappear the second my life stops the maintaining of the structure. The maintaining is to supply the body with what other lesser forms of life collected during the lie they had using more time to collect and it took more life more time to have collected what I eat in order to maintain the body filled with life I call me. That puts me eating them on a higher par of life. If I do not devour other forms of life my form of life will end in starvation and I will tarnish my body structure when I do not deprive other forms of life from what that life took time to accumulate. I would perish because my life would start to devour my body if I do not devour the body other forms of life had. That body is a hallucination life creates to fore fill a function and serve a purpose and then when the purpose is served that body will disappear once more. The purpose that my body is serving is to give life the opportunity to become a second form of time being in movement above and beyond the cosmic movement time instates. My life is movement to extend the movement of time and go beyond time to form a second line of movement. Once life is removed from connecting to the carbon cell the structure of the cell disintegrates totally. You may think that it is possible to electrocute the muscles and have life return but that is not the case. Go into any mortuary and start shocking with low or high voltage, AC or DC, use any amount of amps or just let the electricity sparks run all over the cadaver being shocked and see what form of life comes out of any cadaver or see how much life electricity can replace. What electricity does is electricity proves that life can use a simple thought and in that thought life can generate electricity and the electricity that thought can generate can then propel the muscle. However once the brain is without life that provides the thought, then you can shock the brain material until it is fried meat and in the process it might start smoking but it will not be thinking. In a living person holding cells filled with life it is possible that you may short-circuit the electricity in the brain in order to bypass the thought process and in that then start muscle contraction but that proves that life generates thought and thought generates electricity and electricity generates movement and in the end the movement being established is a distance away from where life has to be situated when providing the thought. Life is not heat and life is not electricity but life does energise heat and life does charge electricity where both is the result of the presence of life and not forming the presence of life. With the above-mentioned obviously present we find that life controls the cell without the control being in any position to activate life.

That makes life the only form of reality while it was holding the cell structure together and that makes the cell structure an illusion brought about in space exemplifying a much bigger hallucination that is without proportions when compared to the cosmic structural illusion called the Universe. The human or animal body does not truly exist at all but is only present in the mind of the person (or animal if you wish) that is holding life. The only part that forms life in a governing or a position of domination has to be not within the Universe where the cell structure is. Then within every cell the singularity point centred in spin also holds life, where that life which is not the I in me that is in the I in me that is in control of the singularity dominating the cell holding life which is in the I in me. The part controlling the body being in a form I can see is also truly what I can't prove because the position in which life is that controls the body is formed by the fact that the body can move and apply movement associated with life and also I can't prove the other person has a body since I only see the light being reflected and never see the body.

What I can use to prove the body is there is that which I then have to use in order to bring such proof by the body being there and teeming with life. The proof that is required to prove the presence of life is in stated in the body being with life. The fact of life being present is that which allows the body to be where it is within the cosmos. Life is not part of the cosmos but being the builder of the body this act of building a body and allowing the body free movement proves that life is in place and the body being in place proves that statement that life is in place while not being in the Universe. But the body is paltry and truly nothingness glued together by life having some thought about it. Then to top this I see light that is discarded by the body that is assembled by mere thoughts and the light is rejected and flowing away from the body. That proves that what I see is formed by an elusion painted with light. I have to use interpretation about some hallucination I think of as being another person to convince me it is the other person I think it is. I can never touch the person to find conformation that it is the other person because I can merely touch a few cells and that would be the hallucination that I will use to form an interpretation that it is the other person that I touched. The same goes for the smelling and the sound. Therefore I have to depend on being fooled and outwitted to believe I see what I see, I smell what I smell and I hear what I think I hear. Every aspect of what I think is I or what I suppose is another person in the Universe is a fantasy that either I or in the case of another person then the Universe creates on my behalf.

If we look at the brain and the way brainwaves circle around the brain and what we see is we see the Coanda effect formed in detail precisely as gravity form in detail. The brain controls the flow of space such as air under the guidance of the Earth's gravity, which is the same thing. We can see how thought controls the space around the brain by controlling the flow of space in a process we think of as electricity or electromagnetic brainwaves as Newtonians wish to put it when Newtonians get technical. The brain matter forms the solid while the liquid just as it does in normal gravity is formed by the flow of atmospheric space that is liquid. However life does not stop there when implementing the Coanda effect to create gravity. The relevancies change and the blood that flows forms the liquid while the brain material becomes the solid and again we find in that line of relevancy the Coanda effect applying. The blood flowing has to somehow be responsible for charging electric current because inside the brain we have electric current flowing and outside the brain we have a magnetic field of electricity flowing while somehow inside the brain we have life establishing the electricity by controlling a shipload of chemistry, which I believe is so complicated that at best we will never detect even a small configuration thereof. When the brain dies everything is present except for the blood that stopped flowing, the electricity inside and the field outside thereby disappeared, and life finding a presence to allow control all aspects of the body through thought shuts down and everything then stops coming together. All of that is time related and I am going to return to time in relevancy as time applies to the body holding life. However, to begin with we have to investigate time as it applies in cosmic related circumstances.

As a result of examining this proposition, I located two principle positions both holding singularity. The cosmos is made up of one type (1^0) that is in two categories where one type moves and the other type does not move. The one is a liquid and the other is a solid. There is life forming a principle time position being other than comic time but in addition to cosmic time that because it is time has to have a position of 1^0 and then every cell forming the body forms that part holding a position in the cosmos because the part is structurally part of the cosmos and that connects to Π.

The condition for the presence of this singularity that forms everything, controls everything and is everything is the centralised to a centre singularity holding relevancy $k^0 = a^3 / (T^2 k)$ that forms by movement $T^2 = a^3 / k$ of space $a^3 = k T^2$ placed in relevancy $k = a^3 / T^2$ that is centrifugally going both ways $k^{-1} = T^2 / a^3$ thereof (**Newton's 3rd law**). In the line that forms when the governing singularity charges a controlling value on space that forms we have the governing singularity 1° that relates to the controlling singularity 1^1 but then 1^1 becomes 1° where it then relates to 1^1, which then becomes 1° that in turn relates to 1° that becomes 1° that relates to 1^1. This is eternity moving across infinity and in that creates the relation of singularity 1° 1^1. This is the spot stretching to become more than the dot and become a line. This 1° that relates to 1° that becomes 1° that relates to 1^1 is time moving forward.

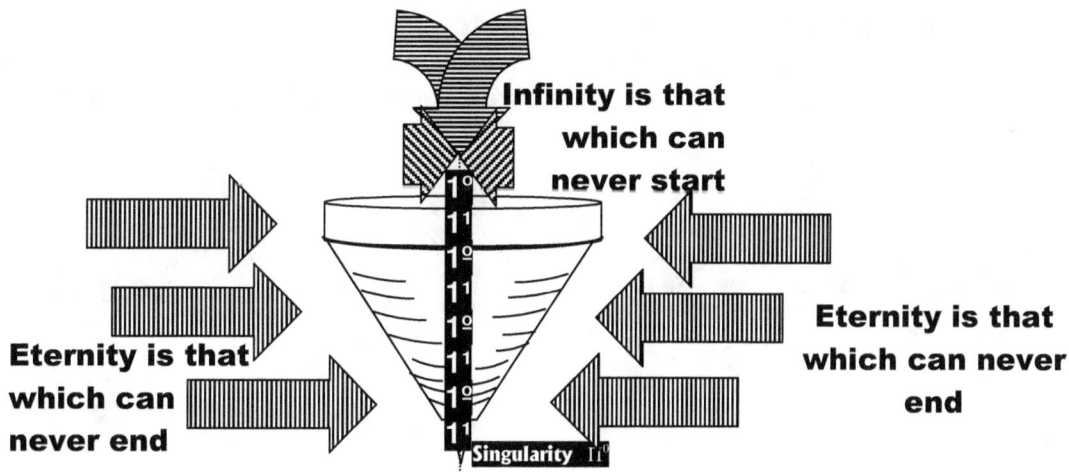

1^0 to 1^1 change 1^0 to 1^1 change 1^0 to 1^1 change 1^0 to

1° = 0.991 going onto 1^1 going back to 1° = 0.991 going onto 1^1 going back to 1° = 0.991

The pendulum arm takes a position of singularity $1º$ and finds the movement in singularity expanding from $1º$ = 0.991 going onto $1¹$, then the singularity moves by gravity towards the earth leaving the pendulum arm to take position again in relation to $1º$ = 0.991 going onto $1¹$ when repeating the process. The pendulum arm holds time to a measured value at the point time forms space to leave space behind as the result of time that left space.

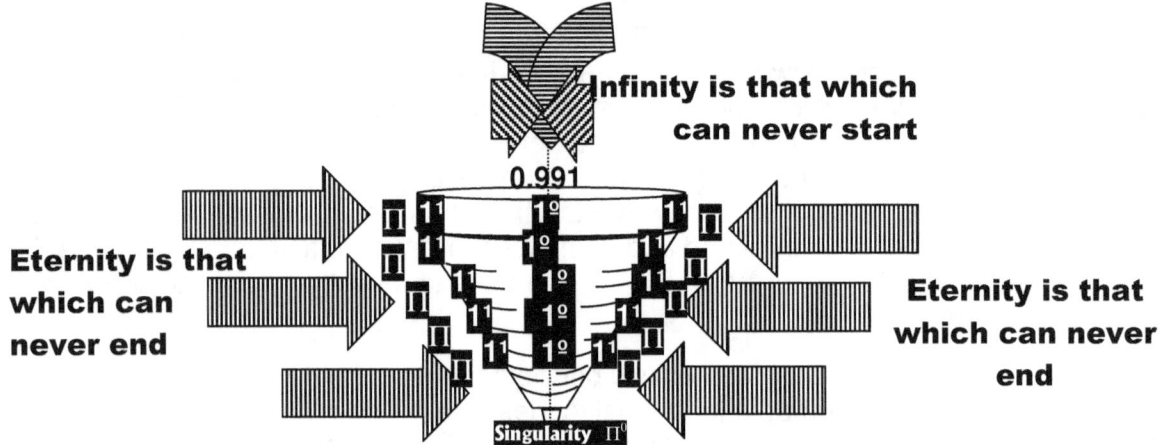

Movement is time, not only is it time related but movement is time. Whether the movement is electricity or wind blowing or water flowing or just mirages on sand dunes imitating water, it remains time discrepancies applying. Time following the line which I refer to as the spot $1º$ forming the dot $1¹$ is according to all accounts delivered by Π the turning of ($1¹$ = (7)+$1º$=(7)+$1¹$ =(7)) = 21

$1¹$ Time coming from the past
+ 1 Time in the present
+ $1¹$ Time going onto the future
= 3 Value of time moving

$1¹$ + $1º$ + $1¹$ = 3 and 3 x 7 = 21, which leaves a value of 0.1416, unaccounted for.

Π = 3.1416 − 3
  = 0.1416
    Movement of material through space is (7 + 7) / 10 = 1.4
1.4 / 0.1416 = 9.86 = Π^2 and that proves that time moves by replacing on e position in the future that then grows to form one to become three and to result in one more future point.

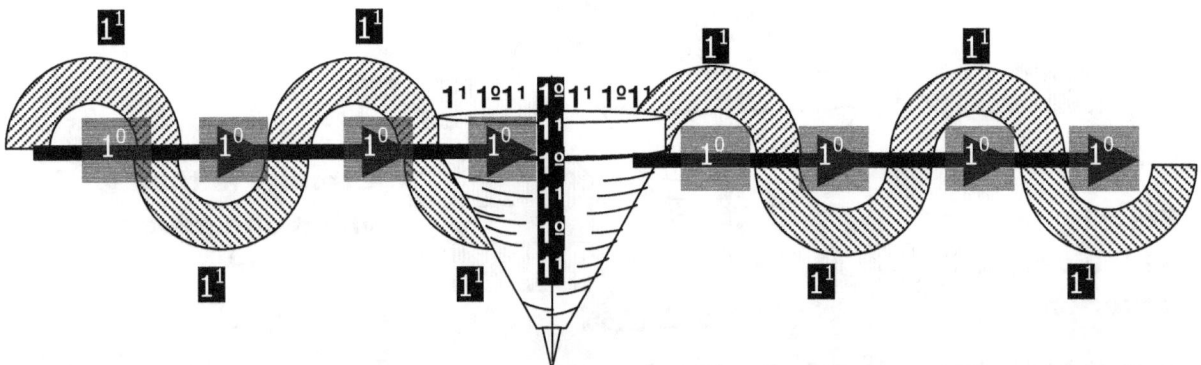

There is a Universe going on we know nothing of. This Universe runs in the dimension of singularity and it is there we will find life. The Universe we think of in terms of singularity the Bible gave another name. It is the called by the Bible the abyss where everything that we think is, is started and from there develops into the one Universe we think is real.

This puts me at the point of the Bible and my audacity to mention the Bible in a book relating to science. Well, I feel very sorry for the atheists controlling the world of science, but since they are the mentally handicapped shortsighted and the thoughtless thinking they are wise, I cannot tie myself down or limit my view on matters to accommodate their disability in thinking. They are the instigators of Newtonians thinking being handicapped by animal constraints and to have my view tied down to their incapacitated level of thinking is the same as me attempting to have this theme aimed at toddlers when writing this book with having toddlers in mind as the reading audience. As much as toddlers will never understand what I say and therefore I write for a developed audience, that much I write for a developed audience without being limited by their disability in recognising what applies in reality and the ability lacking in the minds of the atheist. They can't even begin to think what and where the spot is because their thoughts are too limited. The spot can never be within this Universe because the spot holds no space and still the spot is where everything starts that then continuous to form the Universe they think of as being a reality. The spot becoming the dot controls what is in this Universe also without the dot ever forming a part of this Universe. The spot Π^o becomes the dot $\Pi^o\Pi r^o$ but neither the spot nor the dot has any value in the form of space but also holds absolute validity as substance within the Universe because the spot and the dot is what could be thought of as a pre-Universe universe. It is from this continuous duplication of singularity that the Universe grows, but I am getting to that.

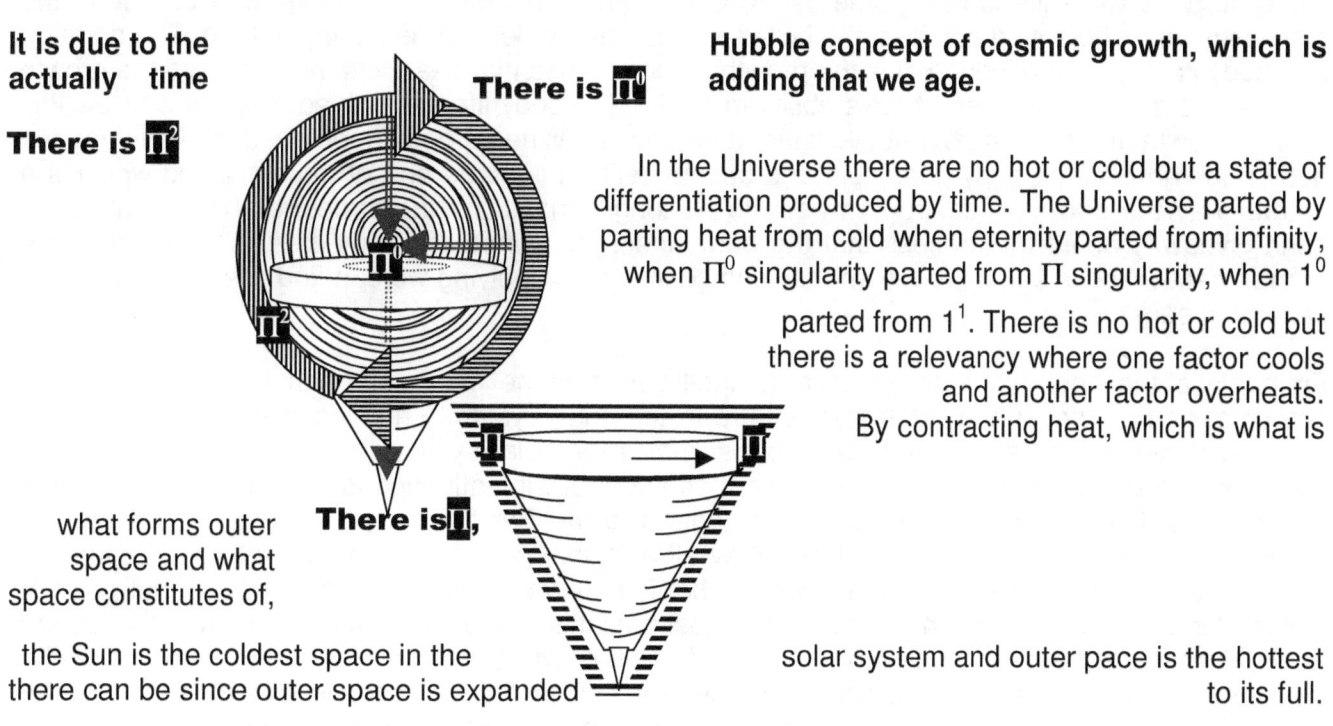

It is due to the actually time

There is Π^2

There is Π^0

Hubble concept of cosmic growth, which is adding that we age.

In the Universe there are no hot or cold but a state of differentiation produced by time. The Universe parted by parting heat from cold when eternity parted from infinity, when Π^0 singularity parted from Π singularity, when 1^0 parted from 1^1. There is no hot or cold but there is a relevancy where one factor cools and another factor overheats. By contracting heat, which is what is

There is Π,

what forms outer space and what space constitutes of, the Sun is the coldest space in the there can be since outer space is expanded solar system and outer pace is the hottest to its full.

I have seen many conversations broadcasted in talk shows on TV where Newtonians (they must be when gauging by their level of stupidity) in the post of medical doctors tell viewers that when in the 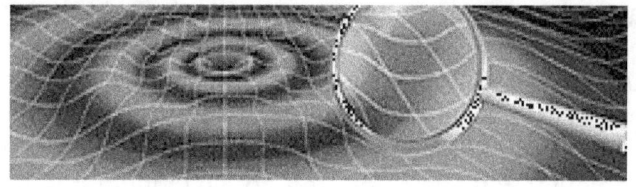 future science would reach the level that if they do this or change that by gaining progress in the future then people would become a thousand years old and people would never age in the wonderful future the mathematician will create in the laboratories. According to then science would become so grand that in the future aging would be a process which man will circumvent and later on finally suspend. Humans becoming a few hundred years would become common practise when injected with some miracle whatever they claim to be a potion that could be invented. The medicine that science are working on would in the future give man a never ending life filled with wonderful ever lasting youth and cures would just depend on the next remedy science will create in the laboratories. All this is farfetched hogwash they rave about while not one of those Super-Educated-lame-brain-Monstrosities sat back to think about what life truly is and what death is and most of all what the process is we call aging. If you listen to them the first thought that springs to mind is how simple it is to be another Educated atheistic simple-minded- Newtonian because then one can hide the lack of thinking behind keeping science simple-minded. Do as Newtonians do best by putting everything on mass and mass will heal everything unknown. To think that when you die you die, you then are dead and then everything that ever was disappears into a veil of atheistic Newtonian stupidity and shortsightedness. If you die you stop living and it is as simple as that! If the lot was that easy to conclude then we would all still have the animal intellect but fortunately there are some that think further than utter Newtonian stupidity.

Life is totally interwoven with the cosmos and also completely independent thereof. If we trace what mechanics it is that the cosmos applies we will find what life manipulates to its benefit and if we investigate how the cosmos grows we will find how the body of man ages. Life uses what the cosmos offers, but life is not part of the cosmos. It is only that few atoms forming a body that humans think they are which life enslaves to use to the benefit of life such as the body composition that life creates for finding something fit for purpose to use for a small while during which time that which life uses is part of the cosmos but that which life uses forming a body remains part of the cosmos after life disengages from the body. The body and what forms the body is left behind in the cosmos as reusable material.

What then remains behind is cosmic but that cosmic part forms only as tools that enable life to for fill some purpose while life is using time as life is on Earth. This part of life being part of time in the cosmos is very short-lived and has a definitive end if it had a definite beginning at birth. The duration of a body is very limited indeed. It is the growth in the cosmos that destructs the body and the growth of cells destroys the purpose of life's ability to control the body life assembled. Let's first follow this cosmic growth and see where that takes life during aging. When a body goes dead the first thing the body undergoes is the body goes cold and become stiff. It is stage one of decomposing which is a natural way the body self-destructs when life no longer maintains the structure. This should be a dead giveaway about the process that the body undergoes when the body dies. If life forms some form of heat which is a time above and beyond cosmic time applying we can find cosmic time to also be heat in some form.

From since the time that man discovered intelligence (if he ever did) man has been with the presumption that the Sun is the hottest centre in the solar system. Later on in the present time, it came to someone's attention that the Sun also holds the solar system in gravity. The Earth by its standard and dominating its sphere of which it can control with influence is the hottest centre in the space of its domain and it holds the moon centred to the Earth. The gas planets are the hottest centres in relation with the most heat and they all hold their satellites captured by a hot centre. All space structures hold in every centre there is that is confirming their independence at that point of securing independence the centralizing of the most heat it is able to concentrate and from that centre holds all material captured or controlled in the domain of what that forms the independence of the structure. I can go on and on but heat in the centre couples gravity to space-time, just as if Kepler said before he was spoken for on his behalf and without his permission or his agreeing to it.

It is very easy to say the Universe holds anything there is contains all and occupies the lot of every thing, but with an Anglo America mentality the practical implications are far and wide. However, understanding this concept proves to be something a little more advanced! Every poster or picture about the Universe indicates a concept suggesting boundaries.

Newtonians are grappling with the notion that one day they would be able to build a computer that will run the earth without the need of human or life in particular. This is of course so Newtonian it is as much hogwash as all the other Newtonian ideas they put forward. The idea that the cosmos stays what it was being what it is, is completely Neanderthal and laughable. Ever point time provides to form space changes as time confirms new points that will provide space with a changing forever. Every instant time moves it resembles the cosmos in precisely the same way it was before except for one difference and that is that nothing is the same as it was before because time moves the cosmos by changing the cosmos. This is where Einstein's flat Universe falls flat.

As soon as science released the world from a flat Earth, Einstein brilliantly came along and gave us the flat Universe. Is that not expanding the Universe into much wider misconceptions than there was before...? To marry our logic with Einstein's calculations of singularity brought about some bizarre ideas. To incorporate a three-dimensional Universe into a flat Universe in singularity is not that simple. It involves understanding time and time forms space while space is the reminder of what time was at any given point in the past.

The list with compelling theories and proof can and does fill many libraries of men with brains more than hair. According to the theory on the Big Bang, there has to be a Big Crunch in order to form a Universal beginning of an end. According to present facts presented, the Universe has an age ranging from between 10 and 25 thousand million years. The overwhelming favourite age is placed **at 13,5 × 10^9 years** and is the most widely accepted number in years or the number of time that the Earth supposedly circled around the Sun from the very start of time. If you believe that you have no idea of what surrounds you and your understanding of the concept involving reality is smaller than a single micron, but there are other books where I go into this total stupidity with much more clarification. Infinity is as small as having no start and eternity is as big as never ending and if that is part of the everyday Universe, only the small minded Newtonian mentality could put a limit on time being measured in circles that the Earth rotates around the Sun. The first scientific formulated theory that was accepted, implied that the Universe was static, going no where as it stayed the same which meant we had a nicely organized Universe that maintained itself uninterrupted and existed in a state of regularity where matter was evenly destroyed as it was created. This suited the Newtonian perfectly since it then was possible to establish constants from where the Newtonian could calculate and re-design the entire Universe to the liking of the Newtonian.

This meant that the Universe would be precisely the same, unconditional to changing the position that matter was distributed through out the Universe. This gave the Newtonians an unbelievable chance to play God with their mathematics and to redesign what they thought God got wrong! As time progressed and a lot of research was done, no concluding evidence was found to substantiate this fact of the Universe being static; and also no proof could be found at any place where matter originated spontaneously. There was no white hole in relation to a Black Hole where material came streaming out. This implied that as matter changed, there was no proof that it was ever renewed. The list of theories born from that confusion is as many as they are bizarre. To make sense of everything I placed the Universe into dimensions and after implementing the dimensions a lot began to make sense.

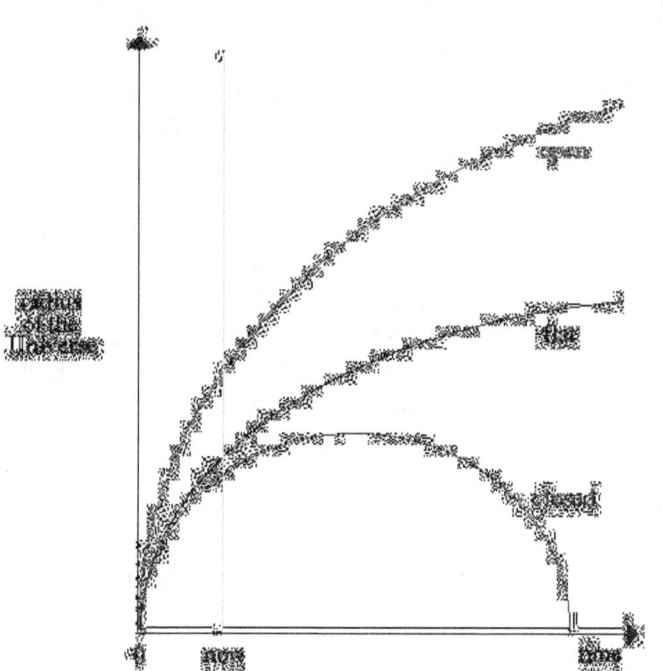

I found it in Kepler's formulae $a^3 = T^2k$.

I placed gravity where there is only a line **k**, moving from somewhere and going somewhere, but never having a defined beginning or end.

The second dimension is the wave, moving in broken or half circles T^2 from a specific point outwards but continuous. The wave can never be in one position, as it always changes position through time.

The third dimension is a combination of the previous two, a^3 of which even Einstein's flatness could not break free.

As soon as science released the world from a flat Earth, Einstein came along and gave us the flat Universe. To marry our logic with Einstein's calculations of singularity brought about some bizarre ideas. To incorporate a three-dimensional Universe into a flat 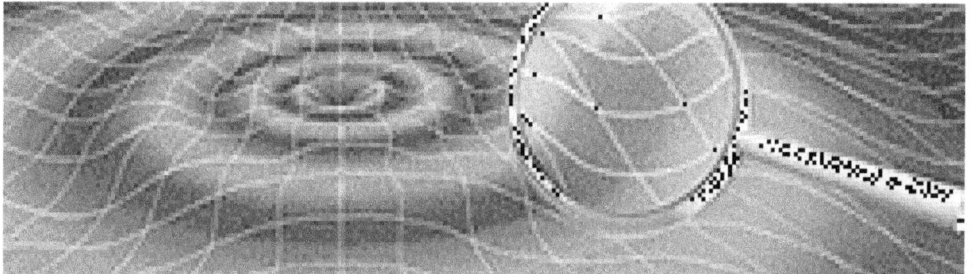 Universe in singularity is not that simple. Whenever I am presented with this picture presentation of a flat Universe I am baffled by the Newtonian stupidity represented by such am image. If there is a top, then a bottom has to be somewhere so it then is not flat. If there are waves, then there is depth and where there is depth there can't be flatness. Where there are blocks, then there is a width and a length and that can't be flatness because the blocks then show variation in size. This picture is as three dimensional as the blanket on any bed.

You might ask what has all this to do with life. If I think of life in terms of dimensions I then have to think of life as being the body. That means that if being the body is the form of life has, then yes, time and life has to end and has to start. But after life ended the use of the body; the form of the body and the composition of the body as well as the structural integrity of the body is terminated with immediate effect. Life is in association with time and is therefore not terminated because life has still got to connect to time in singularity as time moves on in singularity. That is not singular and time is singular moving away by forming dimensions. Space holds dimensions and when I think singular, I think if the triangle as being equal to the circle as being equal to the straight line and this equality defies reality as a dimensional understanding because the shape that the forms represent is totally conflicted in the face of equality.

Every person alive knows that there is a Universe that is not flat and this Universe is chained to time. The flatness k^0 of the Universe is in the space a^3 formed by time T^2 moving between points k.

After I concluded this I then realised I was looking for Kepler's formula that detailed all the factors I was in search of formed as $a^3 = (T^2 k) = a^{3+2+1=6}$

One can draw all the lines you wish, but as soon as you bring in time, the Universe becomes a six dimensional, moving "something" that cannot turn back and cannot skip forward. It is on tracks going in the direction, wherever that may be, through the movement of time. It is $k^0 = a^3 \div T^2 k$

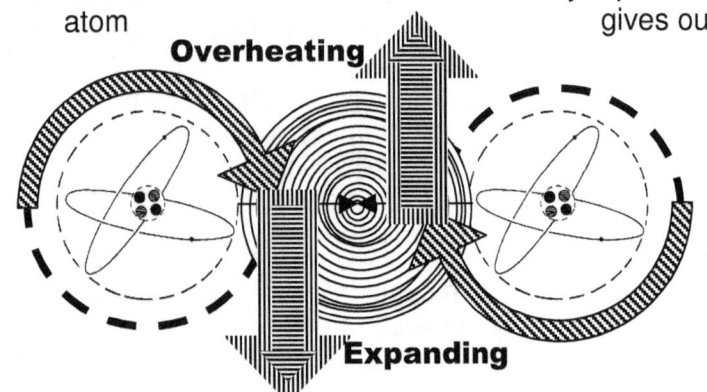

Whenever the electron jumps from a higher into the lowest (innermost) orbit, the atom gives out radiation at a wavelength corresponding to a spectral line of the Lyman series. The jumping down into the second lowest level contributes to the Blamer series. Thinking in terms of space this is a vital clue. There is a release of nuclear charged heat and this indicates to a flow of heat that releases backwards as gravity reduces the flow of heat. If gravity is present then the indicator shows heat released from contraction.

The greater the jump is the greater is the emitting of radiation to the limit of the series, which is reached when an electron enters from the outside of the atom. Outward jumps involve the absorbing of heat and that is inclining to provide space to accommodate the increase heat levels because of the increase or rise in the absorption lines. When the heat level in the atom rises, the electron jumps to a higher band and when the heat reduces in moves down one band. The heat coming about in the surrounding of the atom produces more space because the atom increases the space by applying the electron in a higher orbit ring. The moving of the electron is coupled to the giving out of radiation at a wavelength corresponding to the spectral line of the Lyman series. From this it is so obvious that gravity lends itself to an atom that fluctuates in dynamics by resizing in relation to gained or lost space and space represents heat that releases in concentration.

When the heat level rises or lowers the space within the atom decline or increases. Every atom in every element association shows different corresponding to the heat it is in association with. The corresponding of the atom and the reaction derived from such rising of heat turning into space, is a direct result of the interaction there is in the gravity contracting and the gravity in expanding depending on which sector corresponds to the dynamics if which of the actions of the Coanda principle is in dominance at the time in regard to where the viewer focuses on. The main issue is that the atom forms the Coanda gravity principle as liquid heat interact with atomic solid and the bands are clear indicators the Coanda principle where these lines forming the Lyman bands is adjusting relevancies in accordance with heat intensity requirements.

The rise in heat is a rise in the liquid part that extends the contracting by giving rise to the adding or the relenting of motion. The heat is liquid because the heat is motion and the atom inside becomes the solid since the motion is conserved by the spin of the electron. Today it is the rise of levels that is in focus but this same principle had to be in use when atoms were formed that today is responsible for elements.

If it was true about mass pulling mass by reducing distance then in that case the Big Bang was not possible and individual elements was not possible. With the cosmos down to the size of a neutron and mass confined to that space within the neutron that would form the recipe for the biggest crunch there could ever be. If the Big Bang was brought about mass confining mass by reducing the radius that implied the space within as the Newton formula $F = G \frac{M_1 M_2}{r^2}$ would suggest, then what would ever be more applicable than that moment to bring in all the forces the hell can unleash and destroy what ever was not yet in the Universe.

The entire reduce thought in following as Kepler gravity is part in the heat in which it spins and the Big Bang is the story of heat configuring to form matter in space by gravity.

idea of mass pulling mass to space is a prehistoric and explicitly incompetent explaining science. The Π^2 is a far more suitable explanation and is as true is. It is so obvious that of the movement of the atom

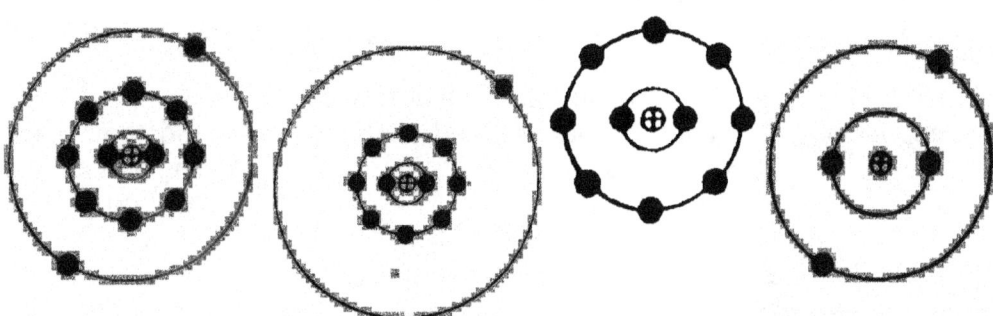

Magnesium (12)   **Sodium (11)**   Neon (10)   **Beryllium (4)**

The most sensible way atoms formed is by the method of the Coanda principle. Those ones that were first formed were most of all very dense atoms where the first had to come in place when time was one point away from still being eternal. With the high velocity of movement applying at the time when atoms formed during the big Bang, the velocity condensed heat into huge piles of protons where each proton attached to a neutron and the neutron formed an electron to serve the chain with flowing heat and dominant and space infinite and one notch off singularity. We still find the liquid time having a vital role in the space the atom uses. At the time when T^2 was almost eternal space was infinite because T^2 will not permit the space a^3 much room to be.

But the opposite is also true that if time was steady then time being so long could pack in large numbers of protons with the accompanying neutrons at the time into the time unit in space that formed. Part of the sage was Kepler's formula applying to the detail of the forming of the atom and giving credence to the density required to form such enormous atoms.

This is most accurate, but this is only in concern of a unit in rotation that is in conflict of its own spin. When time in the cosmos at large view time in progress we find that the factors providing meaningful development are not quite so simple. Time moves by the measure of T^2 but time in progress is by the measure of k. Therefore space in progress of time is $a^3 = T^2 k$ where progress is T^2 but in relation to gravity we find that $a^{3\,-1}k = T^2$. The smaller the space $a^{3\,-1}k = T^2$ that is required to form the atom is, the higher was the rate of spin.

To understand the atom forming gravity it is essential that we understand the top because the top represents everything one can conclude from investigating the Coanda effect and moreover it is an investigation of gravitational principles when we enter an investigation of the characteristics of gravity shown in the attitude of the working principle known as the Coanda effect. By casting a top with a string turned around the base of the top and allowing this string to rotate the top, we do what no other since God achieved. We created a Universe secluded and excluded from other all other concepts forming a Universe. Man establishes a separate identifiable time within the realms of another time. Man presents time by Π.

In the relevancy we find the action and reaction of space-time flow is $a^3 = T^2 k$ and that translates to being $T^2 = a^3 / k$ on the one side and $T^{-2} = k / a^3$. In the times we now live in we can and do produce an optical illusion of $T^{-2} = k / a^3$, but that is implementing the use of a telescope. In the true time we find as a cosmic reality the fact of $T^{-2} = k / a^3$ is rather a mathematical statement and no more than that. In reality we have $T^2 = a^3 / k$ on the one side as time expands and on the other side we find $k^{-1} = T^2 / a^3$. This we know is true because while it is possible by using an optical illusion the reality is that time can never reverse.

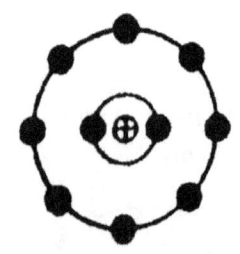

In truth the reality about the opposing actions is that we find normal growth and that which Hubble first saw is the process of expanding space-time by the margin of $T^2 = a^3 / k$ while on the rebound we find the opposing while contracting space-time is $k^{-1} = T^2 / a^3$.

As much as there is expanding coming into the Universe of the atom by the motion allowing …just as much is there expanding coming into the Universe outside the atom by the motion of the atom allowing decrease in density in outer space in material.... Which puts growth in material.... That brings growth in space by density reduction in outer space ...and the growth in heat within material is caused by the reduction of density of heat in outer space and in this the one loss compensates for the other gaining. Material was precisely evenly distributed before half of the Universe started to expand and the other half stared to contract the half that was expanding. In the end the two factors that formed will again be evenly distributed when the unification of eternity with infinity paces everything back into singularity as is happening in the Black Hole at present

The atom becomes the dead giveaway in supplying evidence supporting the idea that spin gradually relented as space grew to become more accommodating. Time is the relation that eternity forms with infinity. Physics to this day never tried to establish where infinity in the cosmos is and physics ignore the location of eternity on a daily basis...and then Newtonians feel surprised that they don't know what time is!

Infinity is formed by one line

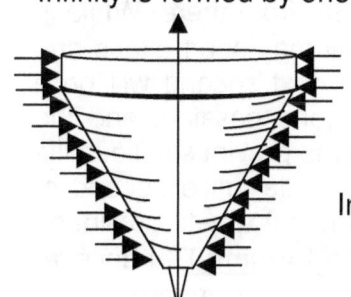

Innumerable many lines form eternity

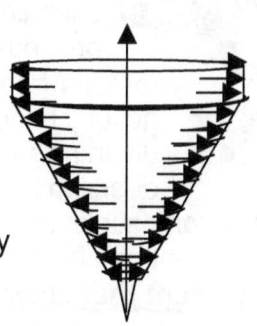

The Coanda effect exemplifies the definition of time or gravity, which is the same thing. Time is the movement of everything n relation to one specific point or gravity is the movement of eternity in relation to infinity. In the centre of the spinning top there runs a line that connect space on the one side to space on the other side without having any space at all. On the outside we have space in the form of heat, which the spinning of the top turns from being a gas to being a liquid. That liquid is eternity because that liquid never ends being present throughout the entire Universe.

In order to understand time we have to return to the top because in spite of Newton's misgiving about the top rotating, it is through the ration of the top that we can learn so much more about the characteristics we find as gravity as it is practised by the atom.

To give the top cosmic viability one must throw the top in and that motion supplied to the top, the throw initiates a time line within the top.

After all it is gravity that keeps the top as it is spinning in an upright position while it is spinning because it is gravity that stabilises the cosmos. Moreover, what is actually in progress from the top spinning is the Coanda principle activating gravity and that happens in accordance with Kepler's formula.

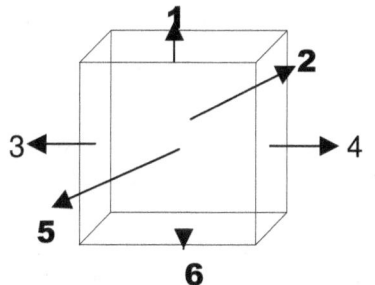
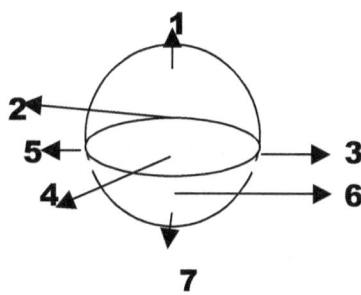

This means that in the cube at the point of contact between the cube and the sphere the cube experiences such a contact point as if the "bottom falls out" of the cube and without a "bottom" to support objects, they fall to the sphere as objects does fall to the Earth. The one side resolves its position in favour of the centre taking the side over. Remember that a body "floats" in space, but at one specific point it starts to "fall" to the Earth. That is gravity and it is a dimension change much more than any force. I shall explain this last remark later on. That too is the Lagrangian system with five cosmic structures holding relevancy to the centre structure where the centre structure stands in for seven positions diverting from singularity and the orbiting structures standing in for five positions in space.

From such a point it holds, when it leaves that point going to every other point, the new point will be opposing any other point from where it came that was not pointing in the direction to which the first point is pointing, whereby it diverts from the direction it should extend and changes by 7° the direction it holds. No matter what the point is or where the point leads, such a point holding a specific direction will be unique in the direction it is rotating because at that or any other specific point wherever, it will be directing not in the direction it travels but spins in the direction flowing from the centre point outwards.

Any point will be it opposing itself within the rotating of 180° changing every aspect of its previous flowing characteristics it previously had or will once again have in 180° from there. While in rotation from the point of an outside observer all may seem static and never changing but to the object in spin every next second will be a diverting from every aspect it was in every second passing, and the direction it held in relation to the direction it held the previous mille, mille second will totally be incompatible with the direction it holds the very next mille, mille second of rotation. That proves no point can be static or constant, all though it may seem that way to outsiders.

Gravity forms by movement that establishes singularity initiating a circle forming Π. I uncover these principles by placing Π within the formulating of gravity and when using Π, I bring clarity to these misunderstood cosmic principles. I show why gravity is there, how gravity forms and what role stars play in forming gravity. I am able to mathematically prove that there is no difference between how gravity and electricity form and that is part of what I call the cosmic code whereby I show how to mathematically decode the cosmos. I prove mathematically when atoms spin they establish Π that forms the Universe. I show the entirety of what there is has to move in spin or everything falls back into singularity from where everything started. Movement drives the Universe. If mass does generate gravity, then mass has to apply Π to do so, or mass does not form gravity. Everything using gravity forms a circle of sorts, which forms the curvature of space-time, which is Π and which curves light. In spinning in a circle, gravity forms Π as a centrifugal force that condenses space. I found a precise mathematical **cosmic code** the cosmos follows by forming gravitational space-time.

When time started infinity as well as eternity had altogether 3 positions, the past, the present and the future.

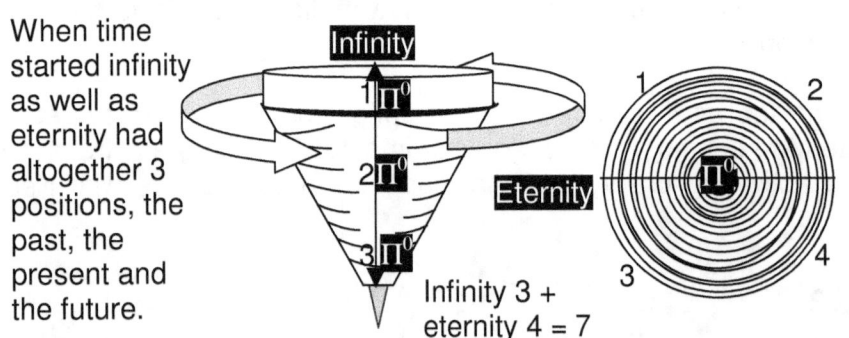

Infinity 3 + eternity 4 = 7

Then eternity parted from infinity when heat separated what is cold from what is hot and eternity formed one more point than before when it had the three points. With infinity and eternity then jointly having 7 the cosmos came into rotation.

Gravity forms when the 7 came from the past to the present 7 and onto the future 7 and this became 21. Not only that but With singularity advancing from infinity to become one it proves that even as we see singularity as one, singularity also is multi dimensional but that ability is beyond our scope we have being in the Universe. That is how the cosmos started. Infinity holding eternity on one spot coming from the past to the present being one spot and onto the future being one spot the cosmos was singular monotonously eternally by repeat. The perfect became imperfect as temperature differentiation brought a partition between infinity having 3 points and eternity always returning imperfectly with 4 points and a cosmos landed between what is a line (3points) and a circle (4 points). If science wishes to find the origins of life they will have to start here because it is here that life still is vested within the Universe…and good luck with that effort because looking at life at that point goes far beyond the ability of the flat minded Newtonian. Π is 3.1416 x 7 = 21.9912, which shows the margin with which time allows space to grow. Time is 3 or in relation to 7 is 21 and time extends Π to a margin of .1416 or in relation to the spin of 7 then 0.991. It is this extending of time 0.991 forming space 1 where life is.

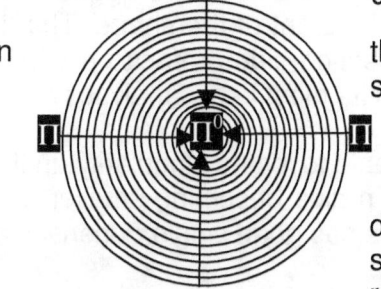

In the very centre of the sphere the form of the sphere dictates that the shape will relinquish space as the line runs from the outside towards the very centre. With this natural state of affairs the sphere is naturally inclined to dismiss all space that it can form in the form as the sphere holds space inside and the form will finally be without dimension. The line shrinking by reducing actually takes place in every sphere as the diameter reduces to the centre. In the centre where the radius line goes single, the form relinquishes the three dimensional form it has inside. Being without dimension in the very centre means that at a point in the extreme centre of all spheres there is a point that holds singularity because this point with no space has a mathematical position although it is invisible since there are no sides to such a point to give that point any dimensions. The shape of the sphere is calculated by using the formula $4\Pi (r^3) / 3$. By reducing r to a point where r is r^0, singularity steps in because only the form remains as Π. Going even further, we find that there then comes a point where Π goes singular Π^0. At that point absolute singularity is

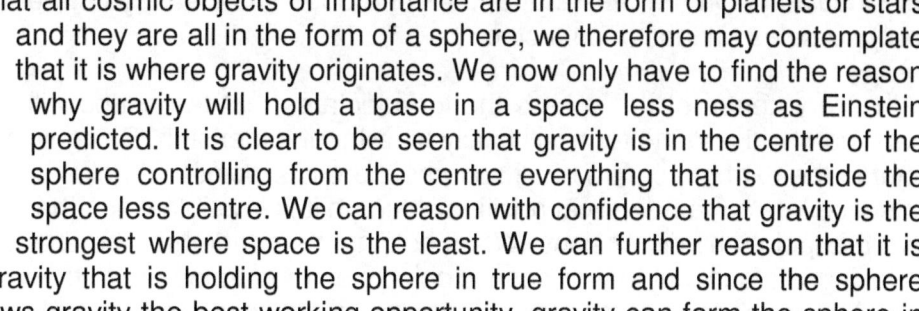

present but so is absolute gravity present at that point. When holding the strength of the shape of the sphere in mind as well as taking into account that all cosmic objects of importance are in the form of planets or stars and they are all in the form of a sphere, we therefore may contemplate that it is where gravity originates. We now only have to find the reason why gravity will hold a base in a space less ness as Einstein predicted. It is clear to be seen that gravity is in the centre of the sphere controlling from the centre everything that is outside the space less centre. We can reason with confidence that gravity is the strongest where space is the least. We can further reason that it is gravity that is holding the sphere in true form and since the sphere allows gravity the best working opportunity, gravity can form the sphere in as strong a shape and form as the sphere seems to have. From every point on the surface of the sphere is where that point connects with the other side of the surface of the sphere by a line that runs through the space less ness of such a centre of the sphere. Such a line also connects by an angle of 180^0 as well as 90^0 to six other lines running from top to bottom, right to left, and back to front, where all join and cross in the centre of the sphere. There are therefore six lines crossing and connecting by a centre from any given point on the surface of the sphere. Such points connect in total six surface points on each side of the sphere while they all support one another through the space less centre. In that absolute space less ness in the centre holding singularity we find gravity supporting and controlling all space within the sphere as well as space connected to the sphere. That is where gravity controls and guides the space, which falls in the parameters as well as under the influence of the form of the sphere. In the gravity centre space goes singular meaning space becomes space less or flat.

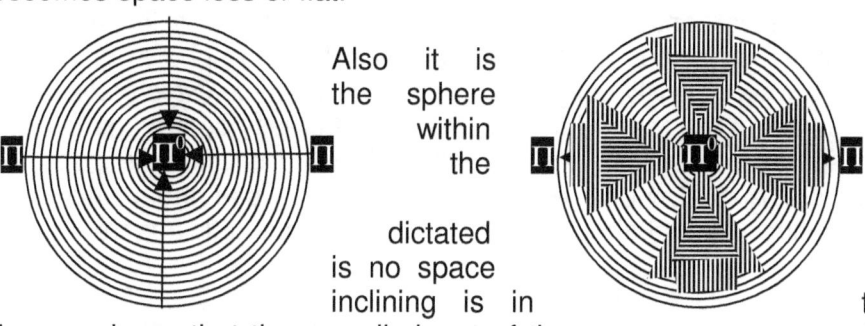

Also it is true that the entire form that is the sphere is controlled from a centre within the sphere. That centre holds the sphere in form and shape. Therefore the strong form is dictated from that space where there is no space and no form left. The natural inclining is in the form of the sphere. It is part of the roundness that the overall shape of the sphere represents and this structural strength is carrying down to the very centre. Because the circle is forever reducing, that reducing which is inherently part of the form of the sphere becomes a tool in distorting space in the sphere and is eventually removing all forms of space from within the centre of the sphere. The very centre ends up as having no space because of the reducing that continuous down to become the space less inner centre. The all roundness is the ingredient that forms the backbone of the absolute strength that the sphere has and that is the component that the sphere is so famous for. The form the sphere has allows the sphere to have a control that is coming from the centre deep inside the sphere where the space vanishes and being without space seems to keep the entire structure rigged. The strength of the sphere comes from the centre of the sphere, which is inherent of the shape. That is why the sphere has such strength in form and the fact that all connecting sides refer to a centre brings credence to the strength that the shape has. How does it work in its most basic analyses?

It is from the layout that the sphere uses as a natural form that we are able to locate singularity. In the case of the sphere the material naturally reduces by measure of the radius becoming smaller to a point where the radius is r^0. At that point the line that will form the radius has gone single dimensional r^0 and that is equal to 1^0, which is singularity.

The cube has sides and the sides form a rather weak and flat surface that connects four corners. The flat surface produces a rather indifferent contact point with no special features on the surface. The corners connect to other sets of corners and those corners form a weak structure without any direct support coming from the other five sides. Without material to fill the body of the cube the cube has no direct connection between any of the sides other than corners connecting at the edges of the sides. Taking the vantage from the point the sphere is holding from the centre out into space there are ten points connecting to the centre. In that are the dimensions of singularity connecting to space where five connect to space in the second dimension of singularity, and five connect in the third dimension of singularity. On the other hand, the cube does show a very different characteristic, which involves only six sides (at least) connected.

In the very centre of the sphere the form dictates that the shape will relinquish all grounds in space that it can hold and the form will finally be without dimension. Being without dimension means that at a point in the extreme centre of all spheres there is a point that holds singularity because this point with no space has a mathematical position although it is invisible since there are no sides to such a point to give that point any dimensions. When holding the strength of the shape of the sphere in mind as well as taking into account that all cosmic objects of importance are in the form of planets or stars and they are all using the form of a sphere, we therefore may contemplate that it is where gravity originates. We now only have to find the reason why gravity will hold a base in a space less ness as Einstein predicted. It is clear to be seen that gravity is in the centre of the sphere controlling from the centre everything that is outside the space less centre. We can reason with confidence that gravity is the strongest where space is the least. We can further reason that it is gravity that is holding the sphere in true form and since the sphere allows gravity the best working opportunity, gravity can form the sphere in as strong a shape and form as the sphere seems to have. From every point on the surface of the sphere is where that point connects with the other side of the surface of the sphere. All other possible points connect by a line that runs through the space less ness of such a centre of the sphere. Such a line also connects by an angle of 180^0 as well as 90^0 to six other lines running from top to bottom, right to left, and back to front, where all join and cross in the centre of the sphere. There are therefore always no less than six lines crossing and connecting by a centre from any given point on the surface of the sphere. Such points connect in total six surface points on each side of the sphere while they all support one another through the space less centre. In that absolute space less ness in the centre holding singularity we find gravity supporting and controlling all space within the sphere as well as space connected to the sphere. That is where gravity controls and guides the space, which falls in the parameters as well as under the influence of the form of the sphere. In the gravity centre space goes singular meaning space becomes space less or flat. That is where Einstein's Universe goes flat because that is where gravity is at its strongest. However my bringing up this statement brings me directly to the point where I get very confrontational about how the brilliant mathematicians treat those they suspect are less inclined to think.

By examining the form of the sphere, we find that there are 6 points on the surface of the sphere holding the form at a specific and equal distance from the centre. Lines run from the centre into space at $90°$ and $180°$ angles of each other from six opposing sides. There then are six lines at $90°$ and $180°$ connecting to the centre from six points on the outside edge of the sphere. As a result of the basic shape that a sphere has, there is a spot in the extreme inner centre of the sphere where the lines in $90°$ relevance cross each other and others connect by $180°$. There is also at that point a spot where all space relinquishes a position and only singularity 1^0 as form remains. At such a point we find the measure of the sphere being Πr^0 with $r^0 = 1^0$. That is where the line that represents the radius as a line disappears, as it becomes singularity r^0. After more reducing continue we get to such a point where we find only Π^0 left. At that extreme point is where space in all form disappears, as the circle providing the sphere the form the sphere has, removes all possible form by going into singularity $\Pi^0 = 1^0$.

Then in that area all form of any possible space disappeared leaving only the dimensions of singularity 1^0. I cannot delve deeper into the argument. However, from such a point there runs lines that connect to space on the outside where six points on the outside points connect to the space less point in the inside. In this book I take this argument much further but for now I leave the argument at that. Those lines carry the structural strength the sphere has. Contact with one point has support of six other points across the whole structure where the other six support every one of the six by singularity and the support runs through the entire sphere including the middle. Where there is no space, there must be singularity 1^0 just because the space filled with material removes zero and only material filled space is present.

That means material fills the lot although in singularity 1^0. If zero was a factor where all space finally halted in zero as the value, then zero would be able to remove the space from the centre and such removing would continue to remove the space until all space was removed. It will finally abolish all space in the sphere and it would remove the sphere. Zero removes all possibilities of anything coming about. Since the sphere is there, a zero factor in the centre cannot be present. Only infinity can be a factor from where space may grow because infinity can extend and grow into and up to eternity.

The implication of this is that following the line down to the centre of the sphere we located the centre of the Universe. That is where gravity is. There is a lot more to that but be patient, we are getting there. In every centre we find a point, which is in truth not there but is the mainstay of all that is within the sphere. The mathematical value of such a point is $\Pi^0 r^0 = 1^0$ and 1^0 is singularity. That is the point where the Universe started and that is where the Universe will finally end. That is the Universe without space-time. That is $k^0 = a^3 / T^2 k$ which proves the Universe is without doubt a sphere…and we just located the centre of the Universe!

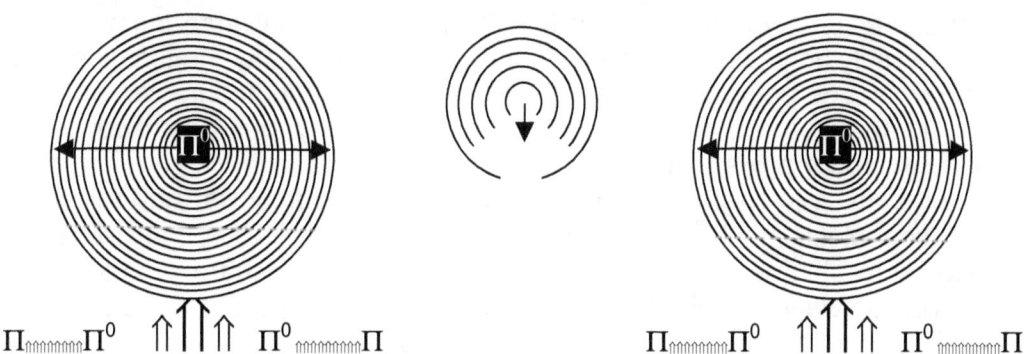

As one can see with the spinning top delivering the Coanda principle, every point overheating can spawn space-time by centralising singularity.

One can see from the top that singularity is established wherever spin occur. The motion generates a position of seven in relation to ten and singularity manifests as 1.9991 as is explained elsewhere. That means any point formed by the sphere spinning can and does start a centre in which no motion holds no space and of which motion surrounds such a point by forming space. Although everything at the time was in the form as a multiple circle, which results in a sphere, the sphere was not the only form present. This too has to do with singularity interpretations. We see a cube, as we know the cube but at first when form came about the cube were not yet a form.

While the one sphere forms on this spot where the dominating sphere secures an edge the dot may be reserved as an edge marker to the dominating sphere. To the forming sphere in progress of emerging heat gathers at that point because the rotation is a result of duplicating and duplicating is the tendency of naturally growing in space-time $k = a^3 / T^2$. In order to find duplicating coming about there has to be heat in order to duplicate what will form heat. The duplicating process is a process of one factor going softer or less solid and therefore more dynamic than the other. To have singularity is to have gravity but to have gravity there has to be a point of motion and a point of sturdiness. The point of sturdy may be in the centre of singularity, but then the solid must be motion. However even today it still apply: what moves forms liquid in the presence of a solid and at that point singularity presented the solid therefore what we might think of as solid was the liquid because it moved around the solid. Where the one factor is duplicating the other factor is compressing $k^{-1} = T^2 / a^3$

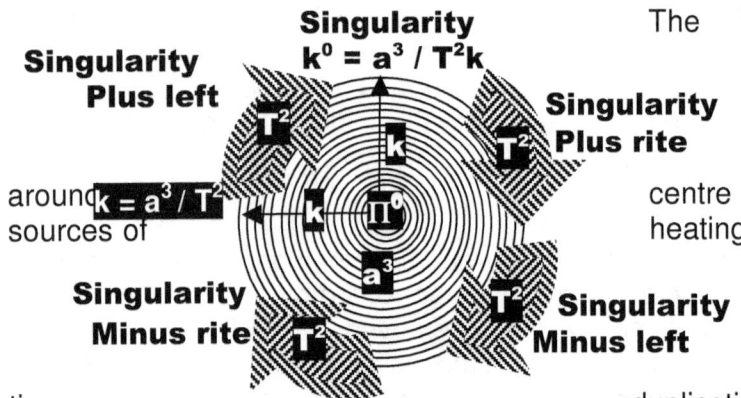

The importance about recognising cosmic behaviour is to see how life manipulates movement by using cosmic movement. The points duplicating is four moving around sources of centre by the square of gravity. The motion is the heating because the heat is bringing about the movement. The heat growth therefore provides the action because the action is what energises the points to provide the motion. The motion is purely is space-time centre from duplicating and the duplicating is feeding heat to the the four points overheating thus the points that shows expanding. But also the duplication leads to the spawning of one point of singularity that provides the installing of the next centre for the next sphere. Because of the principal in which the Coanda works the motion will centralise a new sphere and by appointing six positions around the centre three points will not move while four will move about the three points forming the centre line. The result is that the four points by duplication will reserve the point moving as the next point in singularity because of $k = a^3 / T^2$ singularity will be a natural result of the motion. Then that point will secure a position $k^{-1} = T^2 / a^3$ which will secure six points about such a centre. The centre will bring about four points spinning around three points holding a line singularity. The line in singularity will stand in relevance the contacting factor and the duplicating by expanding points will be four and serve the relevancy by contributing $k = a^3 / T^2$ as space-time only in form. From this the rest of the Universe burst into the next phase of Creation. It is precisely how persons walk by pairing the one side say the left side arms with the other side being in this case them the right side legs. Life can only use what is in the cosmos to use. This example is only but one example of unnumbered many examples I could use.

The gravity is in relation to the spin, which is in relation to the four points spinning which are $\Pi^2 / 2$ and that is the Roche limit. It is the dividing of singularity sharing space-time just as we on Earth share singularity by division between the Earth and us others that is not part of the Earth. The total that forms from the point that spawns is seven plus five plus pi square in division of four totalling twenty one that stands related to the first seven and once again another sphere formed. However this is an eternal relevancy that can never break.

Any object in rotation will have a middle point, a very specific centre point that does not spin. That point once again hypothetical but none the less must be standing still because every line running from that point in opposing directions are also in opposing directional spin to each other. Although the points had the same characteristics only seconds before, they oppose the characteristics it had just before and just after the very second in which they are and to which they relate by similar points also in rotation. Due to the spinning nature of such a point with all surrounding the point very varying second, the value of such a point can only be Π because of its constant changing.

As the top hit the ground after being thrown in a spin it starts to move around in small circles while circling around the axis in a vigorous manner, then it forms am almost motionless stance of complete blessedness as if the top is suddenly satisfied by energetically almost standing still while spinning in a precise circle while adhering to its axis and that is precisely what happens. This surging to find a new dynamic is a very important sign and is of most importance.

With all the excitement of being freed from the depressing of the Earth and no where to take go while enjoying every minute of it, the extending of the drive line runs down the line forming singularity as well as from the edges all the way inwards towards the newly established governing singularity that keeps the whole job erect.

That is why the top is spinning in the first place. The more assertive the spin is in velocity the more reaction there is from the lines running towards the centre and extending as it is expanding outwards. In real terms the space of the top expands as the spin is in contact with more time in space during the same time in period and a bigger unit fills the space in which the top spins. In this the $k^{-1} = T^2 / a^3$ space in which the top spins the vigour and excitement has to allow the structure space in which to expand as well in order to compromise for material relevancy growth the expanding also serves the purpose of room to fit the newly acquired singularity governing the motion that extends and asserts influence to the edge of space-tine.

The support that the spinning top finds in the established governing singularity keeps the top spinning in an upright stance only supported by the singularity that takes charge of the spinning of space-time within the set boundaries to establish time within the Earth's time restraint.

The heat that should supposedly under cosmic law drive the spinning top will come from the governing singularity accumulating the heat in concentration by the contraction or cooling ability the top singularity acquired. But in this case the spin is a result of life's ability to manipulate space-time and lead cosmic events. The heat that would establish such a drive in motion in real cosmic terms would require a lot of nourishing and sustaining from a large number of maintaining atoms that produce a large flow of space-time.

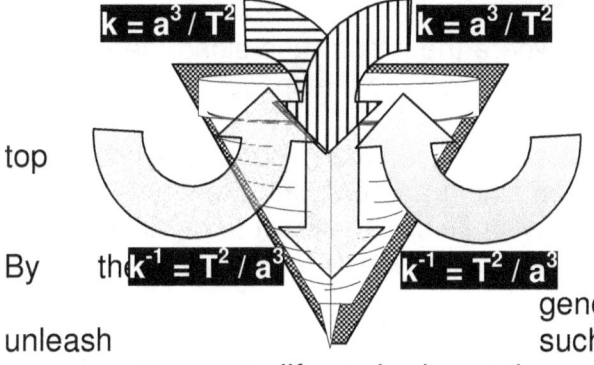

With sufficient energy the top gets into a fighting mood making the top very reluctant to give up this newly established freedom. The behaviour now attributed to the top is normally the manner how a star develops in the galactica cocoon and how the fledgling star gains its birthright to leave the nest of the cradle of the galactica. By the $k^{-1} = T^2 / a^3$ atoms forming a sum total that can support the generating the required gravity to secure the heat that would unleash such a drive c an singularity that is governing the structure movement come to life and release the new star from the blanket of heat that covered the star up to the time of this release from the galactica centre.

The example we can gather from the top shows how desperate a governing singularity can become and how such an exited singularity can put up a fight for life and independence. The top is in a fight for independence while the Earth is restraining the independence. The fight goes on until the Earth finally suppresses the last bit of motion that the top had and the top uses the last motion it has to defy the Earth's domineering control.

When the motion exceeds the level of the Earth gravity, the top shows an eagerness to rise to higher levels of independence in the same manner that an electron reaches into higher rings of energy because the top with motion is in an electron relation with the Earth filling the role as the proton would play its role within the atom and that puts the atmosphere in the neutron role.

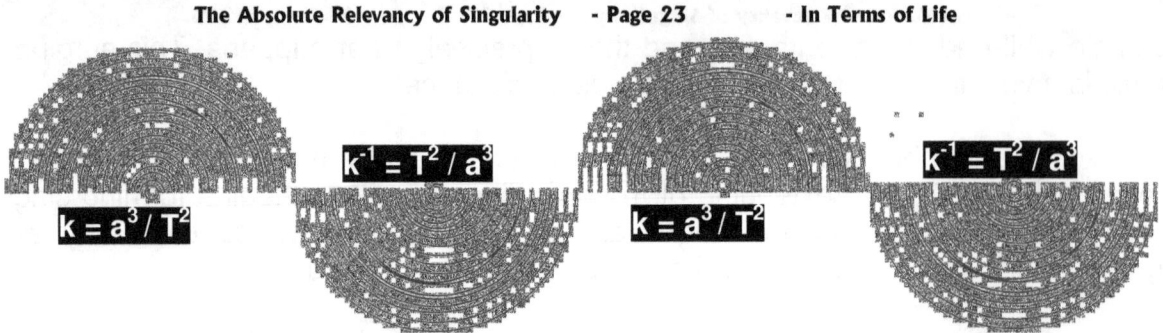

In the sphere centre is a spot that has to be there mathematically by the calculating and measuring of the defining space of any circle we find singularity $(\Pi r^2) / (\Pi r^2) = \Pi^0 r^0 = 1$. In order to provoke the line into action, there is motion requires to excite singularity just as Kepler indicated, where the space becomes equal to the motion and the motion is equal to the space $a^3 = T^2 k$.

Let's quickly establish events as they translate singularity from a dot to a controlling entity that is commanding space-time through the establishing of a separate individual drive. The motion comes about which proves to be that which generates the gravity that drives the individuality in the top.

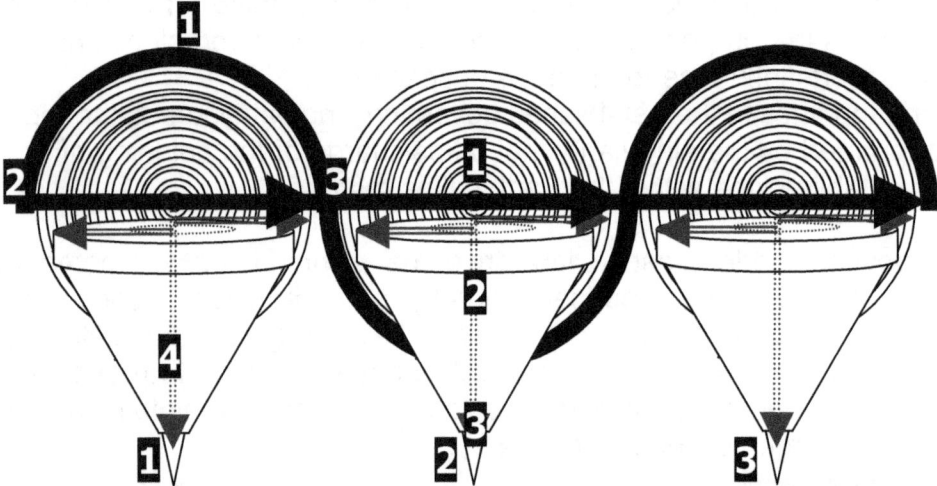

By the motion and the singularity the top evoke graph forms where the graph runs along the line of time. This makes nonsense of Newton's presumptions that the spin nullifies the space.

The balance is a control of motion that is established as a flow of space-time supports the ends (4) holding time while this generates the space (3) singularity containing and creating the space (3) in which the spinning takes place. There is a something (if you wish I'll use the term force although I strongly hesitate to use such an outrageous term) that is generating power to keep the top upright while the top is spinning.

The energy that is charged, has the dynamics stand its ground against the might of the gravity of the Earth that is under normal circumstances controlling the stance that the top has to take, but as if inspired, as the top seems to be revivifying by motion, the top is fighting and rebelling against the Earth's constraining gravity. The top is self-driven, as an electric motor would be. The difference between it and an electric motor would be the origin of the source from where the energy comes which drives the spinning top.

The top stands upright as individual as any self-propelled object can be. Although gravity is retaining the motion of the top, it is not contradicting the motion, all though it still restrains the actions. It is not combating, but is merely suppressing the motion. In this there is no race of evidence linking mass as a factor to any of the above-mentioned actions. What we would think of as air restriction is no restriction because from the restriction comes support that keeps the top standing on a very thin needle edge. The top should tell us so much about nature if we would only listen and learn and not tell nature what we think nature should tell us.

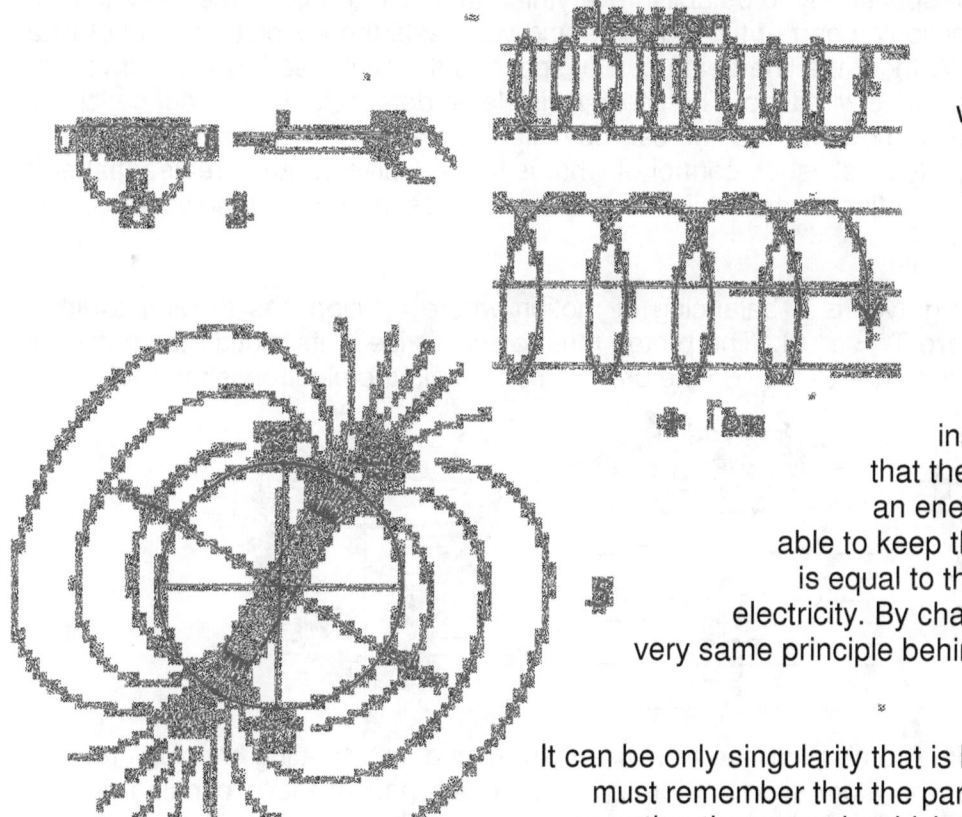

By investigating how singularity interacts with Π when forming space-time the value of singularity becomes multi facet but moreover, the radius takes up the role as an indicator to match and underwrite Π in multi dimensions.

From the motion the top inspires by creating a situation that the top can establish a force or an energy-driven time line, which is able to keep the top upright, and all of that is equal to the establishing of gravity and electricity. By charging electricity we enlist the very same principle behind what is being the Coanda principal.

It can be only singularity that is keeping the top upright. One must remember that the part doing the balancing, that is creating the space in which the top is able to spin, that is establishing the necessary time distance in which the top can spin, that is establishing the time difference that the top can use to apply the motion that extends the time, is pointing to where we find singularity. The line that is evoked by the movement is not in real terms part of the Universe because the line has no sides, can't move at all and is the only substance that is there notwithstanding human failure to witness the line and in defiance of all those wishing to see everything there ever can be and the line is still there while being very much invisible.

That which charges the top to stay upright is the same that charges the generator to charge electricity, is the same as that which charge the Earth with gravity. There is no force but a flow of space-time, which is contracted by motion and duplication to bring about the conducting of electricity. Man may name it gravity and then name it electricity or name it motion and balance but like hot and cold, those names are man-made while the principles are God-inspired and all of the concepts that has different names are in the end, still the same. To the cosmos all these different concept hold dimensional differentiation while honouring the same integrity in the Universe. The lot is still the same thing. It is what started the cosmos. It is what drives the cosmos. It is the engine giving motion as it is giving discipline in the cosmos. It is producing space to heat and heat to material.

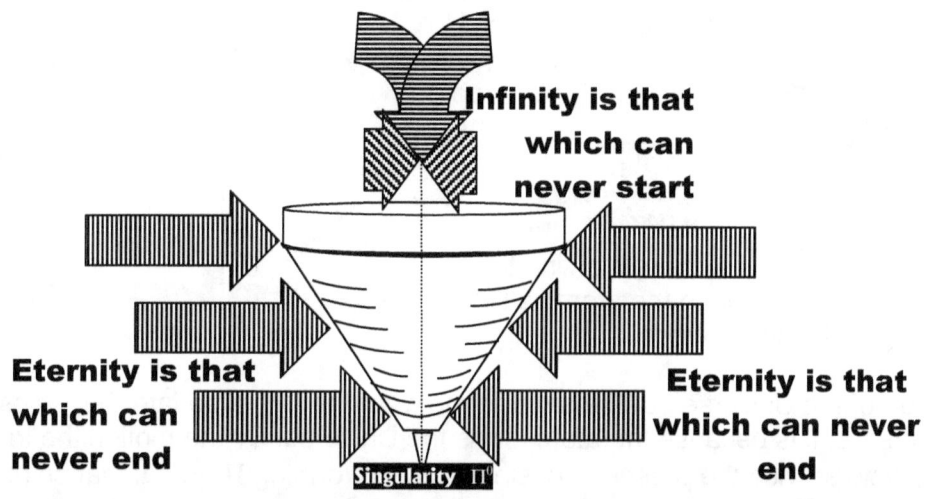

The line that forms has the responsibility to establish everything that is in the Universe while the line by own measure is not a functioning part of the Universe and yet it sets the absolute control of what is going on in our Universe. Without the line being part of space in the Universe it is what drives the Universe in time and in accordance with time. The line is invisible, in detectible, has no space, claims no tangible part in the Universe, however the line can be called into action any place and anywhere from where the line immediately establishes control of what is in the Universe and creates matter in space in time in the Universe in infinite detail while the line is even less than part of the cosmos while being infinite.

We have to recognise that gravity is a balancing of motion where rotation has to refer to linear motion **$a^3 = T^2k$ and therefore $T^2 = a^3 / k$**. The proton must compromise in its motion to enable the electron to move, as the electron has to provide the proton space while the electron is moving.

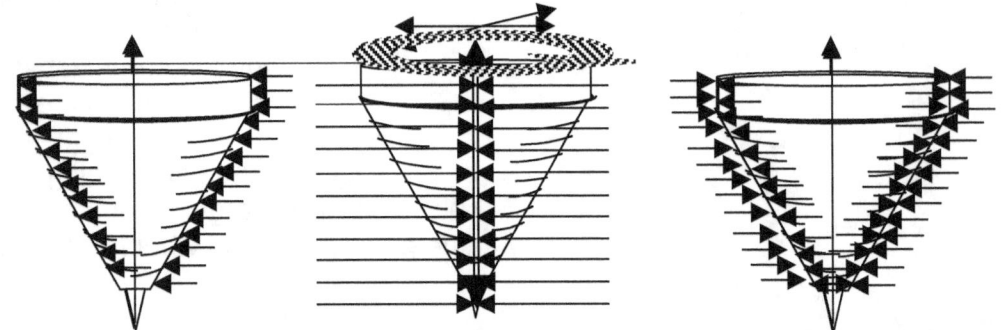

As much as there is expanding coming into the Universe by reason of overheating bringing growth in a space, the atom spinning forms just as much part of the expanding process because of the atom creating contraction which is just forming a directional opposing of expanding, but in essence it still forms part of the overall expanding by the motion allowing growth in material….

…Just as much is there expanding coming into the Universe outside the atom by the motion of the atom allowing decrease in density in outer space in material… …and the growth in heat within material is caused by the reduction of density of heat in outer space and in this the one loss compensate for the other factor's gain. Material was precisely evenly distributed when the Big Bang commenced just before half of the Universe started to expand and the other half stared to contract the half that was expanding. In the end the two factors that formed will again be evenly distributed when the unification of eternity with infinity places everything back into singularity as is happening in the Black Hole at present where the Black Hole is the second last stage stars reach before singularity finally unites infinity with eternity.

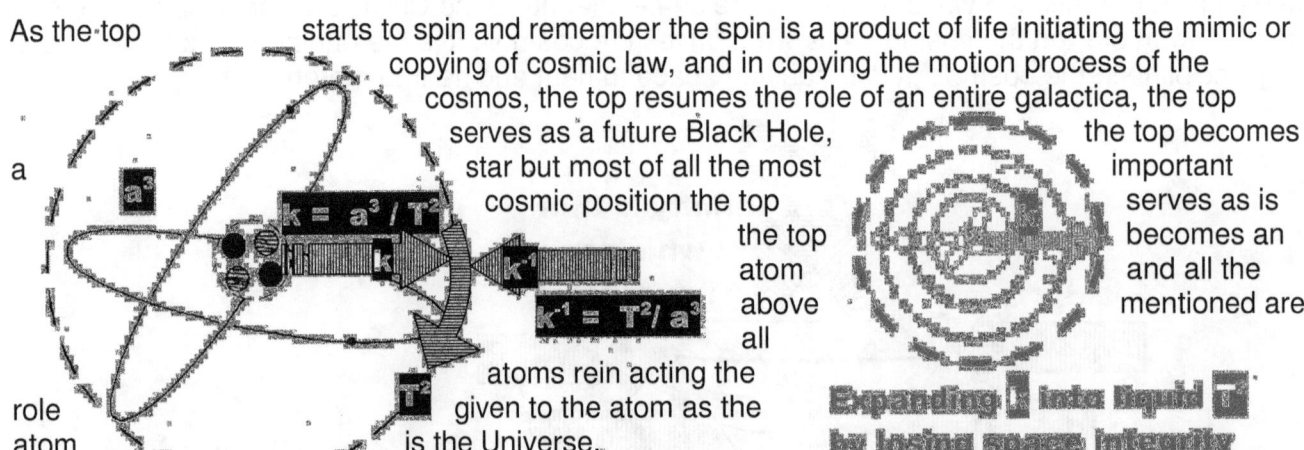

As the top starts to spin and remember the spin is a product of life initiating the mimic or copying of cosmic law, and in copying the motion process of the cosmos, the top resumes the role of an entire galactica, the top serves as a future Black Hole, the top becomes a star but most of all the most important cosmic position the top serves as is the top becomes an atom and all the above mentioned are all atoms rein acting the role given to the atom as the atom is the Universe..

The atom is not just a part of the Universe but the atom forms what the Universe later becomes, therefore what the Universe is, it is because the atom forms the Universe. By the motion the top develops, the top charge into service the presence of singularity Π^0 forming Π and at that point the top forms a self serving Universe as much as any or all other atoms in the Universe forming innumerable many Universes which then are forming the Universe. But with the top spinning one can see the outer space is as much contracting as the inner space is expanding and in that the outer space reducing forms the inner space of the top growing by the margin of reducing.

As the outer space is losing density to the contraction of the top, the inner space of the top is gaining in volume due to the loss the outer space is experiencing. In other words, the atom is growing because by spinning the atom is initiating the charge of singularity and the charge of singularity puts a flow of heat in relation to singularity growing in intensity. The issue that I whish to introduce is that the atom is growing in stature by the sinning motion the top is exerting. That is what gravity is all about. Gravity is the transfer of heat in a process of expanding in a positive direction becoming less dense but more space, and in a negative direction becoming more material and denser by serving movement in order to accumulate heat in order to maintain the role eternity has in connecting with infinity. In the formula Kepler used to introduce space-time $a^3 = T^2k$, Kepler showed that space inside the atom $a^3 = T^2k$ is growing in density $k = a^3 / T^2$ as much as space outside the atom is reducing in density, $k^{-1} = T^2 / a^3$.

Following the direction that this argument brings, it is clear that the Earth as much as every atom in the Universe is expanding by the growth in density in material and the deducing of density of space outside material. In this argument it is clear that the size the Universe is, is today much bigger than the size the Universe was in the past. But unlike Newtonians serving the goal to preserve the integrity of their master and allow Hubble's constant to expand there where it is far away so that Newton being much closer still can contract, the expanding is part of every atom everywhere in the entirety called the Universe.

In the Universe singularity stitches everything into a woven concept we call the Universe. There is the premier singularity that serves as a beacon to everything carrying singularity and that singularity forms part of every singularity charging space-time. Everything is growing but not according to my perceptions because that is growing in alliance with the cosmos. I am not going into much more detail but to say that everything there is (1^0) connects to everything there is (1^0) as (1^1). The network linking what there is to what there is, is not linking on this side where there is no reality but is linking where reality meets infinity by uniting eternity. The fact that a person can view the entire Universe by making contact holds what there is (1^0) in view of what there is (1^1) and this takes place on the side where space is of no consequence because time in eternity meets time in infinity.

On the outside of space there is time that can never end because the time has no outside. Everything that ever will be, will be because it is inside this that can hold everything but nothing can hold this space that there is no outside too. This space ends where this that has no end meets that what there can be no start too. This has everything on the inside just because there is no outside to what this is. Why would I not use names...it is because it is Biblically named? This is eternity.

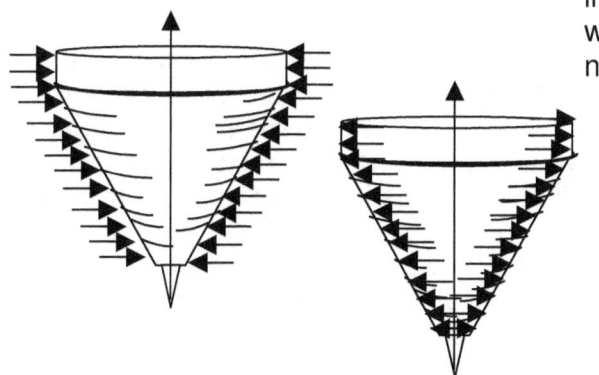
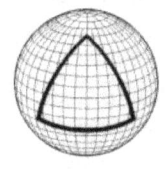
Closed Geometry
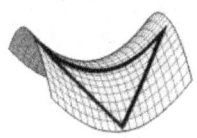
Open Geometry

Flat Geometry

This is infinity. This can never start or begin because this is the beginning of everything there could ever be.

The entire approach to previous cosmic concepts was that everyone including Einstein gave eternity borders and an ending. There were always an edge forming either a flat or a saddle or a sphere but whatever Newtonians envisaged, they saw a border ending eternity. There is no ending to eternity because as much as the line goes straight, the line curves to conclude space.

That is the main issue derived from Kepler's formula $a^3 = T^2k$. There is no outside to eternity because eternity is the entire outside that will ever be available. There is no inside because this is the inside to whatever could be on the outside of this that is representing al the inside there will ever be. The Universe is what is inside that which can have no outside and what is on the outside of that which can never have an inside and reality is where these two factors representing that which can never end unite with that which can never start and moreover, the unification forms reality found only on the other side of where time produces space. This is where eternity becomes infinity and this is where (1^0) unifies with (1^1). This is where (1^0) connects with (1^1)…and that is where we are not because where we are, there is no connection of what has to connect in order to allocate us to where we are.

These proposed Universe formation concepts that Mainstream science proposes prove to be examples that are a sure indication that without mass and without pressures the thinking of the Newtonian boils down to nothing. The problem these suggested Universe concepts have is that it places borders in the Universe and allows the observer to look at the Universe from the perspective God will have. They haven't got the insight to see that the viewer observing the Universe belong inside the Universe because to the Universe there can be no ending on the outside. The fact that this places a picture of the Universe and not of within the Universe as it truly must be, nullifies the entire idea. Here is a far better explaining of the Universe starting at its starting point and also at the ending point.

In the centre of al things spinning a line comes from where a dot first was. The dot came from where a spot once was. The line comes from a dot that extends but just as the dot extended from the spot it once was, the line has no start and has no beginning. As soon as entering the line one has gone through the line. The line has no inside that can go even smaller and yet we know that the line must have some ability to be able to go smaller since our understanding of the concept insist on this reality.

That point without space is where the Universe starts because everything that can ever be, starts at the point that could never start and can have no beginning because it is where everything is beginning and is without limits because it holds no space. That point is in the centre of all material, which puts that point in the centre of the Universe. That point shows where eternity parts from infinity. However to get eternity to part from infinity, eternity needs to move because infinity can never move and therefore will forever stand still.

There was Π^0, which was α^0 or if you would rather have it Ω^0 or it maybe was 1^0, but more correctly it was all the above and the beyond because multiplying what ever constitute the mentioned will bring about what is mentioned to a precise equality. It was a spot that was not. It was a line that ran eternal but because it ran eternal and kept repeating exactly what was before to the precise what came afterwards the line was there and was eternally running, while never changing in the least or growing by any measure. It was not one because before it was one, what was repeated and the process cycled back to before one and before one could be reached. It was such a continuing of the monotony, no change ever once occurred and therefore never did the running produce progress because the progress was in the perfect repeat of what was before. The duplication brought contraction to the smallest detail.

That is where our atheists get one hiccup. Everything that I show is as real as the Universe can be and yet not one point is part of the Universe we see. At the start before the star eternity met infinity and as eternity repeated the past it met infinity at the next point holding the future. Eternity ended on infinity every time eternity shifted. The repeat brought eternity and the repeat was so perfect that the repeat continued. The repeat still is with us as much as we are within the repeat. There was something beyond the Universe that instigated breaking the perfect cycle, which change the institute. There was something that brought a difference and we are within that difference.

That difference was time and that time is what we move through as much as what we see at night. Oh, how stupid and how thoughtless the minds of atheist and other atheistic animals are. Baboons do not recognize this factors revealing the position allocated to infinity as well as eternity because they cannot think and are therefore atheists. Animals are not able to realise that the true value of every point securing singularity albeit (1^0) or (1^1) is without space and is therefore not in this Universe. The location is where space is not and the only value such points have, is the trajectory of time bringing on movement. The points hold relevance and that holds space as a result of time leaving footprints but the true essence is in a place that has no space at all. The only way this can be recognised is by persons recognising another bigger Universe of which we have no part in the present. It is a place one can only reach in having faith and religion. These points referring to singularity is what we see at night when see light or darkness and when we can't see the nothing Newtonians see so clearly. Spiders cannot think and therefore they are atheists, as they do not think what the night consists of.

Reptiles cannot think and without thought they are incapable to see what time is, what space is, what light is and what darkness cannot be. All the animals I have mentioned are mindless atheists because they fail to see beyond the visible into the realms of the thinkable. Because of the incapacity to think the animals are both mindless and they are atheists. Therefore atheists are mindless. The night sky is such a bright light our evolution development protected our vision from the brightness of the night-light in order to give as much better daytime vision. Through evolution development our eyes are protected from the light and we remove the qualities that night give nocturnal animals as such animals see by the night-light. However animals do use dark light and not our light to see by. You can shine a bright hunting spotlight onto an animal at night and the animal will not be able to see the light you shine on it. The animal does not use the light we shine in order to see well as the animal is totally unaware of the light. Then a prowler come from the night and see the animal in the light the night provides and although one shine a hunting light into the eyes of the animals, they remain unaware of human presence. It does not use the light the spotlight uses and the light is not even traceable to either the animal hunting or the hunted. From there we accept that during the day the animals must be using our light to see because the nightlight is inferior to see by. Who says they use the daylight much different from the nightlight because all evidence is there that they cannot recognize our light as light. It is very evident in the manner they go on hunting and grazing while being totally unaffected by our form of light. That which you see at night because you cannot see darkness and you cannot see black is the light the Universe is painted in just like the Bible says. This is not religion and it is not a sermon, it is hard-core and brutal basic science and it the most fundamental basic physics there is. It is the start of the mathematical Universe portraying the only physical way it could ever be.

The size of his top would apply as follows:

 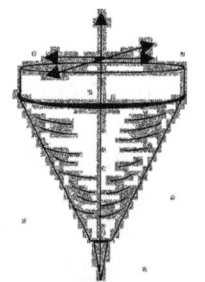 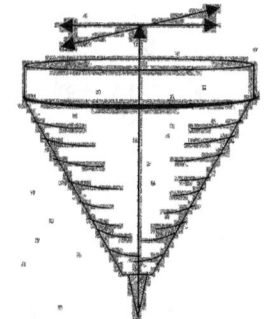

Long, long ago ...   **Some time back...**   **A while back...**   **During the present time**

We accept that we advanced from the Big Bang, just as the Bible says and even Newtonians profess that we advanced from the Big Bang. The big issue about the Big Bang is that every atom grows and as every atom grows, so does every star grow and even the Earth and the Sun is getting bigger all the time, without one Newtonian openly admitting this or recognising this as a cosmic fact.

To save Newton's fraud from being discovered, it is agreed that while the Hubble constant is a fact, it is a very far off fact and while the Big Bang expanding is a reality, it is only a reality that applied before Newton placed contraction as the pivotal force in the Universe.

However, as the Earth is getting bigger all the time, so is the distance between the Earth and the Sun also growing wider apart because the distance between the Earth and the Sun is not space, but is time that is developing by time accumulating and then progressing as space. In that same manner the Earth and the Sun is also not growing in space, but is being part of a time development in accumulating space. Time captured by singularity in the past held heat in space at much denser levels than what we find heat to be in outer space currently. This evidence is apparent from how the Newtonians believe the Universe cooled down and to get that lot to admit to that as they then applauded expanding without renouncing Newton mind you, this compromise was a brave move on the part of the stubborn Newtonian.

According to Newtonian philosophy everything in the past was bigger than it is today. A crocodile that lived sixty million years ago is fifty percent bigger than the current crocodile when compared to the size of the current crocodile. The Hippo of the past was almost double the size of the present hippo and so too was the elephant in the past much bigger than the current one. The tigers and lions that lived millions of years ago was much bigger than the present one and so was the snakes that lived millions of years ago…and the list of giants living in the past goes on endlessly while all the animals in our current age seem to shrink and tarnish to a fraction of what their predecessors were. Every time I hear about a T-Rex that was a long as a rugby field, as wide as a rugby field and stood as high as the pavilion roof of a rugby field and was as agile as a meerkat is today, I get sick to my stomach.

If I hear this nonsense coming from supposed to be intellectuals I gaze in astonishment while I feel like puking when I have to listen to the stupidity Newtonians cover their minds with. To Newtonians everything in the past was big when comparing what was to what is in the present. Yet, the entire Universe including the Earth, the Sun and the Moon is becoming larger…and no Newtonian has the presence of mind to think a little further than what their stupidity would allow? If only I could understand why the Newtonian is so persistent in not thinking! The Universe is growing and that is accepted by accepting the Big Bang. Everything in time is growing and the Hubble constant confirms that. Atoms are growing with time and the fossil samples being that much bigger as they were that long a part of the Earth confirms this concept. If there are fossil bones discovered that dates from back then when they lived several million years ago, the bones by now are stone. They are not bone any more they are stone. As the fossils grew from the carbon in which form the flesh of the animals originally was, it took in more of what forms the atom to become a stone as it now is, and if it took in more heat to become more atom, then the atomic structure grew by becoming bigger. The bones turned from $carbon_6$ to (I suppose) $silicon_{26}$ and by doing that the molecular structure size had to increase in volumetric space. The only way that could be achieved is by the atoms growing in structure and therefore adding to their composition more heat, which is time. The Earth grows and with the Earth that is growing, the atoms that remain part of the structure that reminds of what the fossils once was to be bigger and by now to be stone as it went from carbon to holding much more protons, the fossils within the Earth also feed on the heat coming from gravity and if the atoms within the Earth being in advantage of the gravity feeding the atoms of the Earth, then by the same measure should the fossil's atomic molecular structure also gain in size, if they could gain in atomic proton mass. If everything was bigger back then, then everything back then was smaller than what presently is available on Earth because time is the invert square law by the practise thereof.

I have an idea that when comparing your average modern super intellectual Newtonian with the giant Dinosaur in size as it lived in the past in true reality relating to living proportions applying to both during their lives as each holds the volumetric ratio, the comparison would be about he same as the two pictures depict. I can't see while I try to maintain a non sensational as well as a pragmatic and a stable mind that the dinosaur that lived seventy million years ago was much bigger than a lizard is today. Conditions and land space that was available at the time just would not permit the size of the animal to be much larger in compatible ratio. But reality would not prevent the Brilliant Newtonian genius in going totally Hollywood and fantasize about the fabulous sensation of the size of the fossils without having any thoughts about what pure common sense would insist on. If our Newtonian could get ridiculous while also standing in front of a camera, then their brains become the size of what the so called Hollywood stars have and that is a true Newtonian nothing!

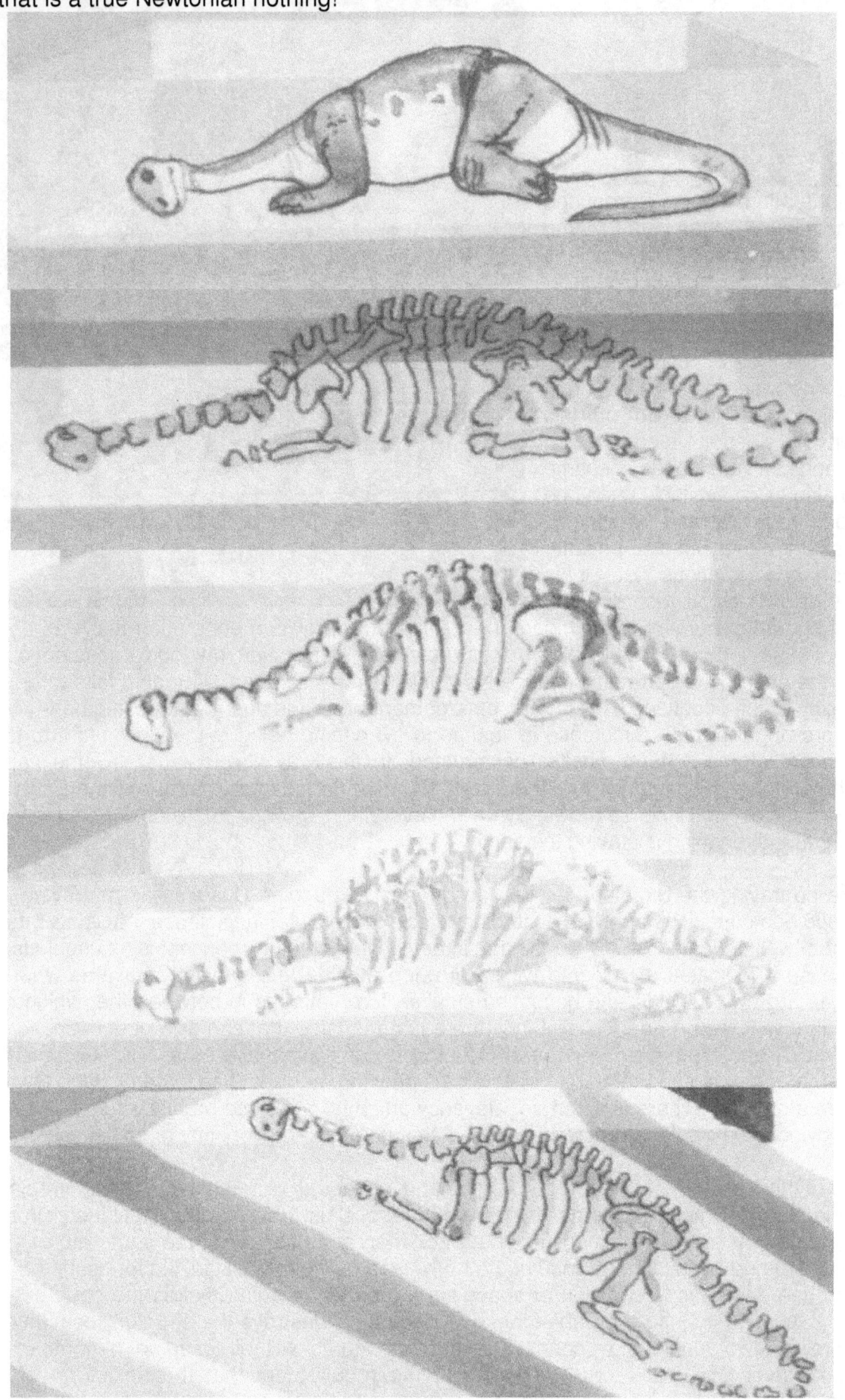

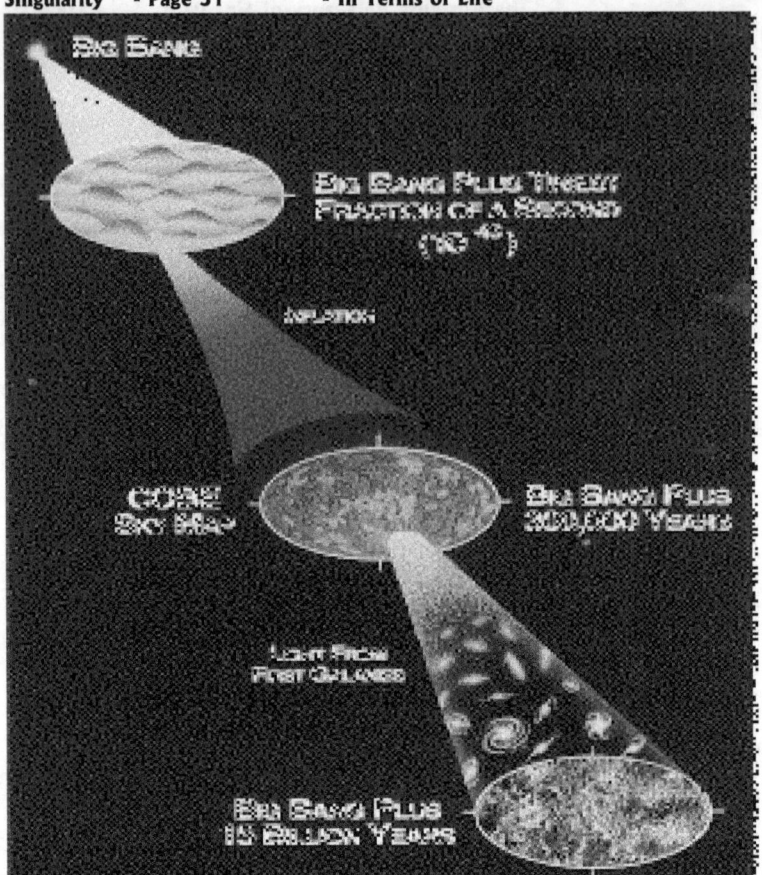

Every generation that produced Newtonians found that the young incoming generation of Newtonians were bigger than the outgoing smaller generations of Newtonians and record kept of previous data about facts written down in the past confirms this statement. The wise and the wonderful, the highly educated Newtonian put this fact down to better foods being available and as the foods became better and also more available everyone eating food, your average Newtonian not only became wiser but also became bigger than the previous lot. It is accepted that ever since records were held, every human generation grew bigger in size than the previous generation that was denied all the healthy fat free foods. But to blame that growth by generation on food consumption is not concluding the case quite correctly because Newtonian simplistic approach again fails to appreciate the complexity of the situation. This human enlargement of size is due to the cosmos and moreover the atoms growing that put an increase in volume to everything being in the Universe that forms by accumulation of atoms to form what is constructed. In this way Newtonians stay the same size because they remain even in growth to the growth found in the cosmos that allows the Earth to grow at the same pace as the Universe grows. By having everything growing, therefore it sets the condition that secures that everything is staying the same.

The picture portraying the Big Bang is showing the ever expanding Universe as portraying a vision about science and conditions applying according to science, Newtonian science that is, as interpreted by the esteem well educated Brainy Bunch and in being Newtonian by concept is as usual also totally wrong because it indicates a Universe with a growing outside and growing outwards whereas the Universe has a growing inside and becoming smaller. Everything is in-between that which can't go bigger and that which can't go smaller. Since there is no outside that can grow just because there is no outside at all and therefore the walls in the picture can't expand, we have to see the Universe growing by the measure and the margin of its atoms growing, and that has to be growing towards the inside where atoms are. There is a certain relevancy attached but since breaks off size by becoming old and en dying it makes life a renewable product that is replaced every fifty or so years.

In the case of the fossil, the fossil is frozen in time and in place of growing by generation replacement of forming larger specimens of the same species, this fossil is growing as a combined structure by using the atoms that formed the monster from back then all is still using the same atoms from the time of death time up to now. In the case of the body used by life to host life, life forms the body in aid of life as to use during its occupation of space-time, then when aging makes the body outgrow its usefulness in hosting life, life then abandons the body as it destroys the structure completely after death where no two atoms stay connected and the atoms will be used in a completely new arrangement forming a total new composition to the requirement that the next form of life will accumulate and in that the dynamic of growth remains with the use of life. This process I explain in much better detail using a volume of book material to explain procedures that took place every day.

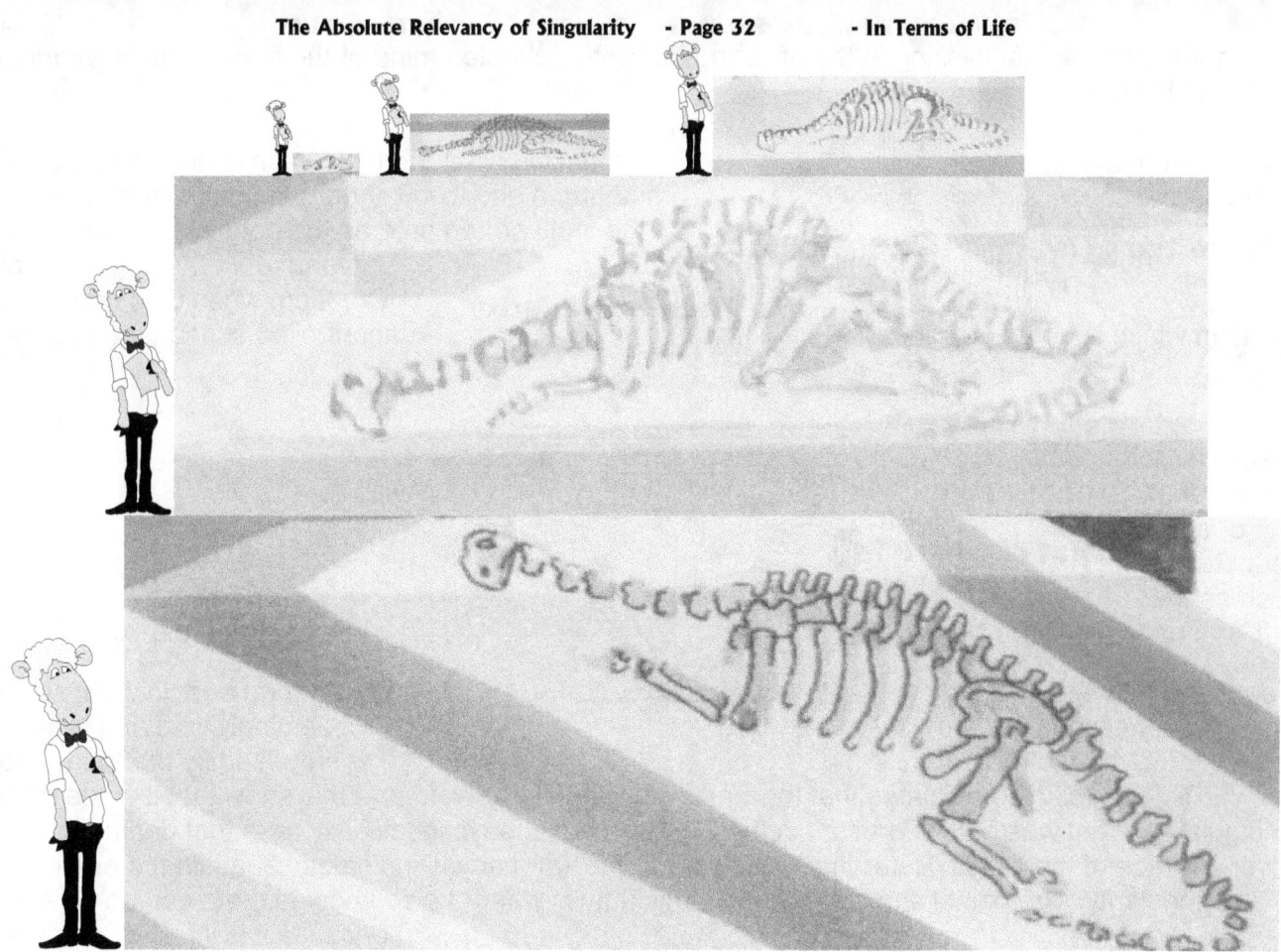

This is a part of **_Seven days of Creation volume 1-7_** or **_Matter's Time in Space: The Thesis Part seven volume 1 to 7_**.

Present date

1×10^2 **years back into the past**

1×10^3 **years back into the past**

1×10^4 **years back into the past**

1×10^5 **years back into the past**

1×10^6 **years back into the past**

The mentally inadequate nature of Newtonians even found a crab that is the length I believe of over eight meters. What this means is that the relevancy of the crab grew with the growing Earth from what it was to what it is and the eight meter crab did not stay the same since then and now we have a crab of eight meters that lived on Earth 265 millions circles around the Sun ago. That mentality is so Newtonian it makes me feel like puking. Is there not one Newtonian that can catch the hint that there is a huge snake in the grass, that their figures are no adding up, that whatever they read they are not reading that information correctly! Are all Newtonian minds so childishly gullible and void of thoughts? The truth is that we have an Earth that grew so much since the crab got fossilised, that an ordinary crab found today might be 10 cm in length but this specie got fossilized many moons ago and since then grew with the growing Earth and is by today's standards eight meters. That doesn't make the crab eight meters when it lived. It does not make the crab a relevant eight meters in relation to the Earth it once lived on. This does not mean the crab was back then 10 cm because species also grew but species grew by generation compilation. The crab probably was in today's terms 10 mm from end to end and the specie by generation growth extended in volumetric size to what it now is. I have wondered on so many occasions what it would take to get Hollywood seeking sensation hunger out of TV science and replace the dramatization factor with a little bit of common sense that will sperm intellectual thinking in the amazing brilliantly schooled mind of the intellectual Newtonian Brainy Bunch.

We start again started because I am about to show how the atom could come about in by looking what the bible says about how creation the multi proton sate as the density of such atoms require. In the beginning of Creation, when God made the uncontrolled space-time and the

There the Π^2 is

heaven which is Π^2

There singularity Π^0 is,

That is where is Π,

Earth which is material or solid substance such as the Earth uses the Earth was without form and void, which means the heavens too was without form and void but since the heavens still are seemingly void and without form, the Bible is not surprised by that part. The Bible specifically indicates that the Earth or material was void and this show that it came from singularity. Light was not yet present in the Universe because it specifically says that darkness was over the face of the abyss. Referring specifically to the word abyss indicates yet again the presence of singularity. The point I showed holding singularity is where I also showed there can be no start because it is infinitely present.

By this it is clear that even referring to a force applying is directly suggesting motion occurring or a tendency of a serious effort to bring about motion restrained that then is a blocking of motion occurring and that makes that mass is the restricting of motion trying to continue in a specific direction to come about. If one takes away the motion or the tendency to bring about motion then it is clear that gravity disappears, and only when gravity disappears would mass or weight relent. This we see happens when aircraft fly or balloons take off.

The Black Hole contributes the strongest gravity since the Black Hole places all motion in space and no motion in the star. In fact, the Black Hole has returned all atomic particles back to singularity or on the rim thereof. Nevertheless, the motion we see comes as a response to a point where that whole ending of space-time centre on singularity which in essence is the result or the product of the motionlessness of the star. Since all stars apply motion and the Black Hole reveals the ultimate form of motion, therefore the Black Hole shows that gravity is space in motion by the reducing of space towards the centre where there is space less ness and motionlessness. In the invisible centre is a point even beyond where space and motion is at its least. Without space, there is no motion and without motion, there is no space. Kepler tells us this as $k^0 = a^3/T^2k$. But that is what the Bible has been saying for thousands of years and the Mindless Newtonian atheists are too stupid to follow the Bible.

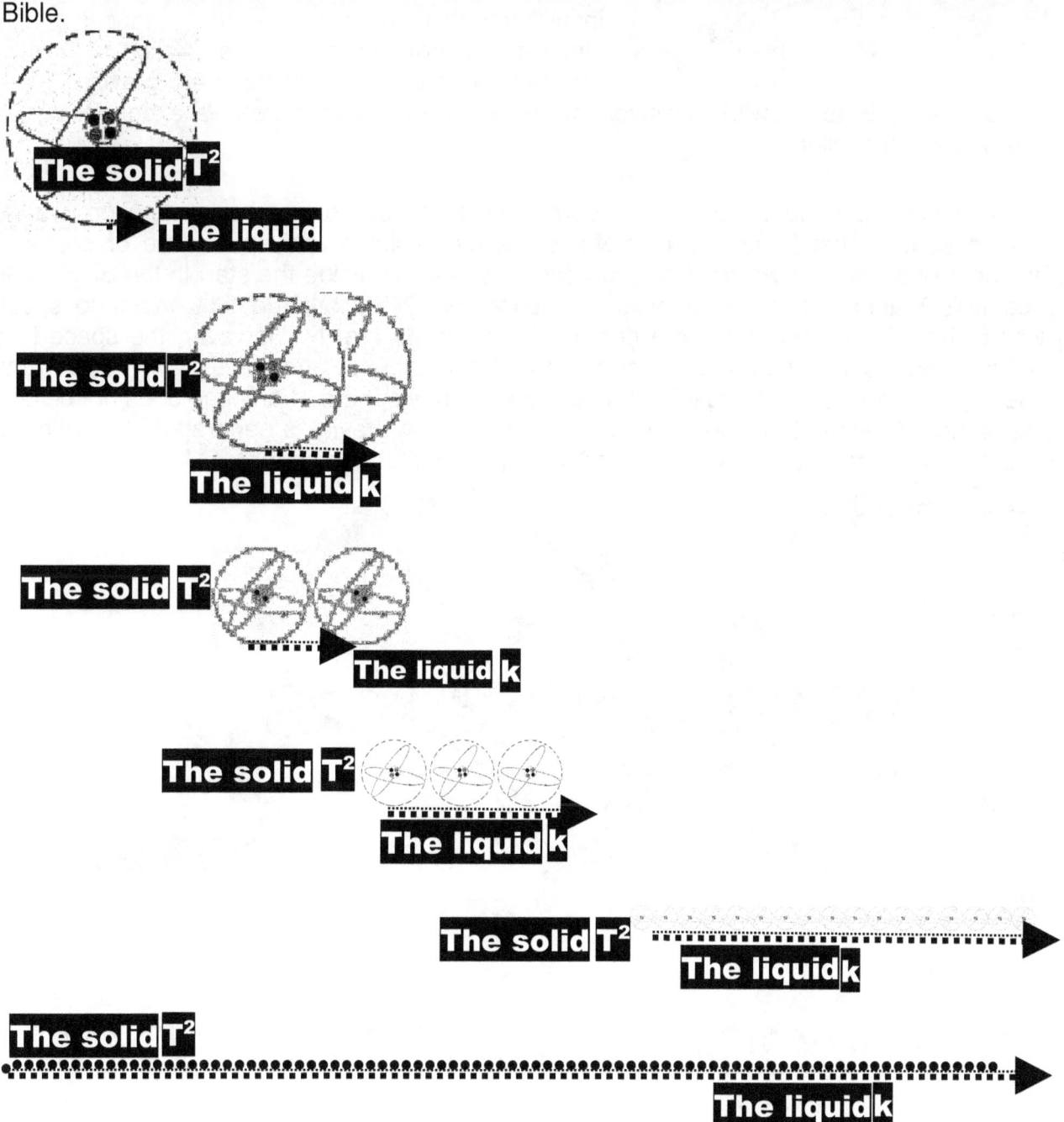

Galileo said all fall equal notwithstanding mass…Newton said mass does all the pulling and Galileo said mass has no influence on the drawing of the object…yet everyone is totally ignoring the fact that mass, be it big or small, proves not to influence the drawing of material that is falling! If Galileo is correct something else than mass is doing the pulling…and for the life of me I can get no Scientist to see what I say in the contexts that I say what I see. The concept of mass being the producing factor in gravity comes across as rather less thought through and more than a bit silly when considering the above.

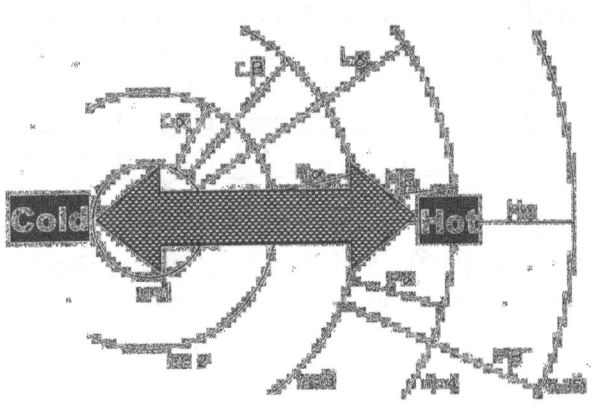
All evidence points directly to the idea that mass can not generate gravity and such presumption that mass can not generate gravity is most inaccurate. Mass becomes a factor only when the Earth restricts any object that is falling, further movement by blocking the movement as the Earth then is "in the way of further descending". The gravity part remains present in the fact that the body still show a strong tendency to move down but with mass applying the Earth restrict or counteract this movement by disallowing further movement. It is completely incorrect to think that it is mass that produces gravity since it is a notable fact that as space shrinks or reduces the stars ability to generate gravity excels. However, space has little to do with massiveness because the mass increases exponentially as space used within a star declines.

If space is infinitely small, then time is infinitely slow. There can be no space when motion is at its slowest possible speed. That forms the ultimate relevancy available and space-time or space in motion is all about relevancy. However, the enormous gravity falls outside the star. In the Black Hole singularity controls matter and space applies all motion that is in fact the time factor to space occupied where the motion aspect is more commonly known as gravity. However, the space less ness of the Black Hole shows that space less ness is the location of strongest gravity. It is in the place that the heat is the most, which is in that centre area of any sphere. If any one does not believe me then test nature. It means that mass has the least say when gravity is generated. According to Kepler, mass in motion within space in motion and gravity is the same thing. $a^3 = T^2 k$.

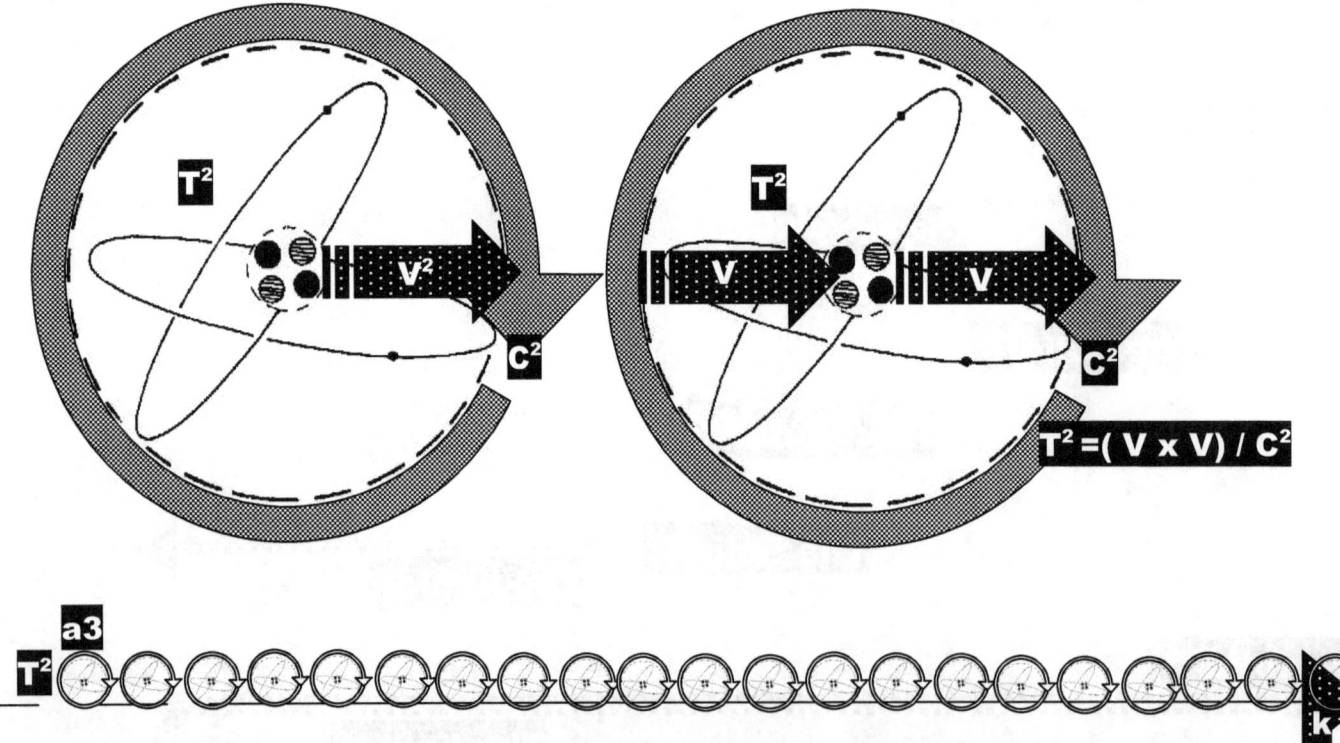

I am not getting in a debate as to relevancies applying and how that operates but the atom applies movement according to a differentiation between what is hot in terms of singularity uncontained and what is cold in terms of singularity confined and contained. This movement is not perpetual but is very precisely defined. What singularity would use as a gradient as to determine the cold to hot differentiation is what brings about a star classification. Time slows down as space decreases. What would this entail if space (a^3) were at a premium then movement ($T^2 k$) is equal to space. The bigger the atom is the shorter would the duration of time be.

In the formula Kepler left us who I might add is the very formula he received from the Universe after studying the Universe he place space occupied equal to movement. That is the Coanda effect. That is in mathematical equation as follows: $a^3 = T^2 k$. The space occupied holds a specific relevancy to the movement applying to the space and in that time determines movement as much as movement brings about time. Newtonians declare that time is

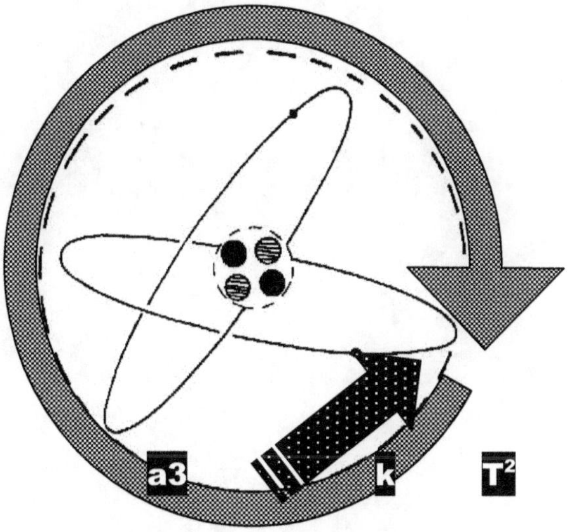

$$t = \sqrt{1 - \left(\frac{V^2}{C^2}\right)}$$ and say no it is not time...but also yes it is time. Let me define what I try to say: Time can never stand still. Gravity is movement and gravity institute time as performing as the movement of everything in relation to one point in infinity. But we know that the movement is in the atom while we think of as space that changes the ultimate relevancy. Why it turns out to be space and not the atom moving is a big issue and going down the road of explaining why this inverse of movement actually forms the change of the relevancy I leave to **an Open Letter Announcing Gravity's Recipe.**

Explaining how this really works takes up the best of three hundred pages and I best leave the explaining in detail to remain part of the book entitled **an Open Letter Announcing Gravity's Recipe.** I mention this in order not to seem to contradict myself in another book.

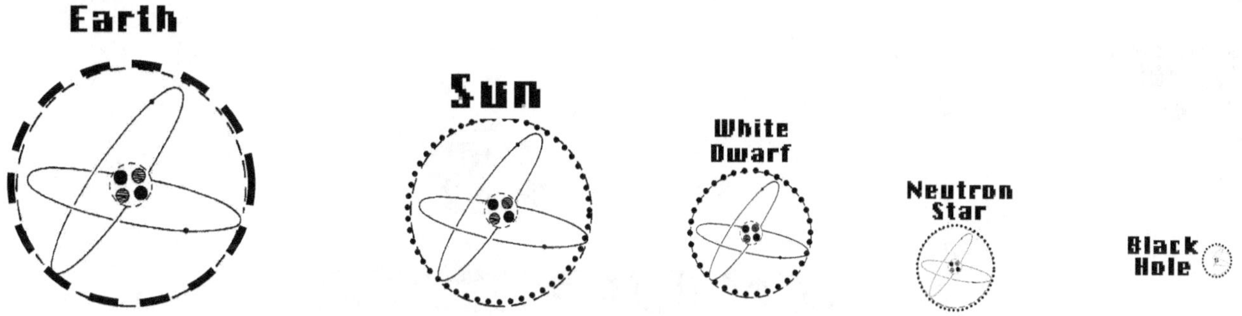

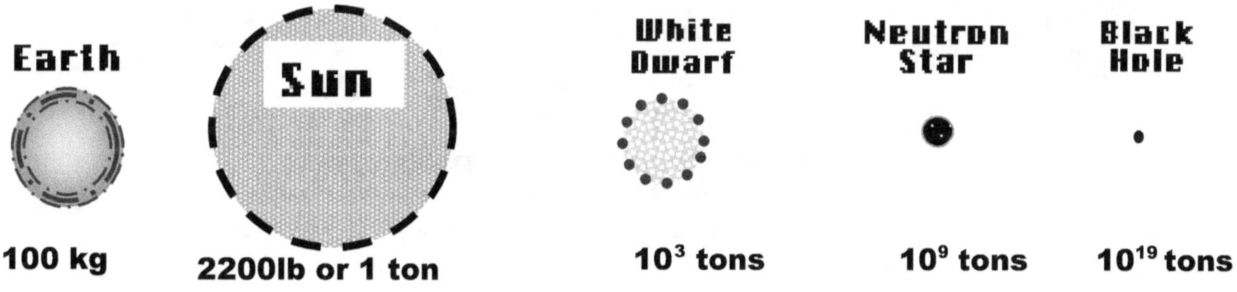

Earth — 100 kg
Sun — 2200lb or 1 ton
White Dwarf — 10^3 tons
Neutron Star — 10^9 tons
Black Hole — 10^{19} tons

The atom is moving. The straight-line movement as the atom moves around the Earth's axis is V gong on to V, which leaves a V^2. The atom is moving in a straight line (circling around the Earth's axis) as well as rotating at the speed of light around the atom's axis by way of the electron spinning. The electron is circling around the circumference of the atom at a rate of C^2. That puts the atom at the value of C, which is the speed of light, which is a standard unit, not the same everywhere but a unit bringing some equilibrium in relation to singularity applying. The atom is always moving except when it becomes part of the Black Hole and space becomes abolished as the entire component inside the Black Hole goes singular (1^0). When the atom stands still as it does inside the Black hole, then movement is transferred to that which never moves in relation to the atom and that is space. Oh, so many smart names were given to the Black Hole in as much as the event horizon and the curvature of space-time and... God only knows what name comes next, but underlining all the naming is one fact: No Newtonian has a vague clue as to what is going on inside the Black Hole except that gravity has gone bananas and where gravity would be intellectual enough to go mentally skew remains an open question!

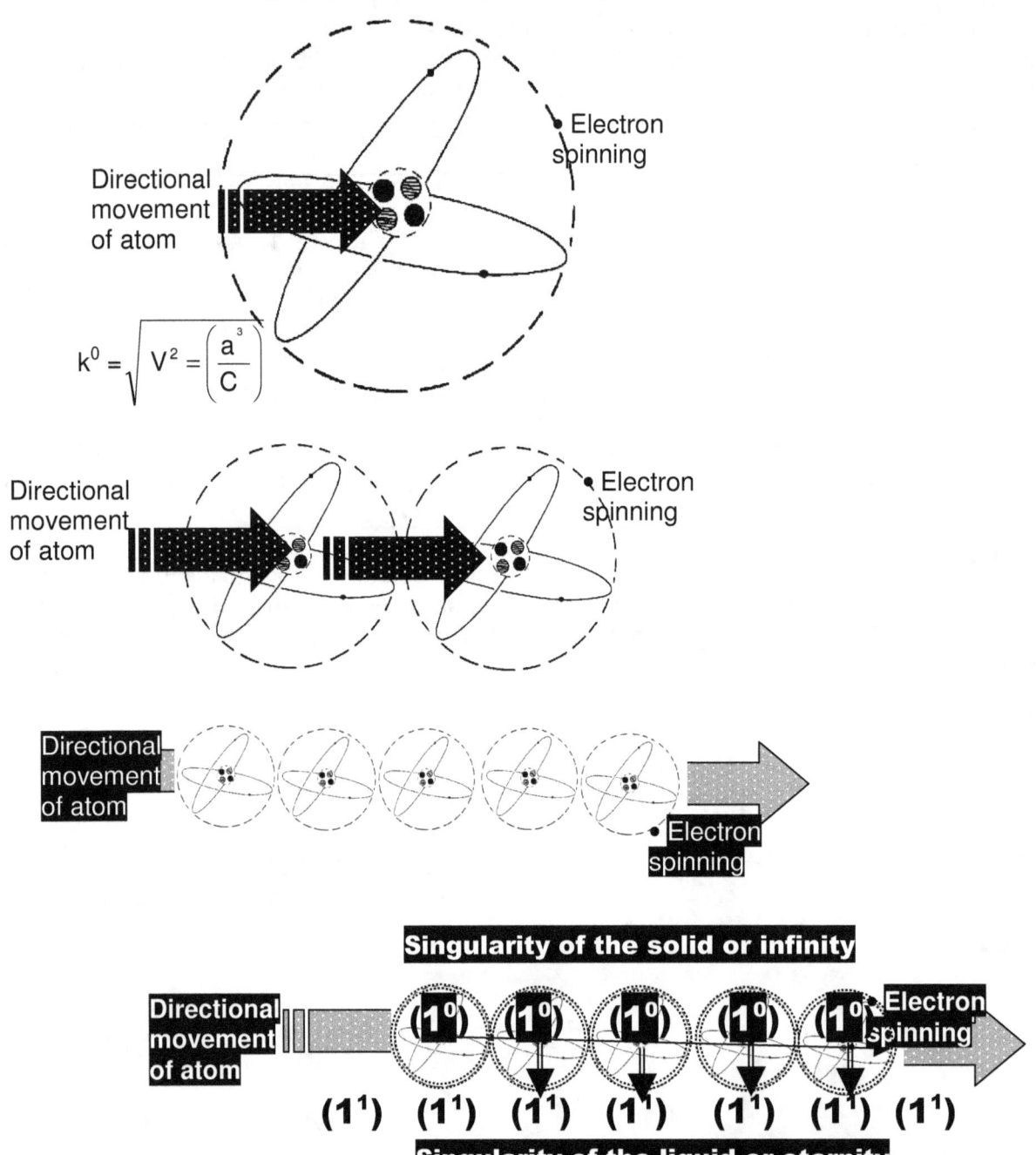

In every atom through the Coanda effect, there is a relevancy applying between cosmic fluid and cosmic solid or heaven and Earth if we go back to Biblical terminology. In every star, it is not the mass as in numbers or weight that sets the margin to gravity but the duplication tempo of material. When a tire of a car spins fast it can have the car run on a layer of water that is an inch or 25 mm thick.

This puts time forming space in a relevancy and the fact of time forming a duplication of space is the reason why the duration of time depends on space forming. Time is not a constant but holds relevancy by movement in space through time. That means time is not a perception as Einstein stated when he said a man sitting in a park next to a girl…Time depends on the movement of the body in relation to the space through which the body moves. With life directly linked to time as life is presumed as time that makes the time life moves through not a perception but depending on movement in time by a body moving through space. Life is time and life uses space during time. That is the function of having a body for a short period. If science are unable to show comprehension about the cosmos and the way the cosmos forms, how can science think they know even the basic about science. With that in mind then life will truly baffle their brains taken in consideration they truly only see nothing with which they try to build a Universe.

They fail to see how gravity works. They fail to see the water is not drawn onto the tire, but is compresses solid. If the water was normal ice meaning it was normal water in a solid state, the ice would have had to be several meters thick to sustain the car. But with the Coanda effect applying, with the velocity of the wheel that high the spinning wheel duplicates it space of the wheel so many times that it freezes the water onto the edge of the tire. The atom within the tire becomes smaller and

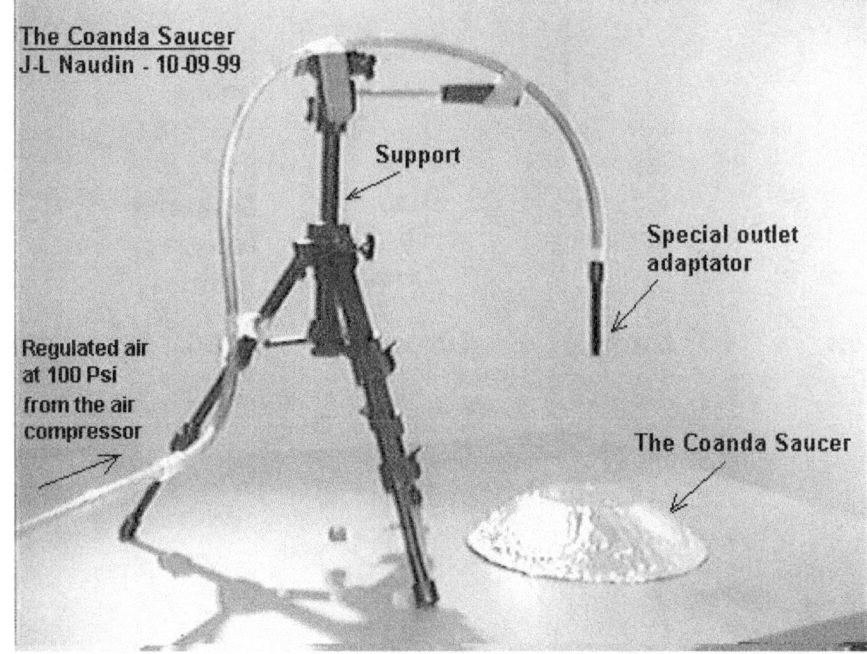

that causes the gravity because the number in quantifiable acting atoms increases although the reality in numbers does not increase. In large stars it is the numbers as well as the reduction that freezes space taken up by particles into smaller areas and by the duplicating of material therefore material occupying space cools and if the cooling is excessive, then it freezes by cooling the liquid into the star. Gravity is a process of heating or cooling. That is why blowing hot air into a balloon brings on the lifting of material into the air, which is anti-gravity applying. Cooling of air is gravity and anti-gravity is heating air. There is a specific ratio between water and air and the Coanda effect experiments conducted proves this. But furthermore, it proves that the Coanda effect is the way gravity works. It proves that gravity is the Coanda effect.

There is a direct link; no moreover, there is a direct ratio between what fluid is and what has for. There is a direct ratio between what came about as that holding no form and that which gives singularity form by which singularity developed space-time. This proves what gravity is and Newton's mass misconception has not role to play because by increasing the ratio in air moving that mass effect that the Earth restricts the saucer is the mass of the saucer reclining.

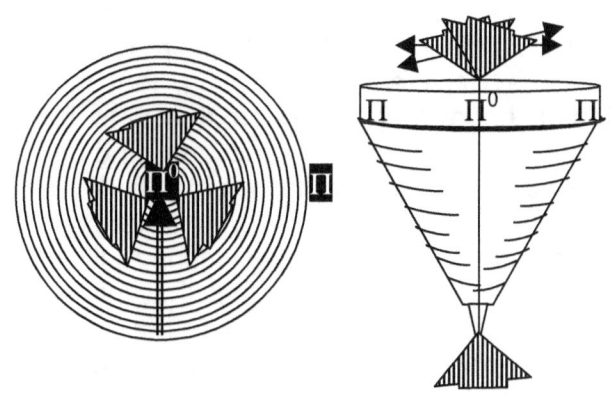

Within the centre of the top there is infinity, a point being present holding no space to the point it can reduce no more. That concept science never realised and with that being so pertinent and so in their faces, yet they missed it, therefore is it truly surprising they lack the ability to understand God and God given life. At the end of the ridge of the top singularity forms eternity as a space that is never ending and that too, science had no ability to realise…then how can science claim they know about science and furthermore it is very clear why given their stupidity, the lot fails to realise God.

Science fail to see God but that is not surprising since science fail to see science. The building block that the entire Universe are maid of science degrades to being formed as nothing and by degrading singularity to a point of nothing, science do not degrade God to nothing but science degrades their ability to follow the true concept that forms science and therefore the Universe. By understanding nothing they fail to realise any position of what could present God and by the Universe being built from something that could never be in the Universe yet it forms the Universe science could never comprehend God because science never could understand the Universe. I do not say god is singularity but singularity represents the form God chose to form all and everything within the Universe.

Time within the Black Hole stands still because eternity consolidated space with eternity making time being infinity as well as eternity. Even this part, which God says represents a part of God, they fail because God says he is eternity through all eternity.

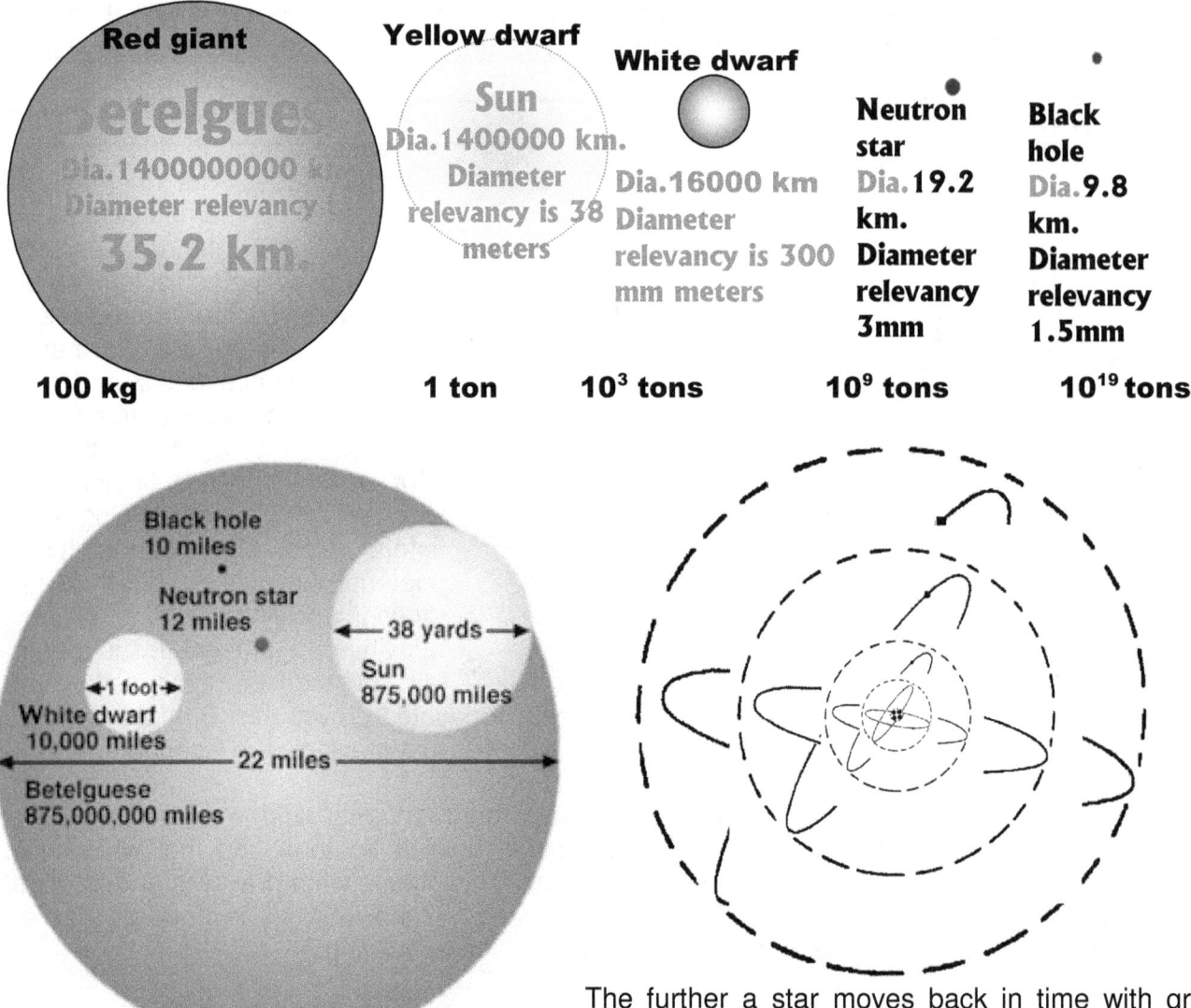

The further a star moves back in time with gravity applying a relevant movement, the more the atomic particles would combine to form the increased gravity while occupying in much reduced space. As much as the Universe is expanding, that much is material within stars also shrinking in time flow. The Moon is not moving away from the Earth, but the Earth and the Moon is growing apart. This has to do with movement and relevancy applying on both sides of singularity. As much as infinity is motionless, we have eternity moving perpetually. If eternity did not move, then infinity could not be motionless because it is the eternal movement eternity holds that keep infinity motionless. Although this is true, with infinity never moving, infinity still has a duration and relevancy of twice that of eternity because notwithstanding the fact that infinity can never start, it has to start and end eternity since eternity can never end, and can therefore also never start. Eternity needs infinity to start and eternity needs infinity to stop since eternity can do neither of the two. For that reason alone infinity has to be twice that of which eternity constitutes because only by ending eternity which is what we think of as the past, can eternity start with a new frame which is what we think of as the future. Therefore the only limit placed on eternity in duration, is how quick it is that infinity will end the one frame and start eternity with the next frame. For that reason we have eternity only holding one frame whereas infinity takes the ending of eternity as one frame and replace eternities ending with the next frame while it draws time into infinity as it draws the Universe into singularity in infinity. That Absolute what Einstein saw but it clearly is not what Einstein saw because the entire Universe we see is time that has gone to the past where time formed space as the past. Once anything is a Part of the Universe, it can never "not" be a part of the Universe, except if the relevancy changes and in such changing the uniting with singularity places the relevancy back with unifying with singularity once more.

Therefore infinity that never moves, also moves twice as fast as eternity that never stops moving. Time draws flat...not the Universe as Einstein said because what Einstein thought formed the Universe is what forms the time aspect of space-time. In the centre of every proton time draws space flat to form the past from what is the future while space-time is forming the neutron where it is the neutron that holds 3D and one Universe in form.

Time draws the Universe flat because the Universe starts where time meets space to form space-time. This happens where singularity ends with infinity starting that starts with eternity replacing what was with what will be and is therefore as much where infinity ends eternity. When infinity ends eternity it holds what eternity was at the ending and the ending of eternity becomes space where space forms the history of time. In every atom singularity lurks as the final control of what applies to space-time and every atom holds singularity directly in association with singularity where one part of singularity is that which never moves forming the relevance with the second part of singularity, which is that which never stop moving. As infinity starts eternity, it pushes the future eternity becomes into forming space that is the past of time. Taking time from the future through singularity to the past it does by taking that which serves the future to the past, which is material. By taking light to form the future of what the past will be light becomes the eternal messenger of space or for that matter, of the past. But the past is also in the future of what eternity takes to the past because as much as I see with my eyes what the past (space) was, I form the future for that space to reach.

Eventually eternity and infinity, the past and the future, is one unit and only a Newtonian (I guess) would wish to place a difference between the relevancy applying because there is no difference in infinity and eternity, except that which can not be, which is a Universe we have because between the two that can never part (infinity and eternity) forms that which becomes the relevancy of what is the same thing and therefore the Universe can never be. By movement of that which can never stop moving, such relevancy places what is the future in time to form the space in the past of what is that moves the entire Universe into the past. It is immensely important to realise the worth of singularity because singularity holds life. The Universe is made up of something that is never part of the Universe but forms the Universe because of space that forms as a relevancy in relation to one point that can't move and therefore can't while it is in place. Life positions within singularity within the atom and uses the same ratio connection that stars use to form a unit. As stars do, so does life accumulate the worth of singularity within every atom sharing the body in space by forming a unit and this life applies to conduct movement that we think of as life. The growth we experience as the expanding of the Universe is the flow of time that remains behind as space and this is what becomes the aging process but also the healing process. If we look at stars we look at how life uses the cosmos.

It is for this very reason that in every star singularity hold a Black Hole captured where the Black Hole becomes the monster in waiting to end space in infinity. The star uses atoms to form the space-time singularity presents as much as singularity presents the space-time to develop the star. Movement is the duplicating of material in relation to the relevancy of liquid holding a moving ratio. The atom connects through the electron all space-time forming to the speed of light, which places all movement in relevancy in some way. Singularity on the other end of the spectrum forms a uniform time component, which also brings a time relevancy. It is not a constant forming for even on Earth time has no constant. It is not an equilibrium forming for even on Earth gravity holds no equilibrium in any two places on Earth. The space that forms a^3 depends on the electron circling T^2 at the rate of singularity k^0 duplicating k space-time $k^0 = a^3 / T^2 k$. Therefore $T^2 = a^3 / k$ and $k = a^3 / T^2$.

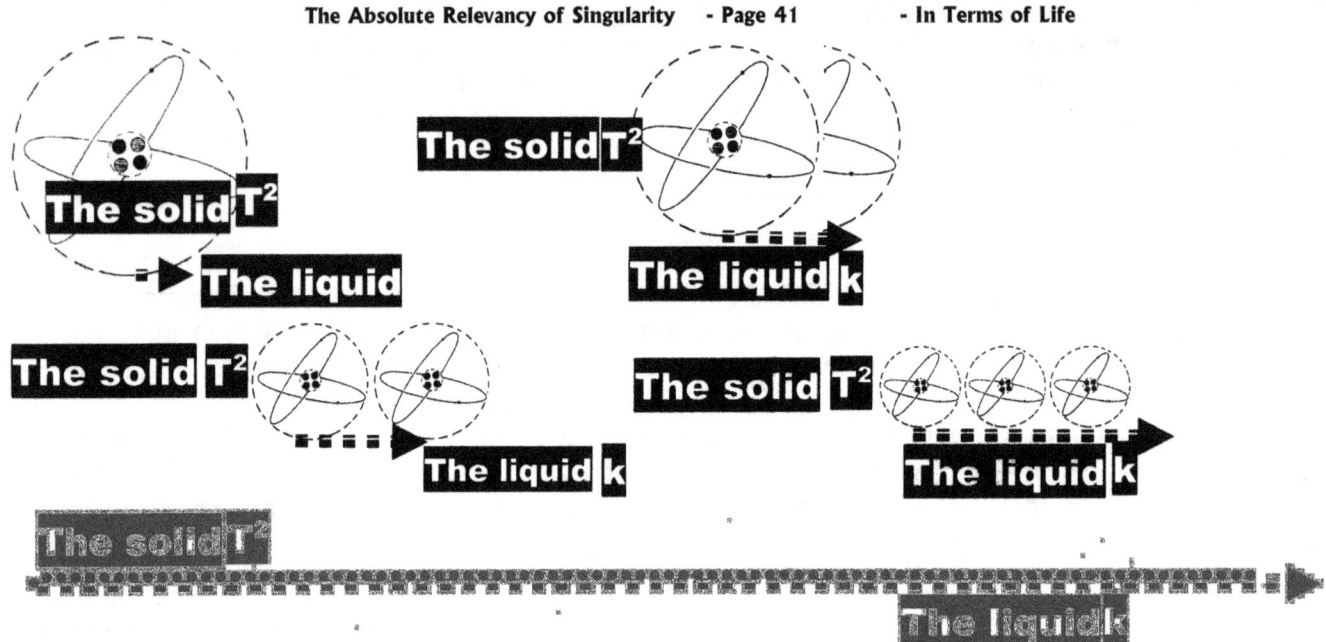

Time in duration is determined by space moving in relation to time.

In order to look at space-time we have to look at the practical implication such movement would have on the occupied space-time that an atom requires. When a body enters the atmosphere from outer space, we are told by Newtonians that it is the friction that rub against the body, which heats the body so much, that the body starts burning. We have all seen a body coming in from outer space, albeit a sand particle or whether it is Mir coming in that we see images of on TV or whether it is the space shuttle getting hot on entry, we always witness a blanket of heat engulfing the incoming object. The oh so wise Brainy Bunch Super Educated tell us in confidence brought on by wisdom that those flames are the result of particles within the atmosphere that is supposedly rubbing against the incoming object and this will allow friction to cause such heat. If it was rubbing and such rubbing resulted in friction, I would guarantee those oh, so clever Newtonians that the entire structure would melt into liquid and fry whatever is inside the body to ashes.

Again I wish to return to the formula Newtonians devote to time being $t = \sqrt{1 - \left(\dfrac{V^2}{C^2}\right)}$. As I said before this could never represent time because time has to be in the square since time holds space to relevancy and that is always going to put time in the square by movement as well as in rotation. When we get to the speed of light it is C that forms the linear aspect and I do not wish to go into explaining that at this venture because it is going to take up too much space.

By moving the atom it gives the electron a specific duration in time to spin around the atom at a velocity of C. This velocity is fixed at the speed of light and can't under circumstances change at will, although circumstances might change and with that the implication of the speed of light might be different in gravity changing from star to star. However, it takes the electron a specific time to rotate the atom while the atom is moving at a specific rate.

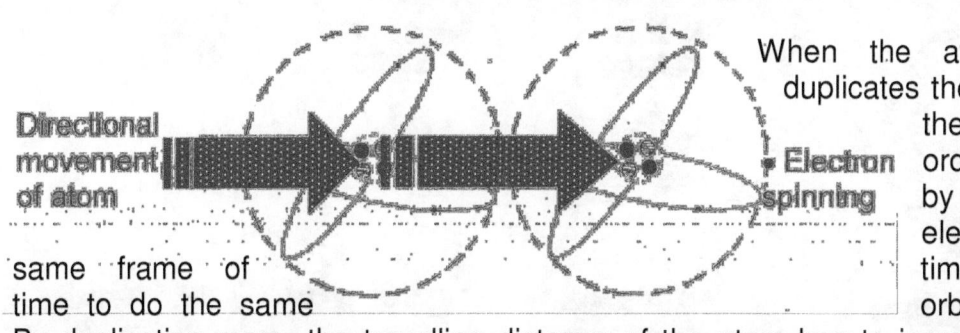

When the atom moves in position it duplicates the entire atom structure. When the atom duplicates more in order to cover a larger distance by moving relatively faster, the electron still has to travel in the same frame of time to do the same time. The electron receives less orbit.

By duplicating more, the travelling distance of the atom has to increase. If the travelling distance increases, this has to effect the time it takes the electron to orbit the atom. Einstein might have been of the opinion that the speed of light is time but if that were true then the speed of light would not have a quantifiable measured ratio of velocity with the space through which it travels to the value of 1^0.

Time stands related to space and only if space was one in terms of time moving through space can time not be effected by space. If the velocity increases by two because the movement of the atom doubles, it would effectively have to reduce the time the electron has which it takes the electron to circle around the atom. If the electron orbit reduces because of time by movement increases and this shortens the time that the electron has to complete one orbit, this will reduce the size of the atom. Since gravity is the movement of space, this is why gravity will reduce or increase the atom's size in relation to the star in which it is.

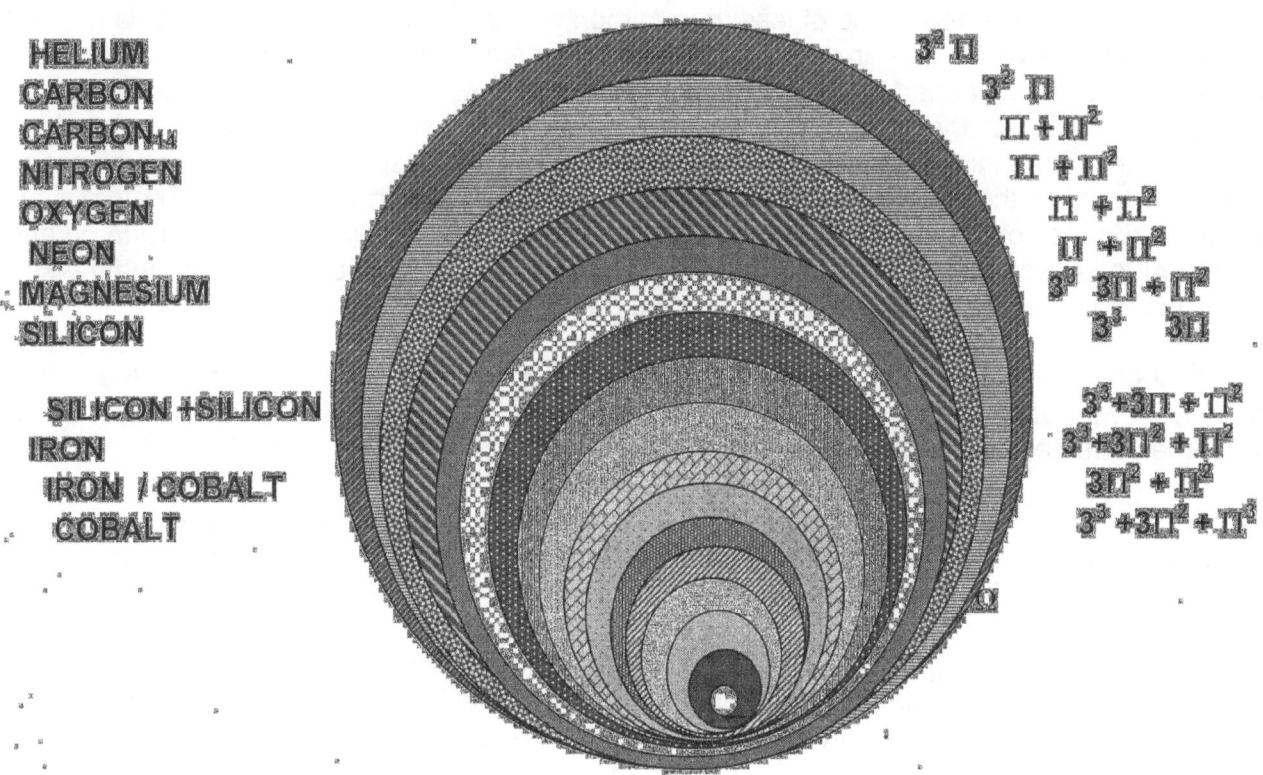

The sketch I show I am not even going to try and explain because it come from one part of the book I named **Matter's Time In Space: The Thesis**. In that part of the book consisting of seven parts I show that as much as every star is the next star in waiting that is in a process of developing it is reducing space as much as time is expanding space. The star reduces while it holds material the size the star occupy reduces as the material in numbers increases to anther exponential level because as space expanding and therefore space is contracting by the same margin, every layer in every star is a star in place to develop the star.

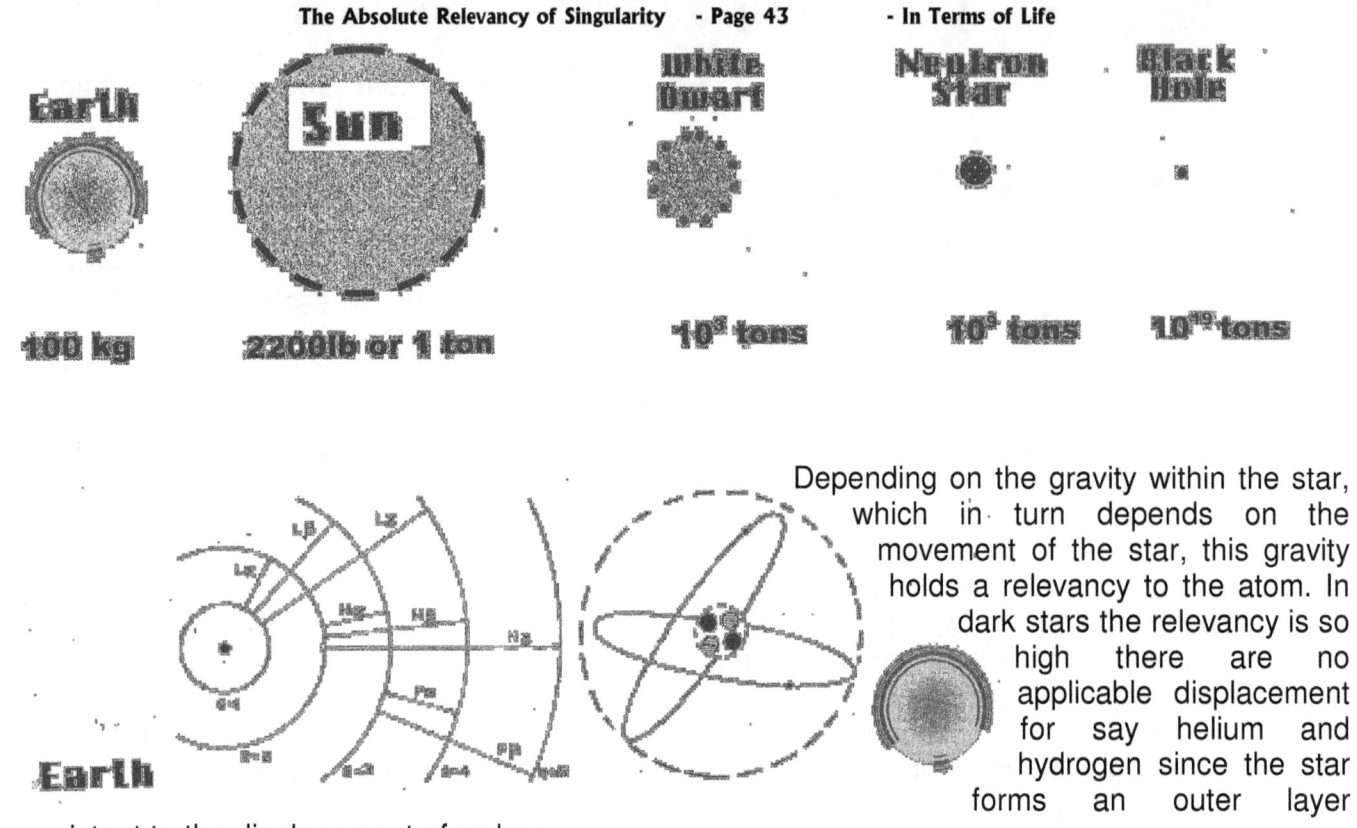

Depending on the gravity within the star, which in turn depends on the movement of the star, this gravity holds a relevancy to the atom. In dark stars the relevancy is so high there are no applicable displacement for say helium and hydrogen since the star forms an outer layer consistent to the displacement of carbon.

Every atom with a displacement value of less than carbon would measure as just more liquid heat tat offers no solid substance. In a star such as the Sun, the movement is still progressive and the hydrogen forms a substantial part of the star since hydrogen allows for volatility to condense gas to liquid. The more substantial the gravity of the star becomes the less space it affords the atom to hold because it freezes the atomic material into a solid state and in some cases there is not enough space left in the atom to allow the liquid neutron to claim space as forming part of the atom. The gravity is so strong in condenses outer space into a more compact liquid that the neutron can afford to be and hence, the neutrons escape into outer space to play a part in later cosmic development and the star freezes gas into an almost solid form the neutron cant be.

There is a specific predetermined relation between the relevancy of liquid and heat and the interaction that every atom plays in this relationship between solids and liquids makes the layer of the star part of the stars development process.

On Earth an atom holds a relevancy of $(\Pi^2+\Pi^2)(\Pi 2\Pi)3) = 1836$, but this relevancy changes as the atom holds a position in more developed gravity fields such as Jupiter and bigger planets as well as in bigger starts as Jupiter is. This places the atom as forming the Universe.

Only Earth with a very feeble gravity would be able to accommodate life because in other cases the gravity will kill life before life could manipulate the atom in order to use it as a host. That also means that the inter galactic space flight Hollywood so feverishly propagate is as much fiction as much as Newton's story about mass pulling mass. If the Sun span fifty percent faster, life on Earth would not have been possible even if everything else remained the same. If the Milky Way span less or more, the earth would most likely not have been suitable for life to manipulate carbon.

There is a link holding the entire Universe quantifiably united; it is the relevancy there is in singularity. Infinity 1^0) stands in regard to eternity (1^0) notwithstanding the space it seems to use from the perspective we have. When the space is less, the duration in time in space being in ratio with time will increase. The time it takes to hold the relevancy in place is time dependent in relation to space depending on time. This comes about in the link every atom holds with C as well as the measured value in infinity in which the atom constitutes.

Betelgeuse
Red giant

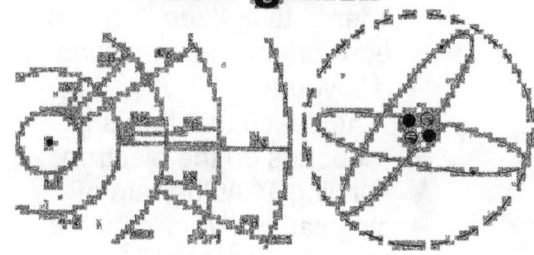

Dia. 1400000000 km Diameter or the comparable relevancy is **35.2 km.**

Yellow dwarf

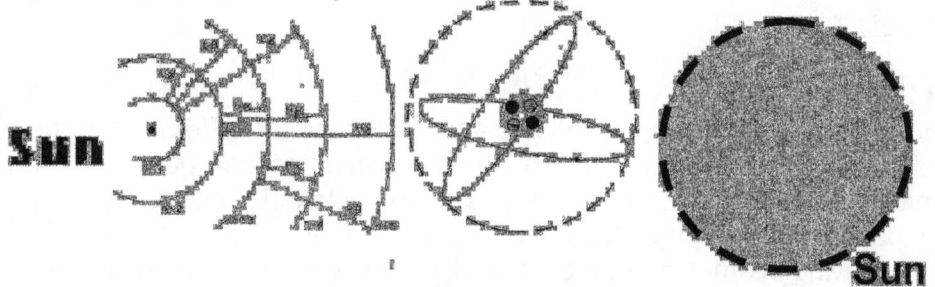

Dia. 1400000 km. Diameter or the comparable relevancy is 38 meters.

White Dwarf

Dia. 16000 km Diameter or the comparable relevancy is 300 mm meters

Neutron Star

 Neutron star

Dia. 19.2 km. Diameter or the comparable relevancy is 3 mm

Black Hole

Black hole

Dia. 9.8 km. Diameter or the comparable relevancy is 1.5 mm

Earth 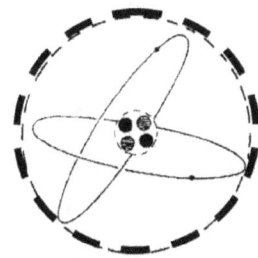   Sun    White Dwarf    Neutron Star    Black Hole

The solid ─────────────────────────── **The liquid** ──────────▶

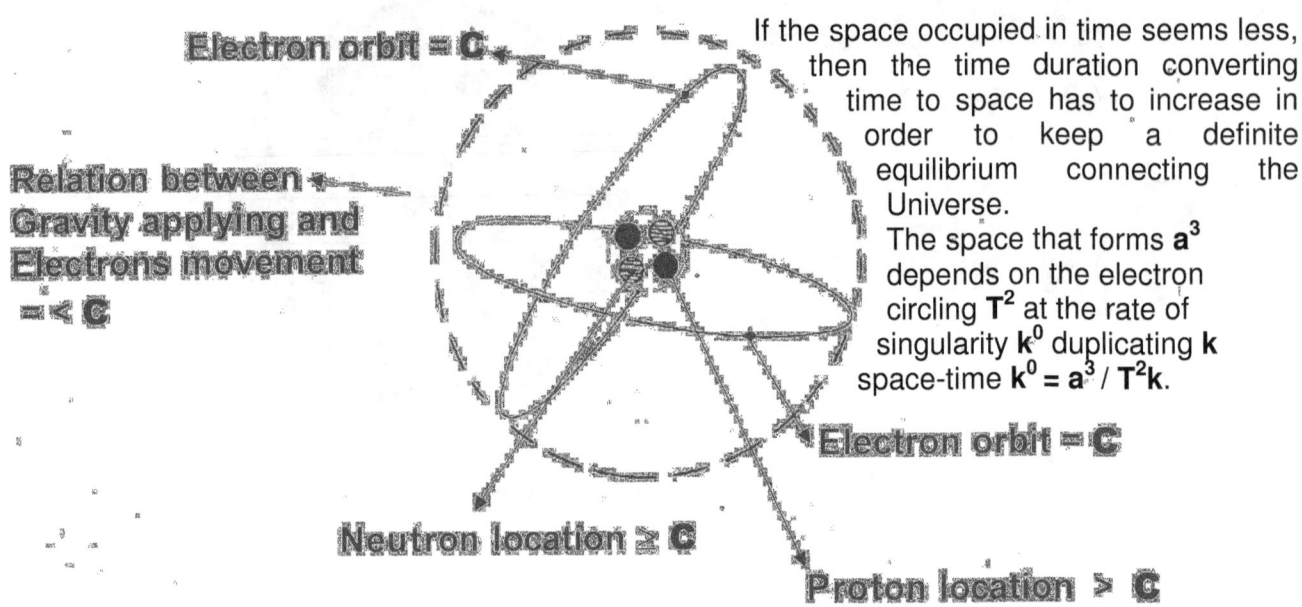

Therefore $T^2 = a^3 / k$ and $k = a^3 / T^2$.

IS THERE DEATH AFTER LIFE?

This article is as much about proving what energy is, as it is about knowing the difference to the state in which a person is that is enjoying life and the state of health in which a dead person is. Newton considered all forms of energy to be the same, and oh boy, was he mistaken. This of course is what all atheists consider to form part of the groundwork that is supporting their belief in atheism. Atheism is the point where everyone believing in atheism then believes life ends completely after death comes to take the person nowhere. That is part of my attack on atheist. Atheists believe they are so cleverly smart while holding an argument that shows how idiotically small they're insight is into reality that they are unable to see their folly. There is a worldwide fashion amongst the very well educated that in order to be regarded by those with the know how that to be a supremely informed person, one must at least be an atheist. The key to find recognition in the field of science is apparently to be completely atheistic. Atheists do not believe in the life after death or a Creator. They do however believe in a Force that pulls by mass without any one ever proving the justification of such a force except that Newton said so. The force they accept does not exist outside the technological criteria of mathematical science where they believe that mass pulls mass. Yet they believe in gravity and no one up to now could tell where gravity is. How smart are they about their self-inflicted wisdom as they only are experts about nothing at all! They are experts on the fact that nothing is a worth that one can measure in something forming a distance between planets. That they believe while that we all know is insane. Are they, them and those in for a rude and brutal awakening since all creation is ruled from what is outside the Universe by something that is unexplainable from the inside of the Universe while it is forming the outside of the Universe? But they think in their eternal wisdom that everything that does exist only exists because it exists in their perceptibility. Any force that might form outside this norm of space value is quite unthinkable and that thought could never present it as to be present in the material universe they measure in units of nothing. The ironic part of this fact is that those well-educated scientists have only one source of information and that is light waves. They think they know what they see. Still they permit themselves to be atheists in their blind state of science ignorance. I do not condemn them, because they apparently know more than I will ever know. However, because I am not that knowledgeable in the field of nothing forming measurable quantities, I must feel my way through the tunnel of ignorant darkness like a blind person. In doing that, I stumbled across a heap of questions that cannot be answered by those who carry the flame of knowledge. That forced me to form my own theories, think and come up with sensible conclusions, which answers all those questions their light of knowledge could not answer.

The explaining of the cosmos must first begin with an explaining about life. To understand the distant one must first familiarize one self the closest aspects and in general move from where we are to where we wish to go. Through culture predating mathematical culture by billons of years we humans see ourselves as bodies of flesh and bone walking on Earth. Our bodies have organs muscles and above and beyond all a brain. Man is so over preoccupied with finding "other life" or life forms in the cosmos that experience has taught me when ever there is so much money going into the senseless search for other cosmic life, with so much eagerness forming the background of this total nonsense, then someone of importance is cooking a stew that holds an evil smell for others of minor importance. But how important is life to the cosmos. I do realize being part of life myself that life to man goes beyond importance and although being part of life I have no idea why life in the cosmos holds that much importance. Why would other life be more important to find than what life is being on Earth? Why would life on another galactica be that important to others except to diminish the God aspect that Created life on Earth?

There is a definition applying to life that does not apply to the cosmos at large and sometimes I get an eerie feeling I am the only one with such a point of view. I can walk to a rock. That is a statement and not a joke. The rock cannot walk to me. I can lift the rock but the rock cannot lift me without having the support of other life intervening and using such a rock as aid to lift me. By itself the rock cannot lift me with all the will in the world. I can take many rocks and build a house. Many rocks cannot build me. All that which I have just said is disregarded by science as science claims there is only one form of energy. I can wish to move and move by myself but the rock cannot even wish by itself. The rock can only move when the Earth aids such movement. I can move with or without the aid of the Earth's movement. By applying some brainpower I can even fly. The rock cannot even move unaided. Water can move unaided and so can wind move on own accord but by doing that they perform in the limits of gravity using the Coanda effect. It is gravity moving water and wind and not water and wind moving. But water and wind cannot move by own willpower. They depend on cosmic gravity to move and cosmic gravity is totally different from the life the body holding life has. There are two forms of life outside the cosmos. The one can form movement by opportunity applying gravity and the other can form movement by applying free will. The plant has a seed holding life and where the seed land it may grow should conditions favour such a growth. Animal life leaves seed injected into other animal of the same specie also having life and from that seed life can grow. But through out animal life everyone normally is free to go where conditions suit its needs best. The one being plant life can manipulate space in time by opportunity presenting the space allocated to promote such opportune occupying. But through all there can be motion but in the case of most plants the motion is in the air capturing space or converting wind to form the structural movement of the plant in question. Other motion than that mentioned by any plant is out of the question because plants generally are stationary and the position the plant is in is placed by nature outside the free will or choice of the plant. I have witnessed plants fighting each other for sunlight and therefore with that in mind plants do have intelligence to some extent by which they manipulate growth and therefore use free movement to their advantage. But the basis of all plant life is freedom in immovability. Plants attach their roots to the Earth and manipulate the air as far as movement goes and the soil as far as security goes. Life in the form of animals also manipulates air by occupying more space but does not apply attachment to the soil security by attaching roots. However, the occupying of space is a freedom issue. That is one aspect of what life is about. Life is to have the ability to claim space by moving and by generating growth with more occupying of space. This goes the same for plants and for animals. Life can take substance other than which was received before and accumulate from that through that. A body comes into this life as small as survival of the specie permits. Then there is interaction between the two different life forms where the seed of the one becomes the food of the other and the waste of the one becomes the want of the other. That is all to do with holding life in some carbon bases body.

It is what we do with what we received during our travel in time through cosmic space that will be of lasting value. The only thing I leave behind is the gene pool that my children have and if the genes are better than those I inherited, then my life may be deemed as being successful. If the life of my children proves to be better than the life I lived I can claim success. On the other hand, if what I leave behind that carries my genes are less than what I am notwithstanding whatever level of success I am awarded, my life was one miserable failure because my gene pool success stops with me.

Another question I love to ask people especially medical doctors are when does a person die. When is a person dead and when is the person no longer part of life? They all come with heart rate and brainwaves stopping and bullshit but the answer is the minute no one knows you any longer or can place you in memory or remember you from history then you are dead. Some persons being tramps roaming the field with no one to care about them is already dead although they still walk and breathe. Others are kept alive in thought and in loving memory for many years after they were taken to their graves.

Life is when an individual re-arrange the protons he has charge over through the gift of life in a certain manner in cosmic space and cosmic time and with that ability to control space-time persons are assumed to be alive. Through manipulating the space his body occupies to his need, he manipulates a specific part of space in time. Every moment that he occupies space, it is a different position that he holds to the rest of the cosmos. Never ever will he manage to occupy that very same position that he held the previous instant because the Earth is spinning, repositioning its position in relation to the Sun, the Sun is spinning repositioning its position in relation to the Milky Way and so forth. I am truly of the opinion that there is no future to tell or to forecast. There is no fortune telling. The future is the combination of the ability to move on with time and the wishes of every one sharing the next specific point awarded to that location during the very next second. Only what happens in that second can be realised by the players determining the outcome of positions of space-time filled and the future is only clear in the second to second developing of the future on that spot. One can see a flow of events leading to a future development. If I pump a balloon by constantly filling it wit air, the balloon will in future time burst. That is not fortune telling but applying logic. However, life does manipulate the working of the cosmos to further its needs and advance the position of individual life on all forms.

The entirety of the cosmos runs on the formula Kepler devised and therefore what life has and what life can manipulate to ensure a position within the cosmos has to come from what the cosmos provides. Take this in relation to Kepler's formula we then find the body holding life (a^3), which is in relation as viewed from individual part life has to manipulate to serve the functions of life (k) extending that control into movement (T^2).

If we put this in terms of singularity (Π^0), which I have shown is where life must be situated; we find the body serving life (Π^3) stands related by different parts (Π) that provided the body with motorised function (Π^2). That secures the three dimensional status life has (Π^3) in terms of a position or base (Π^0) that controls parts of such a body unity (Π) to control and manipulate (Π^2) the movement that secures life holding a space on Earth ($\Pi = \Pi^3 \div \Pi^2$) as space ($\Pi^1 = \Pi^2 \div \Pi^3$) in the cosmos (Π^0)=(Π^3)÷($\Pi\Pi^2$). Life applies space formed three dimensionally to position a point of control by being in infinity while controlling a body moving through eternity. With having the control of a cosmic aid thought as a body in place the manipulation uses the same process that cosmic movement applies in a manner where the relevance (Π) that forms in relation to the present (Π^0) will relate to movement (Π^2) and the movement is circular which ensures that the relevancy forming is circular (Π) by securing that the movement is circular (Π^2) in terms of one specific point (Π^0) in infinity which then secures a roundness (Π^3) that forms an everlasting eternity ($\Pi\Pi^2$) which validates an never ending circle. In this time in infinity (Π^0) secures that there is an everlasting eternity ($\Pi\Pi^2$) in space (Π^3). Every cell in thee body holds (Π^0) and with life controlling from a position in singularity (Π^0), the control life exerts on the body (Π^0)= (Π^0) is equal and this is done by using electricity ($\Pi^1 = \Pi^2 \div \Pi^3$) within the brain to control muscle movement ($\Pi = \Pi^3 \div \Pi^2$) by applying control (Π^2) = ($\Pi^3 \div \Pi$) over space-time (Π^3)=($\Pi\Pi^2$).

That secures the three dimensional status life has ($a^3 = T^2 k$) within the space the Earth provides to the body life control (a^3) within the Universe in terms of time (k^0) which give life a centre to manipulate from other life in a location (k) secured by movement (T^2) that will come by movement where the movement within the body ($k = a^3 \div T^2$ and $k^{-1} = T^2 \div a^3$) secures the control of life. That is life within time uses time in the process of space developing controls movement of space by manipulation and that is how life applies the growth of space-time that forms the Universe and that is the **_Absolute Relevancy of Singularity_**. The fact that a person can move can walk, can love and all that has all to do with the developing growth of space by time (k^0) forming space ($a^3 = T^2 k$) thought of as the Big Bang. Time moves and life being directly linked to gravity and intertwined with time finds the ability to move. If it were not for time moving ($k = a^3 \div T^2$ and $k^{-1} = T^2 \div a^3$) then life would not have been able to move. It is life that controls movement ($T^2 = a^3 \div k$) by manipulating space ($a^3 = T^2 k$) from singularity (k^0). That then forms time in the centre in infinity in relation to space in eternity in singularity where time that moves forms space by holding time that does not move secured in positions in relevance to where every point was in time gone by. Π **Divides** **infinity** from **eternity** where **infinity** can't **move** and **eternity** eternally moves as time.

The **governing singularity** (Π^0) or life holds a **positional validity** (Π^3) the body securing life within three dimensions in terms of any **relevance** (Π) or individual body components manipulated by movement as the **controlling singularity** (Π^2) and life uses the precise process that the cosmos applies as "the Hubble shift" and thus mathematically it equates to $\Pi^0 = \Pi^3 \div (\Pi \Pi^2)$.

If a **relevance** (Π) did not validate a **positional validity** (Π^3) securing a **governing singularity** (Π^0) in terms of movement formed by **the gravity** (Π^2) that produces the **controlling singularity** (Π^2), a three dimensional status, then space (Π^3) would not be obtained and thereby the Universe would not be secured. However, in this ($k = a^3 \div T^2$ and $k^{-1} = T^2 \div a^3$) as well as ($k^0 = a^3 \div T^2 k$) is seated personalities that I am going to explain later in the book as the book develops and information progresses. Life uses time to create space within which it finds an ability to manipulate movement.

<u>Time (Π^0) is life(Π^0), which is the movement (Π^2) of space(Π^3) in relation to (Π) any one centralised point not spinning</u> securing life within time flowing. Everything in the Universe moves in relation to any one single point that forms in any location that then has to stand still to form the centre of the Universe wherefrom that point must be motionlessness to allow everything else movement. In that manner the Universe is constructed and there is no valid solid Universe because the Universe is constructed from singularity (Π^0) that holds no valid space (Π^3) other than being in position (Π) while having gravity (Π^2) that forms the time (Π^2), which is also the movement (Π^2) of space (Π^3).

The flow of time being the present in singularity forms space by moving time in relation to space as much as relocating the present in terms of a past that is determined by the movement of time whereby that action of time moving by the same token is establishing space that confirms the past as it secures the future as time moves on to leave a positional legacy, a footprint of time gone by in the presentation of space. From this we can deduct that the Universe in a three-dimensional form starts at $7/10(\Pi^6) \div 6 = 112$, which is a value forming the start of the element table and that I explain in the Cosmic Code. In the Cosmic Code there are numerous values consisting of Π forming relevancy by which certain rules comply throughout the cosmos. One is 7/10 is the interaction of gravity spinning a sphere (Π^6) within a cube ($\div 6$) and that is how the cosmos forms using Π. In this I prove that for instance amongst so many other things that electricity and gravity is the same thing. This identifies the Universe wherein there is a tiny dot where life can be found. This is where life has to find a place from where it can allocate a position in this Universe and this is what life has to apply in order to be able to manipulate movement.

At this point I can introduce my theory on the ***Absolute Relevancy of Singularity***. Where is the position retaining life? At the point in the centre of the circle a line must start and the line forms a circle only when the circle rotates and in order to rotate the centre point that can't move must therefore reposition the entire circle because the centre point can't move. I am going to explain how singularity links and this way is the very precise way how life finds a way to connect to every cell within the body serving life. By linking singularity, control comes through communication and communication is electricity flowing. Life is not electricity but generates electricity to connect and to communicate with cells holding life and by that process life finds a way to manipulate a body as to move the body in space through time moving space. In the beginning when I explained the way I figured how the line starts, I said a lot of dots has to continue in order to form a line. It would be 1 + 1 + 1 etc. because the line must form by holding singularity. After that point does mathematics begin but in the line that forms representing space as other all factors, then time holds 1. The line can only form when all the points forming the line have the value of 1 being 1^0. In that conclusion one realises something must separate singularity from all other factors because singularity hosts all other factors but is by own initiative Π^0. Only when singularity meets the end value can the end value have Π where the final ring of the spinning circle forms Π. That will be the spot of origin forming the relevance in Π. That will hold the eternal spot…the smallest spot ever because all spots that ever can be were secured in a position in the centre of that spot that must continue as a line that forms. Because of the progress singularity follows from the single dimension singularity only allows mathematics a start at Π^0 progressing further onto Π^0 and from there the line is born as $\Pi^0\Pi^0\Pi^0$ and to $\Pi^0\Pi^0\Pi^0\,\Pi^0$ etc. where Π^0 then may form the concept and value of r. But the line starts at $\Pi^0 = r^0$. This forms because cosmology is singularity based and the value is $\Pi\Pi^0$. This line $\Pi^0\Pi^0\Pi^0$ of singularity can only continue because every spinning atom preserves Π^0 in the very centre and since $\Pi^0 = \Pi^0 = \Pi^0$ the line is the same without finding conclusion except at the end where it forms mass at Π. At the point where Π forms, the movement Π^2 of the circle defines the space Π^3 of the circle and it confirms the centre Π^0 of the circle through the rotation. Let's call this the solid forming or if you wish, let's call it Kepler's singularity. After that singularity forms a line $\Pi^0 = \Pi^0 = \Pi^0$ where this forms another line again as Newton stipulated it by $\frac{dJ}{dt} = 1^0$. Let's call that the liquid singularity or Newton's singularity and the relevance of singularity having a solid base compared to the singularity holding a liquid base comes about by the movement of gravity.

From these conclusions I prove that gravity is the result of four cosmic phenomena interacting to form the value of Π which by movement becomes the value of gravity Π^2 and gravity is equal to cosmic time applying. In order to understand the development of the cosmos and moreover the start of the cosmos and the progress in the cosmos as the cosmos formed, one has to understand the measure of Π. One has to see that Π is not merely 22 over 7 or that Π is a ratio that no one ever bothered to clarify, but Π is the key that unlocks every lock that hides a secret in the Universe. One has to microscopically dissect the measure of Π to find the cosmos in measure. One has to understand where 7 fit in Π. The fact that Π is 7 at the bottom and that 7 relates to a double value of 10 is a key issue. Furthermore, it is very important to see why Π is 10 times two by adding 1.991 on the top part of the equation. In this measured value is what holds the building blocks of the entirety we call the Universe. It is behind Π that we will find the four phenomena, which I named the four pillars performing as gravity as they form gravity. It is by the actions of Π that the Universe develops. The Hubble expanding goes by implementing gravity as Π in the square through the four pillars on which gravity and time rests. It is behind Π we discover the meaning of singularity and how singularity forms the absolute and only building block as a form that forms the Universe. It is in Π we find the Cosmic Code unlocking the meaning of the Universe. Time is centralised in Π^0 that forms Π as space's limit that becomes space by gravity being Π^2.

Space is time gone to the past in which time confirms its presence it had in the cosmos by moving from the present time into space and then onto the future leaving space behind as the past. By forming a present, time is in infinity forming singularity that then has to move on and in doing so it leaves a legacy behind being space. Time is the movement of everything forming the Universe where in time the movement of time relocates everything in space by moving from the present onto the past leaving behind space. As time becomes the past by going to the future it forms space as it confirms the past, and in that space is what time forms by going to the past leaving space behind. Space becomes what time was at the point where time formed the particular space in relation to Π. As time becomes the present coming from the past, time has to move on to the future at the same time and as time moved on it left space that represents that instant in time in relation to other space that was in some position at a specific location at such a point in time wherever that point in relevancy might be. The fact of Π not only refers to form but also validates the Universe by splitting infinity from eternity. By forming space when creating Π, time is using Π^0 in establishing movement Π^2. It is in the process of relocating Π to new positions by establishing Π^2 and connecting this as it forms a network consisting of Π^o by forming space Π^3 in relation Π that establishes infinity Π^o that always stays motionless. If not for movement, the Universe would be one line holding time by repeating singularity Π^o uninterrupted and it is in the diverting of eternity to a position away from infinity that the Universe comes about. This is what happens in a Black Hole where no movement within the Black Hole places eternity that always moves in a standing position to infinity that never moves. Without movement the entire Universe will fall back into and onto one point and everything we thought is real and solid will disappear into that one point holding infinity onto eternity where infinity and eternity then reunites. The Universe is an unreal concept with nothing being a reality but for the movement whereby Π confirms everything in a location in relevancy to all other things in a specific time slot or space.

When I, as a person forms a part of the Earth by the virtue of having mass that connects me Π to the Earth Π^2, stands on the EarthΠ^3, my position in relation to the Earth gives me a specific positional relation to time Π^0 and the Earth. That gives the Moon a future of say one point five seconds being the past in relation to the Earth and that gives the Earth a past in reference to the Moon's future of one point five seconds. Where I am at any specific point in the present, that point I am holding is that which secures my present point in time. The Sun is eight and a half minutes into my past with all the space being in between the Earth and the Sun and by my view of the Sun I have a present time slot, as it also gives me a past of eight and a half minutes in relation to the Sun since the light travelled eight and a half minutes through space to confirm my past during that present instant. That secures my past by eight and a half minutes at the point of giving me a present location in time. However, that also secures my future I have from the point I now have in the present by the margin of eight and a half minutes because that establishes a flow of light that would last another eight and a half minutes of filling a presence worth eight and a half minutes while travelling through space by moving with time and every spot filled on the way would secure a position that I will have in a future presence for the next eight and a half minutes, which then becomes my future as it fills my past. Looking at this scenario in a view from Alfa Centauri the allocated position Alfa Centauri holds in space relating to the Earth, gives the Earth a past of say four point six years while this secures the present and having that present secure the Earth to a future of say four point six years by forming time as space between Alfa Centauri and the Earth and this is confirming time to the tune of four point six years. By securing movement it forms time in having a past in relation to the present that by the same margin also secures a future in relation to a definite past. This is how the Universe builds space in establishing time. This applies to all allocated positions of rotating objects throughout the Universe. This means that every point away from Π^o serving as Π, wherever that might be, secures my past I have by giving me a future in terms of the present Π^o. This then forms the way that life manipulates space to establish movement and that is done not by the body but by thought controlling the body through which life established validity in movement. It is thought that moves the body and the mind controls space by controlling a body holding space in the flow of time. But if the Universe did not "expand" or grow by forming an input of singularity as time movers on, movement would not be possible and the cosmos would be static and life within the cosmos would be static with no means or measure to progress. See the image of a person thinking about his future?

Not only does he manipulate the space his body occupy, but with thought he is able to rearrange other protons in his immediate surroundings and by movement of the protons he control in his body. Even in having a thought, he still maintains an irrecoverable position in space during the time he is having that very thought.

The Earth might provide his following location in relation to the macro but in the free will that life allows him to manipulate he holds the ability to direct his next position within his direct confinement where he normally has the absolute say in where he will be. The movement requires a ratio of space (a^3) going through time (T^2) by the distance (k). By having singularity refresh relevancy of movement in association ($k^0 = a^3 \div T^2 k$), the next position that singularity (k^0) has' also puts life in context of the new allocated position (k) of space (a^3) in time movement (T^2). That is life: life is the ability to apply motion to cosmic structure life claims control over where the cosmic objects already in cosmic motion receives additional motion from life on top and beyond the motion of space-time. Life is the ability to control; and manipulate space-time, securing (for a brief while) his own destiny (from one instant to the next). With the blessing of life, life within a person is able to manipulate the future position he will hold, but not only that, that person is able to plan the future arrangement of other proton positions and have the way and the will to occupy space-time in future. No one can sit and predict the future because there is no future to predict when bringing life into the frame. The Sun will have the planets in orbit, but that I might be at a place tomorrow is highly speculative and most probable at best. That is because what man controls is feeble given the cosmic picture and the surrounding can alter the destiny of the person to the extent that within less than an instant life can lose all control of the planned future he thought he had. This unpredictability does not extend to the cosmos but only applies to life having some allocated place within the cosmos. With life every aspect may change and such a change will have a knock on effect eternally because it will change positions in the one hour leading into the next that the day following has any chance to be anything and all possibilities are available to come out. This shows that life is an adaptation of cosmic events and life being unpredictable has no part in the cosmos but only has a rental space in one specific cosmic place at any single time. Life is not part of the cosmos and only idiotic atheists with no sense of reason or connection to reality would try to place life within the cosmos as just another cosmic factor.

Man control destiny by thought…it might sound also dramatic but little is truer. The next thought that comes to mind is the person's destiny, which that person will follow. If it is within his limited capabilities he will manipulate space surrounding his thought to establish the future to the way his thought find a need to manipulate the space going with time to the future. There are no rite or wrong, but only future paths with consequences derived from any choice to the person and to others that share space-time with such a person on Earth.

It is what such a person thinks that determines the future of the person where the future of that person is created by thought. Only at the point where he seizes to find the ability to supply motion to the cells under his control above and beyond the motion cosmic space provides then only will he seize to live or then die. When the blood odes no longer flow giving the body life-support, not oxygen, does the life fail to find the body sustainable in control thereof. Life is not electricity and neither is it heat. It is a drive compiled of energy that is not part of the cosmos and not found anywhere else in the Universe notwithstanding the misinformation atheists try to instigate. Life is in carbon linked to singularity finding gravitational feasibility on the Earth.

Newtonians whish to place life all over the Universe but fail in their ability to explain life other that performing in the way electricity will perform. On what grounds do they wish to distribute life if they are unable even to declare the most, most primitive basic function of life? Once motion is no longer part of a cell life has left the cell.

Every instant of the Earth's spinning through time and in doing so, relocate to a certain position, it is relocating the persons position to time in space as well.

There is no geodesic null in space, in time. At the speed of light time through space to cover whatever distance there is between whatever objects there is. Once the body loses the ability to locate the next position of the body by choice provided by life, will life no longer be part of the cell or be part of the Universe Life is motion of cells in motion within space-time.

By the way since stars are part of the cosmos therefore stars can never have a free choice in movement but stars are on a set path leading to become Black holes. That is why the star moves from one minute to the next minute as time drives the star through eternity where eternity will one day match infinity. This then rubbishes Newtonian science, which atheistic came to preach, proves once again to be incorrect in their assessment that stars are able to die. Only that which has the ability to apply motion free and independent of cosmic space-time motion but in sequence with that of the cosmos can a body lay a claim of life on life. When a person is in control of the vehicle he sees as his body can he claim to have life as long as such a body is attached to a larger object and by finding security in the motion of the larger object can such a person control the body he claims to be him. Every atom holds in its innermost circle a waiting potential Black Hole. The cosmos will not allow death while we connect the ability of life to lose control over cosmic material as a process we associate with death. Death is where life loses the ability to manipulate the body by establishing movement while it is the gravitational movement driven by singularity that drives the star and the star are forever unable to stand still. Everything that is in the Universe is part of the Universe because it moves. The Universe is formed by applying singularity which as a single substance that cannot be part of the Universe. Only when singularity by duplication moves by a line forming that spins, does singularity forms space. When singularity does not spin, the Universe will fall into one spot uniting infinity and eternity and that then is not part of the Universe at all. How the hell can that then die, except in the misunderstanding of the mind of the Newtonian atheist. Life is the substance that has the ability to accumulate material in response to needs and requirements in order to maintain manipulation. However, that simple not is not either. In the Universe no other cosmic enterprise holds such ability.

The body not only holds more space as time rolls on, but time by input is not only determined by space occupied alone. Time has a much more prominent role to play. The most important device connecting the cosmos in its entire is also that which is what Newtonians altogether misses while they hold blindly on to their vision with mass. Newtonians with their unbelievable stupidity doesn't even know what time is while time is what keeps us alive. Time is life. Time is what I have to live with and to live in. Time is what I use to keep me motoring by life-energy on Earth. Time is life. When my time runs out, life itself runs out.

$R^3 / T^2 = 1$ $\qquad$ $2R^3 / T^2 = 1$ $\qquad$ $4R^3 / T^2 = 1$

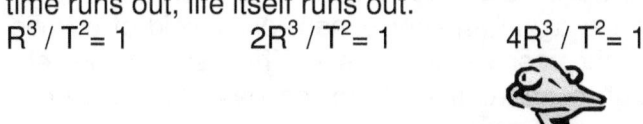

A bird grows by occupying more space. Its relation to time will influence the space-time value affecting the chicken in occupying space. That process we call growing (up?). However, the reasoning is rather complicated and intense in explaining I shall leave it at for now until I have done a little more explaining.

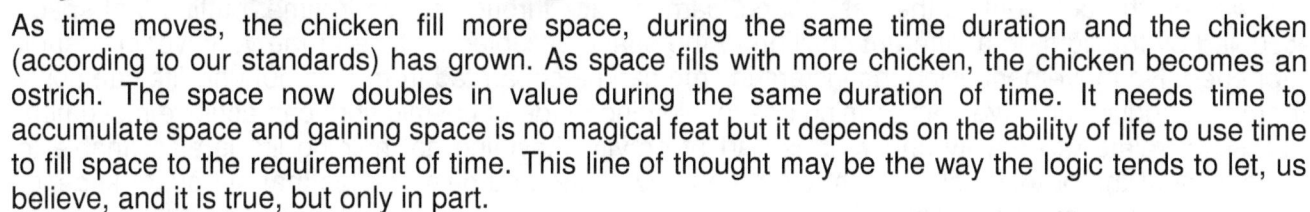

As time moves, the chicken fill more space, during the same time duration and the chicken (according to our standards) has grown. As space fills with more chicken, the chicken becomes an ostrich. The space now doubles in value during the same duration of time. It needs time to accumulate space and gaining space is no magical feat but it depends on the ability of life to use time to fill space to the requirement of time. This line of thought may be the way the logic tends to let, us believe, and it is true, but only in part.

I can feel the anticipation of everyone that is waiting for the climax, as all of this is a build up to the climax coming. This cannot be the climax because this, which was just said, is general knowledge and every one knows about it! To them I ask: does science know about it. Why does science confuse life with the cosmos? Why are life considered by science to be part of the cosmos as something found everywhere at a dime a dozen…and this is sustained without bringing the most miniscule bit of evidence to support such a mindless and baseless presumption? Do not fool yourself. With more than eighty percent of Scientists being atheists also, Science do not know about any difference in an energy called life which has the ability to manipulate space in time by choice and the cosmos that works on an unseen cog of gears. They are unable to see any difference between the energy used by life and the energy that other factors in the cosmos has to generate motion through gravity. Above all, can those atheists not see that one energy form being in life is totally incompatible with another energy form driving the cosmos at large? To those and to them whom has lost the ability to realize that energy is forever I say start thinking and stop feeling superior about your insight. Stop looking at what you know and start thinking about what you don't know! The rock has not the energy of life therefore the rock has not the ability to manipulate space-time but the body life holds, has the means to manipulate the space and the time on Earth as long as it holds in the carbon structure the energy of life. Life can be but life also cannot be and once life cannot be then the body left vacant of life and the rock has the same abilities. If other forms of life does take charge of the corps and treat it as waste by devouring the structure, the structure will mummify and become the same as the rock is where this is fossilisation. The body is part of the cosmos, and remains part of the cosmos even to the point of total destruction where not one atom finds a location next to an atom with which it once formed a body holding life. I have most likely one or two atoms in my body, which Julius Caesar once had. But I sure as life itself do not have the life in me which Julius Caesar once had. In that is the difference between the atheists with no insight and being without such basic understanding and being one that has the ability to believe in a divine power being outside the Universe. The mentality of the atheist is the same as the animal it shares the Earth with. Me, whom are a believer in the Greatest

Power of an Absolute Creator have the ability to see there is a big difference between the body holding life and the body not holding life. And to those that say one can shock movement into muscles and create life I wish to say: wake up, you can manipulator the muscle while life is part of the muscle because one of the functions given to life not attached is the ability to generate low currencies of electricity as part of the deal through which it finds the ability to manipulate space-time. The life energy itself may part with the body but can never part with existing because energy and life in that is indestructible.

Let us use a tale to try indicating time in future present and past. Let us also with the same explaining show how life can influence its surroundings and even its future by manipulating the space-time it occupies and intends to occupy. But there is space it does not occupy which it also will manipulate by mere thought of free will. Life occupies space in time, with that no one can disagree. Time is the immediate, the present, and the position of unawareness. It is where singularity (k^0) moves (T^2k) space (a^3). Even the thoughts one have, and which we consider beyond time, is in time in space. However, I must admit the space we refer too is once again another dimension devoid of cosmic space through which light travels and therefore not in the cosmos we can see. The instant I can see, that which I see is an image carried by light through space coming from the past. The present I cannot see because the light is only leaving at that point and therefore being away has not yet reached me. I am the future of such light and such light is bringing me images of factors in the cosmos gone by. The distance and the time it reflects on is a relevancy I wish to create. The fact is that in the instance time forms in singularity light becomes a reality as space and in space light becomes involved in space through which light is travelling in space. The light in space also come in as a part and combined that then is history...How much history, well that depends on personal interpretation. The very, very, instant light becomes part of the picture, time has moved on, and that part we see through light, is the past space, carried back through space in time, putting that space occupied by the light, in a different position to the space occupied by time at the very, very instant it happened. No movement using the instant in time can be separated from time but only life can apply control as additional movement forming space. Nothing made of cosmic fibre can achieve movement that is not related to gravity and wind is part of gravity. Anything not holding life is also unable to move without the granting of freedom by the energy of life, and best of all is that our super-superior atheist never got that far in thinking!

Man can plan to manipulate his surroundings and his future by thought but also man is very dangerous through thought, making man a disaster more to himself but just as much to his surrounding...the surrounding which I may add has not received the ability to be or to be somewhere else by choice of the free will. The surroundings have to take what man thinks and for that reason man is destroying this planet at a grand scale.

Present

A man realizes an idea at the one instant. The space-time he occupied during the time duration he had the thought, vanishes with light to the past, and the thought is stored by means of light or electrical brain impulses, which is light because it is heat, somewhere in his brain. Even the storing time, is no longer part of time because light carried the massage through space to another part of the brain and that fibre is in space standing apart from the instant where thought takes place. The position the man held in the space during the time of the thought is carried simultaneously by light, to the greater cosmos, away from the space he held in time during the thought. The reflection on the thought through the body carries through space by electricity charge, which is heat, and heat is also light. The moment the massage carries it is history repeating thought within my body. My arm might be a part of me, but my arm is not I. My arm is an attaching part holding life on my behalf as part of my command and under my control but that life is not part of my life. I can do fine without the life in my arm although it will amount to being uncomfortable. The fact that my eyes translate a massage it receives in the form of light to the form of electricity makes my eyes merely another life supplied to my benefit but not being I. Being blind does not culminate in my dying so my eyes might be fibre with life under my control, but although it is part of me it is not I. I am a very small part of what I presume I am when I presume I am my body. I cannot become separated from me by time or by space and that is a reality. When I send a message to any part of my body that message crosses space and by crossing space it becomes apart from me while forming a part of me, but it can't be me with space between me and the other point. This is exceptional important to understand. There is general life and then there am me, the person I think of as me being in the place where I generate thought, emotion and a will to command the rest of my body.

No man has yet come to realised the manner in which man establish thought in the very explicit specific and in the very detail that which changes him and her from me. Although I have tried in many occasions to come to realise just what it is that may allow me to think of something and to allow someone else very different thought, but I have not achieved an answer or even a hint of a conclusion.

This person now got his idea, judge it as brilliant through emotion mostly. Then the thought helps him conclude an idea. The moment of realization of the idea, the idea was already in the past, carried by light to another part of his brain, which is his memory. (It is not HE; but now stowed in his memory, because he and his memory separated by space in time.) In the simultaneous event, the space he occupied during the time he realized the idea disappears into space, carried by light. The space went to the past tense of the position he held during the realizing of the idea and the idea went to the past tense he held to the position his body held in the duration of that time. Both occurrences went to the past at the speed of light, however, the speed of light is not equal to time, because it takes time for light to move through space in time. Time is a much higher dimension than the speed of light occupies, because light carries the value or image of time gone by.

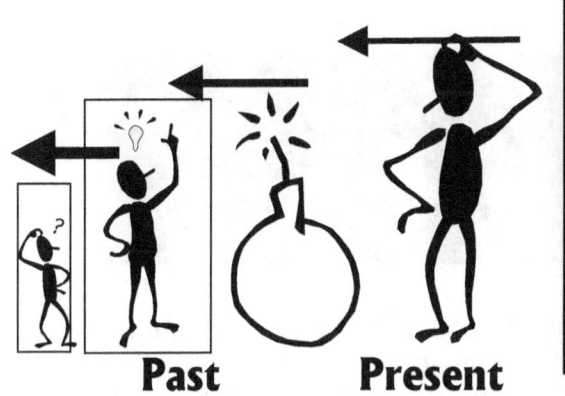

Thought is mostly a process where the one mans genius is the other mans foolhardiness. There are no rules to say which should be where. I am totally and utterly against war, but warring is as necessary to man's survival as his breathing is. What can be worst than raping a woman, but at one time rape was as much part of every culture as was cannibalism and back then cannibalism and rape became the spoils of war, the achievers achievement and a treat to success.

In realizing the situation of a threat, the man contemplates about the problem in hand. If the man's body was in the same time-instant as he was, and his thought process was in the same time-instant as his body was, no time in space was needed to decide on a form of action, as he then was able to move his thought and mind in the same duration while moving his body in time occupying the space at time present instant. As his body is not in the same time –instant as his thoughts are, it takes time occupying space to establish the manipulation of space-time in order to adjust to the situation in hand and have his body respond to his immediate thought. It will take time occupying space in duration of a specific value of space-time to manipulate occupied and unoccupied space-time held by his body which forms the densified space-time to establish a new cosmic relation between matter in occupation of space-time and space-time not in occupation of densified space-time. In more understandable interpretation of the English language, it will take time a period while the watch is running to access the situation and then decide where to run too and on top of that let his body know about the decision he took. If his mind and body were in the same time instant as his realization, he could move all of his controlled space-time the very instant of realizing the problem he faces.

Past   **Present**

In realizing the situation of a threat, the man contemplates the problem in hand. If the man's body was in the same time-instant as he was, and his thought process was in the same time-instant as he was, no time in space was needed to decide on a form of action, as he then was able to move his thought and mind in the same duration in time occupying the space at time present. If time was the speed of light, he could command his muscles to take action at precisely the same instant that the man realised what was happening. All things would happen at the same moment during the same instant with no parting of time by frame, if time was equal to the speed of light and time could stand still in the fraction. But light uses time frame by frame to move and therefore one frame is without light and the next the frame is with light and the next instant the light moved onto the future. Light would have taken all the time frames it used to travel in one go if light was travelling at a rate that was equal to time moving.

There are two possibilities in which the future can develop. If he can act by space-time manipulation quicker than the fuse burns he would be able to apply a time in the duration in reacting faster than the fuse will burn. If his space-time manipulation is slower than the fuse are reacting to its space-time reaction, his chances of enduring life in space-time diminishes drastically.

> By releasing so much heat, man is taking that space to a time, which was very far back in the Universe development. It takes the heat to a concentration where life still had no part in the cosmos, no role to play or a position to be and with man being in that spot is not very conducive to life as such.

Present

Time is when the person has a thought he thinks he gets. What he thinks is a bright thought about presenting an explosive idea and bringing this explosive idea into reality turns out that he is massing with time. The explosive idea manifests as life that manipulates space-time by thought formed in time as motion. That is life. Life is manipulating space–time by applying movement as time, at the will of life. As the speed of light is not in the same time – instant as time in the person is, it takes time occupying space to establish the manipulation of space-time in order to adjust to the situation in hand. It will take time occupying space in duration of a specific value of space-time to manipulate occupied and unoccupied space-time through densified space-time to establish a new cosmic relation between matter in occupation of space-time and space-time not in occupation of space-time. In more understandable interpretation of the English language, it will take time while the watch is running to access the situation and then decide where to run to while ordering the muscles by brain command that assessed the situation while running. Since the space-time in which the gunpowder is igniting and therefore increasing the space it holds, the space the gunpowder is occupying will enlarge dramatically, and with that will the duration of time extend dramatically because it is forming more space than the space previously occupied by the time in duration compared to the surrounding space occupied by time. The situation will lead to the loss of time synchronised with space and then should the person share the expansion of space in time with the gunpowder expanding space, what he then afterwards would suffer from, is **_bomb shock_** or **_POST-DRAMATIC-STRESS-SYNDROME,_** or what ever cockamamie name the doctors may come up with, just to legalize the enormous amounts of money the doctors will suck from the persons already victimized.

Once the space expands, the duration of time will inversely react. Since the space becomes to small to hold the time within it, the time has to extend its duration in order to establish the appropriate space to occupy the time in duration as to match the same value as the surrounding space-time. The space the released heat creates through the explosion immediately intensifying the heat balance and in that specific space a^3 the volume of space will increase which is space-time depicted in the Kepler ratio ($a^3 = T^2 k$). By increasing space a^3, time will transversely become effected but while k indicating the limit of the space effected by the sudden increase in heat concentration, k then responds to a^3 increasing whereby the only other factor that can react by equalising the ratio maintained to the changes brought on in the situation is time T^2 that depends on the spin of the concentration of heat in that area. The time factor T^2 will have to pick up the difference to equalise the formula.

Past   Present

Now the position is reached where the mind matches the body as the mind now must take full charge of the body and suspend the future to gain advantage from its ability to realise the past creating a future

Life's ability to manipulate space-time results in time as the tool allowing reoccupation of space by matter during any segment of time, leaving the previous time in space to the past of the position held by time in space- time. This continuous movement of time through space is in a frequency of ($a^3 = T^2 k$). Since space only occupies the value time holds during the duration of time in that particular space occupation, space holds no value but for the time in the space. With this in mind, we can mathematically discard the space, and value only the time as $\Pi^2\Pi$.

My criticising of the Physics Academic intelligence proved very expensive to me as it was this far enormously costly on my part of the intellectually influential divide because I declare to be of average intelligence and like millions of others on Earth, just like all these millions are, I am also a believer in my faith as others are some or other faith. As I am just another believer in my faith, and are confronted by the same questions these super intellectuals are seemingly incapable or unwilling to answer, instead of trying to sound intellectual like they do when putting forward the most idiotic arguments just to sound clever while coming across sounding like a total moron, I truly try to reason the why, the where and the how. I don't say outer space is made of nothing and then measure that nothing that forms the Universe in astronomical units. There are an untold number of these total incompatible facts that I do discuss in other books and I started making sense when I stopped accepting the Newtonian way of thinking as being brilliant.

Then I realized the super intellectuals only have one source of information that lead them to their conclusions and that which they use as information is measured light. Through light they see the cosmos but they do not even understand that which they see because what they see is not reality but it is a hologram in space. I am not criticizing those Intellectuals because please see this as my way of praising them. Through the light they see they don't see that what they see is history and not truth. To underline their misunderstanding they wish to map the cosmos. By such an attempt they truly indicate their level of not understanding what light is and what light is about. By not understanding the language the cosmos speaks, then how can one understand the writing. They declare that light is energy...well yeah, no fine and by declaring that, that says it all...My question to them is name anything that is not energy!

When a human body walks without life, it will not manage to do so because the energy left the body. Apparently science is too ignorant to see this. When a body of a human is without life the body has as little ability to walk by own initiative as a mountain can. Without life the body becomes just another cosmic object. Science this far failed to realise such simple detail. They are unable to see that there is a massive, no a cosmic difference between the ability that a person has holding life and a dead body has that has lost life. To their inability to recognise the simple detail as humans would, they see the cadaver without life being equal to the moving human with life because science apparently fail to differentiate. That is clear evidence of total lack of understanding the most basic physics because they are unable to fathom that the cadaver remains a human body minus the human energy called life and that it is life that drove such a body to have independent motion where that motion that life provides only brings about independent movement when life fills the human body. This obviously proves that life is not from the cosmos and not being part of the cosmos, while the body that carried life is part of the cosmos.

I challenge every person to come with a better explanation than the one I give, because those that were forthcoming up to now were as backwards as the rest of cosmic Newtonian science. We are what life decides and the body is what life accumulates. When life starts out life accumulates what body structure it fits and it is not the body that determines life. I can be what my father and mother was because from them I received life and that life gathered material to form a principle that reflects their form. Life collected the material and it is not the material that formed life. If life abandons the material at any stage of development, the material stops growing into a developed body but becomes very cosmic. Any minute life leaves the body, the body stops being. My parents did not supply the body…I did, but they supplied life, which is what I am, and life is situated not in the material but it is in singularity that holds no space. I am not a reptile because I with life accumulated the cosmic debris that forms the body I recognise as mine forming a human. I cannot be my body since when I leave my body, my body distorts into more cosmic debris. There first is life by thought of life being possible. That principle we call emotion. Then from the thought comes the deed we call sex. The urge to have sex is the urge to procreate and multiply. From a thought a few cells align to form a distinct and specific form of life. The life decides the form and not the form deciding the life. All the cells forming

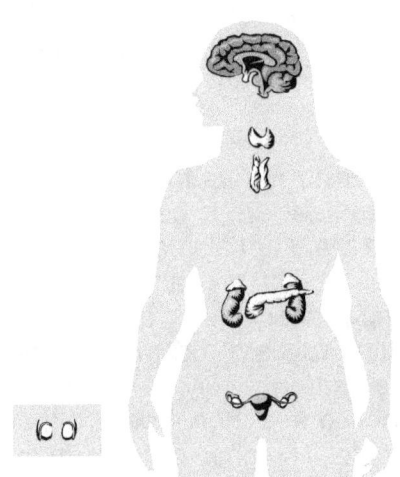

the semen holds an equal potential value of 1 = 1 in singularity making the cells being apart but also forming one unit and the complexity coming from this unity, man could never understand. When changing the form of the cosmic structure, we have what we call the human body, and then what we change about the human body does not change life afterwards. Life is what I am and not the body I hold. I can't be an ape because the life that selected me did not select the body an ape holds. I can't be a donkey because life created genes according to the prescribed formula life representing me chose and not the other way around. The classic Darwin atheist tries to put relevance on the body as the mentor establishing life, but the fools have to put everything they can't link onto some missing link, which I have to accept if I wish to accept their fantasies. Life is eternity condensed into a short time frame, but that is what only life can be. Life is the accumulation of every life holding control over the cell that compiles the complete person. Every cell confirms the person's ability but the person confirms every cell.

Let us look at their definition of energy. Energy is, as I understand it, indestructible, which means it cannot be destroyed. Energy can only transfer from one form to another form. Let us look at the example, which is used to teach scholars at school. We take a rock and move it from a ditch and role it up a hill. With this rock on top of the hill, we have a lot of potential energy that was transformed from static energy by means of the kinetic energy as they transferred the change in energy from one type to another type by repositioning the rock. One thing they forget is that when relocating the rock it used life as a source of energy and only life gave the rock the transformation in the type of energy. Their changes made to the energy and the relocation of the position of the rock came through life's effort of manipulating space-time. Science fails horribly in recognising that life brought about the manipulation on behalf of the rock and as a cosmic occurrence the rock will never roll by itself to the top of the hill.

The rock by itself within the abilities allocated by nature to the rock, the rock finds a total incapability of such manipulation forming movement of space-time. The energy transfer came from life to a cosmic object and by reading half of the writing they seem to think they understand the full picture while it is clear that they only see a quarter of the writing. The Atheist not believing in God but in cosmic forces cannot see from where the energy came. If it came from a human body filled to the top and overflowing with energy, which is what we named as life, they donate this energy to the cosmos blindly. In the vent where our bright and super intellectual atheist are of the opinion that life and the body is one and the same entity, then please take any fresh cadaver and let the cadaver repeat the effort what it did when life was still present. What ever the atheist claim to bring life to a body while life is part of the body and also holds the opinion that which brings life to the body is so natural, it is a normal comic phenomenon, let the atheist then produce the natural substance and force the cadaver or the rock to get to the top of the hill without using the aid of life. If they say it is electricity that establishes life and then with life and electricity being the very same, then shock the cadaver until it does achieve the feat by completing the repositioning of the rock from the bottom of the hill to the top of the hill.

The human effort is much more involved than calculating the effort it took to get such a heavy object up a steep hill. They can calculate what the body supposedly lost in energy by rolling the rock up the hill but no one ever challenged them to prove the truth of the energy measured in the calculations, or to prove their accuracy in the calculations. How much energy went into rolling the rock to the top and what were the losses lost in the effort alone. In the transformation of energy by relocating the rock, other losses occurred, like heat, sweat vapour and friction losses. Those losses they can't calculate but life is just not that simple nor is it as simple as they see it. The science apparently does not take into account energy losses brought about by anger flaring up with arguments when conducting the process. The never try to calculate the human temper raging by fighting caused as a result of frustration born in incumbency through personal inability to achieve what ever goal without the effort draining emotional energy. The heat rising when the body fights emotions is as much losses and is part of the totality of losses. These are also energy losses and being losses, it somehow must be accounted for.

After all the sweat and wrestling, the rock is on the top and we have a situation with potential energy from which we can derive kinetic energy when the rock is rolled down hill. This was solely achieved with life giving all the required energy. They now claim the rock will produce the same effort going down the hill as what went into the rock by getting the rock up the hill…well then our atheist better do some farming and find what effort it takes to remove rocks from any location and reposition those rocks to another position. By saying that I admit that I do not agree with any of the above mentioned, and will later state my point of view. I will however declare at this point that Newton's statement of energy and work being the same thing is utter nonsense, knowing very well many are reaching for a knife in punishing me by castration. This I say as I have seen the look in their eyes before and in the past. This nonsense is the simplest example we use to teach children in school. I too had to teach the children this nonsense, in the period when I too was a teacher and I admit it is total misinforming the youth.

When a body is with life, the body can move independently from the rest of the cosmos. In this there is great prominence because this is where Newtonians detour from reality in every sense and on this, Newtonian atheistic stupidity feeds like a parasite feed on blood. When a person is with life the person holds something nothing else in the cosmos has. The person requires no special talents to walk about and do many things nothing else in the Universe can do. We with life, take all such abilities so much for granted that we think of life and everything concerning life as the cosmos. At one point I was asked what would the purpose be of the rest of the cosmos where life can't intervene or exist. Then the other thought just as ridiculous as this idea of life being most prominent is putting life in equality with the rest of the cosmos. When I am with life, I have every form of ability there is available to life to move and that I can do as much as my free will, will allow my ability to perform as if there are no limits.

From this we take the scenario of a person's life. A person is born, and after that momentous occasion he continuously move about on this planet for the best part of sixty or seventy years. In this, period a great deal of energy is used to walk, run, laugh, cry, think, produce and reproduce. By doing that, he would from time to time state that he feels tired or without energy. What form of energy will this man be referring to because there are many forms of energy the man can refer to in spite of Newton's views? I have once heard a scientist making himself such a fool. With his PhD and all it would have served him much better to keep his trap shut and let every one wonder about his intellectual competence instead of knowing about his intellectual incompetence, the PhD notwithstanding. That scientist declared that if God was energy, God could be coal, because coal is energy as well. How bright can they be when they are that stupid? Now I would love to invite him to a meal and see him devour a plate of coal since they maintain that coal and food both are energy. According to Newtonian wisdom food is energy and coal is energy therefore it is energy we are talking about. If coal is energy, he can make a meal of it, and then live very cheaply. What he does not seem to grasp, is there is many forms of energy, which differ totally as we remove the heat by burning the coal and in doing so one can tap the energy. That does not make coal digestible or eatable although we use sugar for burning energy. However, coal cannot walk, run, jump and laugh, as life can. Life definitely being energy is able to jump and laugh. I can jump and laugh when I am full of energy and with the coal also being full of energy and energy being energy according to Newton, then why can coal not jump, laugh and play? I cannot even begin to imagine one brick crying and moaning because his friend was thrown into a fire. Coal cannot have sexual intercourse producing an offspring and then caring for it afterwards. Life, on the other hand, does have the very qualities as energy. This means that there are different values and forms of energy, of which life is one. If life is so different to other forms of energy, God could be another total different concept of energy. This is the problem that I have with those super educated geniuses that lacks all ability to think. They can make the most bizarre statements and by superiority they hold in an academic position they can release the biggest idiotic claims mentally possible and through their class they could get away with such stupidity unchallenged. They, the most learned after all, have the PhD with which they can prove by showing such a diploma on paper and with such intellect and qualifications that should also supposedly represents their superior intelligence.

Then life removes the control it has over the body and all the abilities associated with life vanish. That which remains is just more cosmic rubble called flesh, muscle and bone. The body remaining without life becomes as much a rock as a rock is a rock. The body without life is only more cosmic material. It is life that gives my body the ability I attribute to being my person. The body only holds what I contribute to being me and when life is no longer part of the body that served me a while, every aspect that contributes to being what others and I may think of my person, disappeared with life disappearing from the body I had. My body is lifeless which means it can't move. It has the energy ability of a piece of rock that fell down a mountain. That means that it is sad, stupid and mentally impaired to think of the body as being me while I know that without life there is no aspect of the body that remotely resembles any characteristic I had except what light conveyed as being me. If that rock is equal to what I think I am according to the Newtonian atheist, then it is little wonder their small intellect tell them computers might replace them.

The world contains a wide spectrum of different people. Of these people each one is in an arrangement and a wide spectrum of occupations, in which, people earn their livelihood. Seen from my personal occupation, there are two types of livings to do. Since I am a farmer, there are farmers and then there are not farmers and in between there is nothing else but farming or not farming. Some of those that farm produce wheat, corn, barley, nuts, sugar cane, vegetables and many other produce. Cattle and sheep farmers produce meat that is used by others to convert into energy for their own personal use. These farmers are providers of potential energy products. They produce food, for the other group of the human population that uses this energy product to maintain their strength to apply it to other methods of occupation.

All people have one thing in common. They devour one form of energy, which is known as food. They have to kill off life to attain life. Plants and animals are only life in another form and to live a person must kill or die. Such killings is needed to maintain a life cycle, and the consuming of food must be done on a regular basis, to enable a human to live and reproduce for a lifespan of seventy of eighty years. The only precondition is that life would parasite and vampire on other carbon-based forms of life, whether it is plants or animals.

Living off other life forms this person maintains his way and means of life, thus transferring energy from a form of food to a form of work. Then one day he collapses and becomes still. That person becomes unable to move. We call this state that the person is in, being dead. Being lifeless means that even if I take a shovel of food and force it down his throat, he still would lack the ability to transform that food energy to movement. Why then, would he lack the ability transforming energy to energy when food is energy and life is energy and all energy is alike according to Newton? The dead person received energy in the form food and he should then have energy to convert it to movement because when he was with life he was able to use for such conversion.

He can't perform the converting simply because he does not breathe any more. Now there is the solution! There is no oxygen to burn the food to produce the life to perform a task of transforming one form of energy to another form of energy. So we go and pump his lungs full of oxygen while shocking his heart with electricity using such violence that he hops around after receiving every jolt but what good does that do... it does no bring about the energy called life in spite of all Newton's correctness going around. We are inflicting all energy that life removed and that which we think that went vacant by injecting adrenalin and insulin and liquid protein and all the acids life produces but still the atheists energy the atheist claim is used to produce life in the body does not result in eternal life and energy coming back to the body where all energy being the same does not replace the same. It should because Newton cannot be wrong, Academics will have you locked up for thinking that Newton ever could be wrong but realising it too we then should be surprised because according to the superior intellectual atheist there is only energy and this the person did receive in the form of food and now oxygen and electricity mixed with a cocktail of medical life producing chemicals. Still above all odds given by the esteem of the atheist the cadaver still does not breathe, leaving us with the question: why isn't he able to breathe? Because he is dead, he kicked the bucket; he has reached the point of no more suffering, his mortality came about or whatever way one feels to state his position of eternal rest. When a person is considered dead, he lacks energy to such an extent that he cannot bring his own body to the grave. Others like me having life and therefore movement, and I still with the same abilities have to carry him to his grave. We, that are alive, and maintain the process of translating food into life, have to carry him the dead (he who is without life) to his grave. He can't go to his grave notwithstanding whatever energy I give to the cadaver.

The only difference between him and me is the energy form known as life. However, life is not the same form of energy as food, oxygen, heat and electricity. Even if I force all the food down his throat, and pump his lungs with air, while I heat his body with a blowtorch and shock him with electricity up to where he jumps around like a ping-pong ball, he would still find himself unable to walk himself to his grave. That means the one form of energy is not the same as the other form of energy. Where is Newton now that I need him too explain...where has the atheist's intellect gone?

It is widely accepted that there seems to be a generator in the brain that generates electrons which enables the body to function. We know the flow of electrons is due to the process called electricity. On the other hand, do we? In a later chapter, I shall indicate what difference there is between the flow of electricity and common heat. However, for the mean time I would stick to this accepted fact that electricity is conducted by the flow of electrons. Now, you can shock the cadaver with electricity until it hops about like a funny-putty ball while frying him with heat and electricity in the process, if life has gone absent, conducting a flow of electricity would not reinstate life. But medicine found a way around this state of affairs.

Medicine put the cadaver on life support, with a heart machine a lung machine and all kinds of other machines and denying the longest living partner to become the longest survivor by letting the other partner dying first. By the way, this method of prolonging life has nothing to do with life being precious as the medical paternity claims, but it links far more with the money paid by his medical aid, being precious. Once the cadaver's line of financial support dries up, his life instantaneously becomes worthless as they switch off machines faster than you can say help. Life is very precious while someone can pay for the maintaining thereof. Once the money dries up, the precious disappear into a coffin to be forgotten as part of the greater past. Then the cadaver finds the problem that it seems unable to live which means it is dead. Death means the brain is unable to send electronic signals by means of amino acids to the muscles, which would enable those muscles to continue with the normal functions. The cadaver finds itself without the energy called life. Once life has departed life has departed with no possible returning.

At this stage, I think that I pointed out there is a difference between a body filled with energy called life, and a body that lacks energy and called death. However, the energy that I pointed out called life, is miles apart from the energy that consists of food, air, the burning of it and the destruction of it. There is a broad difference between the food process and the actual form of energy called life.

Now I would like to ask those Super Intelligent Atheists and consumers of food and air to explain where the energy form that is called life, has gone. No one can destroy energy, but can merely transform energy from one form to another energy. This is scientific gospel. Life as I pointed out, has a different value to heat. Life cannot be destroyed, that means it has to be transferred from one form to another form, and life itself is not heat, electricity, or food, because applying all those other forms of energy cannot raise the dead. The fact that energy must be transformed and cannot be destroyed is proved by science to be unquestionable.

The only answer to which I can conclude is that science is applying a soothsaying method whereby science is ignoring the findings they concluded to prove their religious fashions of practising atheism. With life being an undisputable form of energy and energy that cannot destruct, it seems very unscientific to propagate atheism as a fact. Atheism says the end of life is the end of life. Science show life is energy being another form than what consumable energy is and science shows energy is eternal. Then where did the energy of life go because it can't destroy or vanish or disappear from existing and therefore defecting is quite out of the question?

From these facts, one has to conclude that there does exist another form of life after death. Life continues in some other form maybe but according to science it has to continue. The sickening cancer of atheism runs much deeper though! One has to consider energy in the state science recognizes it, to truly appreciate the flawed manner which Newtonian Atheism trips over their own reason by their cleverness.

In space or as I prefer to call it Universal time $T^2 = \Pi^2$ and k is Π
An explanation about the relativity of space-time and space to time can be as such: Time is a flowing image of frozen minutes, fragmented to seconds.

The following scenario would better describe it.

Let us presume a person is standing, looking at a moving object that he wishes to capture on film. Time in space is passing him by at a rate of sixty minutes to the hour, sixty seconds to the minute, and there the human concept of a valid perspective of naming time fragmenting ends. When a shorter duration requires naming, we simply divide the second into tens, hundreds, or thousands. To this day, that suited most of man's requirements. Going into cosmic time as time develops into space, some problems arise about the way we establish time by means of applying historical, cultural and mostly other accepted norms.

In Newton's era, establishing time to accurate detail was of less importance. The camera was an invention that was still waiting its discovery centuries to the future, and since science is stuck in Newton's era about the way of practising physics we have to remain there. In Newton's era time measuring to explicate detail, still remains relevant to the requirement. Modernization brought with it, some focus on measuring time with more accuracy, but in the cosmic relevancy, it still lags eons behind any true requirements. In this, science still has no idea what importance time has. A good example is the reasoning in the riddle of throwing a ball at an oncoming train, where the ball has to stand still after hitting the train.

Did the train also stand still or did the train move forward at an unbroken time interlude. This silly riddle forms the very essence in misunderstanding time. In order to move the cosmos have to use time because only by using time does anything including everything there is, move. To move space $a^3 = k\,T^2$ distance measured in space is a factor $k = a^3 / T^2$ and the duration $T^2 = a^3 / k$. There is not one single thing in the entire Universe as large as it goes or as small as the object goes, that could stand still. Everything moves in relevancy with everything else while such movement will forever involve one point, not object, but one point in every object that can't move.

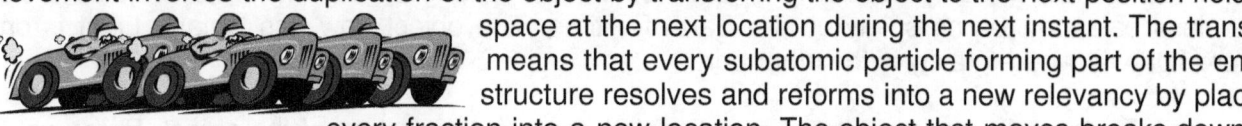

Movement involves the duplication of the object by transferring the object to the next position holding space at the next location during the next instant. The transfer means that every subatomic particle forming part of the entire structure resolves and reforms into a new relevancy by placing every fraction into a new location. The object that moves breaks down its entire structure and replaces the entire structure to fill the next position. This it does during the next singularity faze time provides as time takes he body from the past moving thee body through the present and shifting the body to the future. Time always has three valid positions of which one is in singularity. Now use your mathematics the Newtonian is so proud of and calculate how long does it take to break a car down atom by atom and then to rebuild the car atom by atom. Take this frequency down to the level where every subatomic particle shifts and the number of times it must shift to move one micrometer during one billionth of a second.

Life is the free movement of occupied space within the realms of cosmic movement thought of as gravity. To understand the relevancy of movement better we have to look at time per space in the smallest possible unit. Speed is relevancy ad gravity is speed. A craft have to move straight up at 11.21 km / s to beat gravity hands down and escape in to outer space. This puts gravity in a relation between distance (km) and time (s) so what the hell has mass got to do with the entire affair of what forms gravity, but that is how the Newtonian mind works.

Newtonians are fixated on the part no one can explain, which is how mass forms gravity while they ignore reason completely. Gravity is about speed and speed is about movement. Understanding gravity is the same as understanding something about movement forming speed.

Travelling is converting space in relation to time coming from the past time or to from a history factor and moving on to the future. Speed is about an amount of space being pushed to the past during a specific period of time. It is the number of cubic units a^3 that the car k fills during a specific duration of time T^2

Freezing a sailboat on film will require something like an hourglass. Anything more will be over estimating the limited requirements. Any effort by any photographer with any equipment will be sufficient because frame per second will be wasting film.

To freeze a vintage plain in flight travelling at hundred and sixty kilometre per hour, will require more skill as the photographer would need to slow time down to a pace that would allow the plane only a few frames of movement in one second. Capturing a Dakota flying at full speed would need more skill and better equipment. The Dakota is moving at a far more rapid pace per second. This is a captured instance in an instant in time where thee instance represents space and the instant represents time.

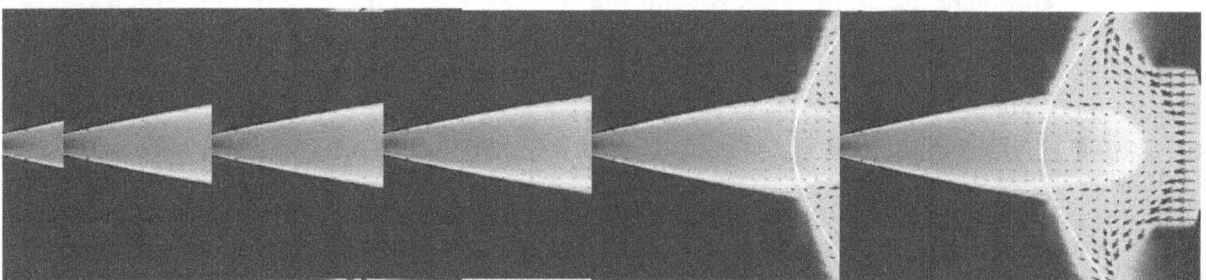

Freezing a photon would require skills that are more professional as well as involving a lot better equipment than what man can ever build. The problem is in the equipment, not in freezing light in time. To freeze a photon in an instant of time would mean fragmenting time to a fraction of a second, and yet doing that will only freeze light, and not time. Standing on the Earth, all objects are in motionless balance, relying solely on the Earth for time. The object now has weight. Light is one sector of gravity where gravity is the other part.

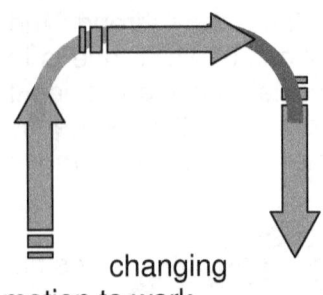

Going up takes time to reach a point of directional change and the return of the object. But to go up high is no chance of applying cosmic energy because cosmic energy does not allow such motion. Cosmic energy is gravity and gravity is moving objects towards and in relation to centralized singularity. By going up and fighting gravity is only done with the aid and by the intervention of life. This happens due to life inflicting influence by manipulation the movement in relation to the surroundings. Gravity is not manipulation but it is about cosmic control of movement according to a precisely determined destination. Life cannot control the cosmos in terms of gravity but life can manipulate the flow of gravity to set independent motion to work.

changing

At the point of return, the object in motion has exhausted the energy supply life supplied and therefore it is on route back to Earth. That is gravity taking charge and applying control to defeat life being the manipulator. The going up life supplied some energy to accomplish the task but in the end life may not sustain the gift of energy to defeat gravity. At such point, the only supply of energy is the Earth using its concentration of cosmic time.

I focus on this issue to show that life and the cosmos are not the same and life is an addition to what the cosmos normally is. The cosmos might blow out some gas but will never have the ability to overcome the gravitational draft going down while when this happens it then is life intervening. This is what the atheist tries to ignore in his rambling about the possibility that life came from Mars and landed on the earth back when life originated. Life in the very beginning had to use gravity to escape from mars but being almost cosmic at the time it then did not have the ability to overcome because only with the intervention of intelligent life can anything escape to outer space.

Gravity stopped the motion going up, which the going up can only be the product of life manipulating movement in an attempt to defeat gravity. Gravity has the ability to freeze the movement by redirecting the flow through which time rearrange space and that is the way that time influence the directional flow of light.

However, when looking the very detail the question is how does gravity accomplish that. It takes time to send an object into space taking time and it is taking time for an object to relapse back to the ground. It takes time to allow light to travel through space and since time is much faster than light flowing though space, then time has the ability to change the direction light flow. It is not gravity bending light but the flow of light is redirected.

> **Life in the body forms time in space**
> **All else outside the body is in space-time**

As I said before, Newton considered all forms of energy to be the same, and oh boy, was he mistaken. It is not surprising he formulated gravity the way he did. Life is as much about proving movement, which is what energy, is as it is about knowing the difference to the state in which a living person is and in which a dead person is. There is a worldwide fashion amongst the very well educated that when any person wishes to be regarded as a supremely informed person by those with the know-how, one must at least be an atheist. The key to science is apparently to be completely atheistic because those boffins in science can't think of a way that science could prove the presence of God. Therefore having such well-respected wisdom and clear insight into nothing, atheists do not believe in the life after death, a Creator or a Force that does not exist outside the technological criteria of mathematical science. That is your local Newtonian. The Newtonian is Newton addicted except for one part, where Newton was very religious and God Fearing the Newtonian became so animal-like in thought, Newtonians nowadays fails to see the presence of any Godly Entity, but in doing so they are just as wise as all other animal minded life fails to see.

The Newtonian atheist share a lot in mentality with the sheep but I will not go as far as calling the Newtonian a sheep because I feel it is rather unfair...to the sheep that is. The sheep does not present its image as one being superior intellectual and therefore the sheep does not wish to be seen as very brilliantly-minded in order to fool all other life forms on Earth, whereas the Newtonian tries its best to hide its stupidity by concealing its status of being mindless. This state of being sheepishly mindlessness Newtonians have an all out effort to hide behind a fashion they present in the name of atheism. But on a mental level, the sheep is as big an atheist as the Newtonian ever could try to be, and when saying this I have to apologise to all sheep that may feel hurt by the truth in my mentioning this. To them (those called Newtonian-atheists and not those called sheep) there is light and there is dark and the light is something other than dark. When they are asked what dark is, they will tell you there is an absence of light. What brilliant mindless genius they are! One can only see light and one can only use light to see and therefore the darkness one sees must be light one sees, because one can't see nothing that is forming light. In keeping true to this mentality, the Newtonian-atheist thinks when death takes life, it is because death took life away and destroyed life and not because life has departed only from the body life used for a very short period.

This brings us to time and life. When one places the duration every cell will have in time being alive and living in the body, it will surprise me little to find out the duration of the life expectancy of every cell put in continuous and not simultaneous, life duration will hit eternity. If every cell starts life the instant the previous cell died off and all the cells go through this process one by one, I think the body of the human may just never die. But that is not the purpose of the cell. The purpose of the cell is to aid the full compliment of human life preserved by countless cells that form a compliment of one human body. The human cells do form a structure of life in a human being, but life requires so much time in the total time it takes to culminate in forming one human life fully, there has to be innumerable cells all running together to form one human life living in a total compliment of seventy or so years. The cells have to put time in one piece called the human life to establish what life of the human requires. The cells all have to join in one active body in order to find the ability to support life as developed as the human is. The reason why we sleep has to do with different life entities holding time at different levels all joining forces to aid human life and sleeping is the rectifying of the different stages where each time frame of different cells requires another sleeping period as a stabilizing mechanism to equalise the duration in the body again.

I do elaborate on that in the chapter I call Man-in-Motion I show what part physics play in the development of life of the human mind, as I interpret it. The human body is not reality but is only a manifestation of life's input. Once life disappears, the human body disappears. The human body can be or can't be at the whim of life, which means the human body is a reality at the will of life and only because life wishes it into reality. Other than that and if not with life, the human body is imagination, which makes the human body imagination and not a reality. If life's imagination is not present to support the imaginary fact of the human body, then the body seizes to be. It starts decaying and corrupting the very instant life no longer supports it being reality. On the other hand we know life is the energy that fills the body and since energy can't ever disappear, the fact of life is eternal and can never disappear because it forms the energy that drives the imaginary substance that we see as the human body. Life puts time into the body by manipulating movement and fills energy levels in the body.

In short I wish to touch on food and why all forms of life need food to live. As I said life is time and to live one needs time. The human needs a lot of time to sustain all the time, the cells forming the body has. Again I state my argument that time and life is the same and that makes the Universe part of God's eternal being.

We see the food we eat as carbon fibre of whatever variety we take in. That is not what we should see. We should see what was accumulated by life in the time life was in time within that which is accumulated and remains in a form of time we are able to devour. Every blade of grass holds innumerable cells where every cell is a product of life while life is serving time. It took life to put the cell in position in forming the blade of grass. It was life that filled the cell with occupied space presenting the time it took life to fill the space. It took the blade one season of growth to form the blade of grass. In a patch of grass holding one hectare there are innumerable blades of grass all representing time filling space where life took time to hold and fill the space with energy. Does the cow coming to graze eat grass...well it depends on what we think of as grass. The grass is the memoriam of what was present when that which was present filled what we see as the grass. Life took time to fill the space we see as grass. Life deposited time in the form of space to produce the product we see as grass. The fact that the grass can no longer grow once the cow detached the blade from the tuft of grass supports the fact that life is no longer part of the grass. Grass is life but the life grass represents are miles away from the developed form of life that the cow represents. The grass presents the cow with the time the grass served to collect the filling that the cow uses to sustain life. Plenty of grass has to serve plenty of time to sustain a well developed life form as the cow has and the cow with such high development can't have such high development and still waste that much time serving space. It needs what the lower form of life leaves as an inheritance to the cow. Life is not just more of the same life but there is much discrimination of how much developed life is. The cow with the developed form mammals have being as developed as the cow is in a form holding life, represents much development and being so much more intensely developed, the cow needs much more time to uphold the status of life the cow represents. It is the time needed to sustain the life that the eating takes and not life as such. The fibre the cow takes in represents life in time and the time is accumulated by all the grass the cow eats to sustain the complexity in time, the cow as a form of life holds. The low form of life that the grass represents has to spend time to deposit space as space-time, which then is available to nourish the cow so that the cow can find time to be more in life and offers more, produced with time while the cow's life is representing time. It is the space-time the grass as life that is forming space used that the cow now uses to sustain life by extracting the required sustenance from the space occupied by which is what life left on Earth. The cow does not devour grass ass such but instead the cow feeds off the time that the grass in life represents.

Then the human comes and collects the time the cow spent collecting the time the grass spent on Earth in time to fill space. The human is so much more developed that the grass would hardly do to fill the required time accumulation by feeding on grass alone. Some apes and Baboons spend all day feeding on grass but intellectually they have not progressed that much further than the cow mentally has. Eating is just reaping the time the next form of life spent accumulating space and that is all that there is to eating. It is taking the time another form of life used to fill the role it had to play on Earth. Now we get to the question about what plants use as food.

Newtonians try to sell the idea that plants take in a few minerals farmers put at the roots of the plants and together with water from that a tree of many tons grow. The plant takes the minerals and the water via the roots and then grows from those few grams of compacted minerals an enormous tree of many tons. Those plants use a few grams of minerals, which plants then use to fill the space that the plant eventually occupies. The connection between a few grams used and many tons of material gained eludes the thinking mentality of the Newtonian. Then with this mentality of not comparing facts and putting in grams to receive numerous tons, they extend into what the product will be of the contamination of the air that is polluted by waste of fossil fuels. They go on about global warming but never, not once... try to speculate what sustains plants. What do plants use to build fibre and where does the fibre as building material come from that plants use. This is global warming and the idea of global warming and the release of these gasses are forever connected by Newtonians that don't use facts to support claims but as Newtonians so well do, Newtonians only use suggestions to cause hysteria.

What we will find with the carbon gasses is that heat in the form of radiation will become more and that will affect human life, but plants will remove the carbon from the air. But it is the same politicians that give the same Hoggenheimers and Mammonites as industrialist's permission to cut down the trees and make a fortune in the process. They always destroy with the aid of the politician-criminals sitting there as our democratically elected representatives and pay those that are supposed to serve the public but is only there to serve their money master and to ensure their money making masters make a packet of money with the political deceit. Global warming is coming and no one can do a thing about it because it is part of gravity cycling around in relation to whatever centre. The release of the gasses by harvesting the fossil energy is connected to life, and is not a cosmic process. Life is as connected to the oil being there because the oil is a remainder of life that was. The mining process of the oil is life connected and the oil and other fossil fuels are just the remains of life that harvested energy from the sun in the same way life now harvests energy from them. The fossil energy we harvest is a historical process that was in process of life acting its principle but is now the history, repeating to again benefit life and release energy to the advantage and favouring life. The fossil fuel is a dispensary life back then froze to benefit life at present. Releasing the gasses is life connected as much as the oil being there is life connected as much as mining and using the fossils, as historical Sun released energy stored for billions of centuries and now forms gas that will again feed plants in the future. The plants remove the carbon-whatever oxygen from the air and remove the oxygen to use the carbon as plant material. The carbon they remove from the air with leaves is the food they use to nourish and support their life requirements. They use time to fill the space with energy we take from them as material which is time masked as plant material. It is the time they used we now claim to support our needs because time is life. We, with more advanced life do not have the required time to use in collecting food and therefore we have to use their time they spent to accumulate and store energy to fill our needs.

Everything holding space in the cosmos grows and that includes mankind as well. Everything prehistoric that is fossilised was big. They were so much bigger than whatever still is today! The elephants back then were a size and a half bigger. The rockeries were twice as big and the sharks were three times as big. The dinosaur was bigger than the ordinary housebroken garden variety Elephant. It stood almost the length of a rugby field and was just about the breadth of a rugby field but with that enormous size it could outrun any modern rabbit and still it was so nimble that it could devour an Elephant before breakfast for breakfast and remained more agile than your common meerkat. When those Newtonians get carried away, they get high and far on their imagination and whatever were, had no limits to what was possible. I have witnessed the tracks of a centipede that had the breadth of over a meter. No one thought that comparing this lot with modern reality was getting ridiculous. No they were Newtonians and were prepared to clown and cheat and swindle without thinking as long as they could cash in and make money. Even today they are getting more absurd without trying to fault their comparable findings.

To every human hearing about global warming, it is synonymous with carbon in the air and gasses that heats the Earth and all that brainwashing the Newtonians so spontaneous share with all that wishes to hear is backed by the hellish bankers, oil barons and money lenders because those representatives of evil and money have one interest and that is to make more money. The Newtonians get five seconds on TV or an article going on about saying nothing in some paper or magazine and that is used to let a statement loose like a fart and what ever that statement informs how delusional and incomprehensible that may be, that information as it comes, everyone believes without testing, because it is in such a manner in which Newtonians always commit deceit, even in classrooms at school. Those Hoggenheimers will support any and every lie told by Newtonians on behalf of the oil barons and the Hoggenheimers as long as their filthy rich banks and vaults are evermore filling with money. They fill the air with the carbon gasses coming from the oil they sell. Now they want to be paid to remove the dump that is a product released by the petroleum they sell and results from what they sell when selling their petrol to the public. Then they want the tax paying slaves to pay them to clean the air their petroleum filth filled in the first place. They pollute and become rich beyond comprehension and then the tax paying slaves (us again) have to pay them to clean the mess their money making left us with in the first place. This they wish to do by spreading lies and unsubstantiated rumours through the devilish gospel that Newtonian atheists are renowned for and thus are used to sell and spread lies. Never does any Newtonian think where plants get their food as plants give us food. The fact that plants take the carbon the carbon gasses fill the air with, as a food, has never been published and goes unmentioned as if no one ever thought of it. The petroleum they sell as energy is just remains of what life deposited billions and trillions of years ago…and yes it is trillions upon trillions of years ago because this garbage that Newtonian believe about the age of the planets are as if the planets are only all just four point five billion years old is as big a lie as their mass that is responsible for gravity. The energy released from the fossil fuel was deposited by life billions of years ago and still the energy that was gathered by plants that lived then and was then taken from the Sun back then, still today sustains modern life in so many ways. Our cosmic energy we use comes from the Sun, but only fossil life could store it in a manner we can produce work and gather from it any ability to use it again. The Hoggenheimers tell the Newtonians to tell us those gasses is a problem, which in a way is a problem but not in the manner that the Newtonians tell us they foresee.

Global warming has been in progress since the end of the previous ice age because if not we would still be enjoying the previous ice age and only criminals such as Hoggenheimers, Newtonians, Mammonites and politician will be evil enough to find yet another way to bleed the taxpayer into paying for the biggest farce the lot created since time began. Global warming is a fact but it is part of gravity. The carbon in the gasses was part of the Earth before and still is part of the Earth. It is not additional all of a sudden that now suddenly poses a mal balance. If it was part of the Earth then it could contribute to tipping some part of a balance but not tip the entire balance. Global warming was coming and going on a regular basis in the past long before these gasses were released and with the release of gases it will come again notwithstanding gasses released or not released. Global warming came and went as many times as the magnetic fields swapped and ice ages came and went. Global warming and global cooling is cyclic in nature and is a result coming from the rotation changing directional relevancies through gravity changing directions by circular movement of the Earth going around the Sun that is in turn going around the Milky Way and as it is going around this is a natural flow from amongst other influences rotational directional changes also going around whatever the Milky Way is going around. There is nothing those evil bastards I call Hoggenheimers and Mammonites with all their money can do with money to stop the process as much as the carbon is heating the world. Being Hoggenheimers has nothing to do with being Jewish or any form of a religion faction or being of any specific religion because the biggest Hoggenheimers I can show sit in Christian churches and while sitting with their sanctimonious expressions on their hypocritical faces as God fearing prudishness trying to hide all the love they have for their god Mammon. They pay the politicians to be even more pharisaic and false in attentions. Honesty is another scam the criminals called politicians spread as pharisaic gospel because they're masters, the Hoggenheimers and the Mammonites pay them with money and position to spread such false double – Dutch gibberish jargon as that global warming can be stopped by human intervention. I have not come across one politician that is not the worst form of criminal one can find and only Hoggenheimers, which are bankers and Mammonites, which are the industrialists and the insurance broking firms beat those politicians as down rite disgusting criminals. The Hoggenheimers and the Mammonites pay the next elections propaganda to brainwash people into believing this politician that is a professional layer is God sent to bring salvation. America invades Iraq to get funds to shoot the living crap out of Afghanistan so that the war can be paid by Iraq's oil and to accomplish that America and Britain tell the people that Iraq sponsors terrorism, while not one shred of evidence can be found to support such a claim. What does the Anglo American public do when all this falsifying of evidence becomes known the Anglo American public do what they do best, they vote for Blair and Bush and give support to the two monsters because they are Reborn Christians. Those mass murderesses should get the death penalty but instead they are rewarded with another term in office because the Hoggenheimers and the Mammonites reap the oil reward and pays for the ticket to get them back in office.

What we will find with the carbon gasses is that heat in the form of radiation will become more and that will affect human life by promoting cancer in some cases, but plants is the only effective way to remove the carbon from the air. But it is the same politicians that give the same Hoggenheimers and Mammonites as industrialists the go ahead by law to cut down the trees and make a fortune in the process. Hoggenheimers and Mammonites always destroy with the aid of the politician-criminals sitting there as our democratically elected representatives and get paid as those that are suppose to serve the public but in truth they serve Hoggenheimers and Mammonites and only the money strong as if politicians are there to serve their money masters and to ensure their money making masters make a packet of money with their political deceit. Global warming is coming and no one can do a thing about it because it is part of gravity circling around in relation to whatever centre that forms the circle centre. Global warming is a cosmic institution. It comes and it goes as ice ages and as hot ages and as magnetic field interchanging and swapping. Releasing the gasses on the other hand is life connected as much as the oil being there is life connected as much as mining and using the fossil fuel as historical Sun-energy that was released as energy and was stored for billions of centuries and now forms oil we change to gas by using the heat released in the process and as carbon gasses it will again feed plants in the future. The plants remove the carbon-whatever and oxygen-combinations from the air and remove the oxygen which then is released as pure oxygen while using the carbon retrieved in the process as plant material. The carbon they remove from the air with leaves is the food they use to nourish and support their life requirements. They use time to fill the space with energy we take from them as material which is time masked as plant material. It is the time they used we now claim to support our needs because time is life. We with more advanced life do not have the required time to use in collecting food and therefore we have to use their time they spend as plants and lower forms of animals on Earth to accumulate and store energy to fill our needs. This is one fallacy. There are more…

The cosmos is without life except for one small spot that in all cosmic laws should not be, because there is no reason for planets to be. Notwithstanding all the fraud Newtonians apply to make their criminality and deception legal, putt8ing life in the Universe in any other place that Earth remains a concoction of corruption and disinformation. Life is on Earth and until concrete shatterproof and visibly confirmed evidence is found it will remain just that Atheists love to give corrupted half baked ideas. Give evidence to support suggestions as I do and not distortion of unproven facts or proving madness in reason by allowing only vague thoughtless ideas to surface. Until the atheists can come with solid undeniable proof, as science should be based on, life has only a place being on Earth. Those that say life came from Mars must indicate what made life leave Mars and what enabled life to escape from the gravity with which Mars would arrest life. If they say life came from some galactica or some far off cosmic dust cloud as Chandrasekhar tries to portray, then don't stop where the suggestion then requires proof. Tell us at what speed did it come, what energy propelled it, what directed and controlled the movement, what determined the destiny in which it would travel and if by chance this happened, then please give the odds of this happening in the face of the facts that only the Earth can possibly accommodate life. Tell us how long the journey would last and what odds were in order and in place to have life come from some cosmic dust cloud to land at this spot called Earth. If the mad atheist wishes to be believable in the outrageously bizarre, the mad atheist must then put reason behind his mindless babble and become scientific by providing substantiated facts, as I do! It is always us, the believers that have to prove there is a God. I prove the entire Universe consist of one single spot by many where if not for movement the spot will disappear and in all accounts the spot is not part of the Universe but because of movement forms the Universe. The point is just like the presence of God, invisible but it forms an entire Universe. Try to disprove that! I think the time has arrived where atheists better start proving that there is no God! I will challenge anyone any day anywhere to any form of debate on this matter of life being anywhere else in the Universe than how it is found on Earth. Even locating a distant message coming from afar life form is far fetched If we would send such a message it would reach no one ever because cosmic space development will fade the impulses to become just more singularity with no signal ability even before the signals leave the solar system and remember the solar system ends anywhere measured in light years past Pluto.

It will be an opportunity for the atheist to show how brilliant his mind can reason and explain how the cosmic dust allowed and even spurred on the huge travelling that such a venture would undertake. How long did it take for the dust to reach the Earth and remember the coincidental factor, as the Earth then is the only place suitable for life to grow? How did all that carbon realise that there is an Earth and took off precisely in the direction of the Earth to land at this point. What drove the carbon and kept propulsion going all the while the carbon travelled across millions of billions of trillions of astronomical units. Take the chance the dust had to land on Earth in volumetric space held in relation the space the Earth holds and try to think of how small the chances are that the direction will pinpoint the Earth considering all the other space involved, not forgetting the quantity and density it must have had to propagate the Earth in the quantity that the Earth has. Life on earth is in abundance. Life cannot create carbon but can only seize the amount of carbon on Earth. Try to calculate the chance there is to have the carbon spread across the Universe in evenly density and still fill the Earth in a quantity as it did. If those odds became completely insane, then try and calculate the odds there is for the carbon to decide on the Earth and move across such vastness using time to fill the Earth with the measure that does fill the Earth. After these odds are established, then challenge me on my accusation of those atheists of being madly corrupt and in effect be criminally deceiving, to even suggest such madness using the level of mindless reasoning. Can everyone now see why I say the reasoning ability of the Newtonian atheist matches in all respects the intellectual capacity of the atheistic animals…there is no intellect to find between the ears!

We have life using space with the flow of time to fill more space as time progresses and all this becomes eventually solid proof of the Big Bang being correct. Life can only use what is in the cosmos and what is part of the cosmos. If life can grow in space then it must be a characteristic that the cosmos lends life the laws to apply the conditions that life can manipulate. Life cannot create anything but can only redistribute what is already present, which includes cosmic growth we see as aging. Only because the cosmos shows a growing in space can the life factor use this phenomenon to accommodate the growth in life stature.

Space (a^3) will disappear when not validated by time (kT^2) therefore life can only be present when motion applies (kT^2), In order to hold life in space (a^3) time (kT^2) or motion of space has to validate that space during that time. As the chicken grows every fragment of time (kT^2) filling space (a^3) and space (a^3) disappearing as time (kT^2) fills the next space (a^3) in time, while space (a^3) carries the movement of time as time (kT^2) forming the space (a^3) of the movement coming from the past into the present and onto the future originally filled in time (kT^2).

The space (a^3), which motion as time (kT^2) filled during that specific duration in time (k^0), disappeared as time (kT^2) moved on to the next instant (kT^2) and by moving on, it leaves the space (a^3) it occupied as the space (a^3) that is filling the present in the past. The cosmos is time in the past acting as space (a^3) becoming time in the present acting as singularity (k^0) to move on to the future (kT^2) as relative movement. That is what life manipulates because when life remove from singularity and then the movement life provides will also leave and what is left is cosmic occupied space (a dead body) moving at the rate cosmic gravity allows (being still and lifeless. This is not however, where the bus stops. In this life aging plays a major part, as life is extremely sensitive to time. Why would any one age? This simple thought baffles the greatest Newtonian minds into obscurity.

Every thing holding space in the cosmos grows and that includes mankind as well. According to Newtonians, everything prehistoric that was and now is and are fossilised, were big. They were so much bigger than whatever still is today! The elephants back then were a size and a half bigger. The crocodiles that were alive back then when everything was big in relation to everything being small in the present, were twice as big back then as those now found and the sharks were three times as big. The dinosaur was bigger than the ordinary housebroken garden variety Elephant. It stood almost the length of a rugby field and was just about the breadth of a rugby field but with that enormous size it could outrun any modern rabbit and still it was so nimble that it could devour an Elephant before breakfast, for breakfast and remained more agile than your common meerkat. When those Newtonians get carried away, they get high feeding on their misinformed science and far away from reality on their imagination and they wish us to believe that whatever was, had no limits in size or to what was possible by reality standards. I have witnessed the tracks of a centipede that had the breadth of over a meter. This doesn't provoke thought from the Newtonian mind but only feeds bizarre mentally insane explanations. Not one thought about what would explain the phenomena when comparing this lot with modern reality and seeing how much they were getting ridiculous. No, they are Newtonians and are prepared to clown and cheat and swindle by crooked facts without thinking as long as they could cash in and make money on the fantasy they use as the religiosity called science. Even today they are getting more absurd without trying to fault their comparable findings. Their presumptions go beyond the bizarre and compares well with the insane! From this one can read that space-time is time that forms space and space is coming from time.

Time forms space. Time gone by leaves its legacy as space filling the Universe. That is the reason we find the expanding we named the Big Bang and we are part of the Big Bang in every aspect. The growth is not only constrained to the "void" but also to the atom. The atom gains heat by expanding into space and the space expands by losing density, as material flows from the space to the atom. Every atom there is grows and that includes every atom in our body, which also is growing in size. This growth is what is behind aging and that growth of material taking up more space is a cosmic reality that kills life. However, this growth is immeasurably small but is throughout the Universe because it is time. If not for this growth movement would not happen. By allowing the body life, life occupies to outgrow that which life can sustain. That is the ability that gives life movement. That too is aging. He body grows to a point where the size of the atoms reduces the number of the atoms used in the body and the restriction of overall capacity forces the body structure to fall in on itself.

The atoms forming cells get bigger and with bigger atoms the human body can cope with a certain size. When atoms get bigger in growth by holding more material, the body will entertain bigger but fewer atoms as time goes on. Because fewer atoms being bigger are not supporting the construction of the body that well, the body becomes overgrown and frail. This principle we named aging and the process is cosmic. There is sweet nothing humans can do about it because as we can see with fossil bones, the process carries on in growth and the fossil becomes bigger as time proceeds. Time forms space and space is what remains of time as time went by. Whatever is in the Universe can never leave the Universe and that even includes time as a cosmic factor. But this goes beyond the conception Newtonians can gather because it doesn't include or provide for mass to be calculated!

Every generation human this far was bigger than the previous generation. The size of man is on the increase in space used by the body. Sons under normal growth get taller than what their fathers were by about from about half an inch to about an inch. This is used by life to grow nails and to grow hair and in the children faze to grow and develop overall. The atoms forming cells get bigger and with bigger atoms, but being genetically restricted by birth, the human body can cope with a certain size the gene restricts the body. When a person is born, the atom is at a specific size. Then as a child, this growth is used as the child grows into an adult. By the time the person becomes middle aged, the atom begins to outgrow the initial required size the human genes allows. To compensate for the gain in size, the body reduces the number of atoms used. When atoms get bigger in growth by holding more material, the body will entertain bigger but fewer atoms as time goes on. Because fewer atoms being bigger are not supporting the construction of the body that well, the body becomes overgrown and frail. The body begins to bend over while at the same time shrinking away. This is because as the size of the atoms grows, this growth increases the body while the genes using gravity restrains the size by the initial limits set at birth. While growing, the limitation deforms the human body into a bended posture and while the actual size of the body reduces, as the number of cells in use becomes less in order to compensate for the growing cells, the cells used to maintain the body are growing bigger. Aging is the product of the Big Bang growth and the product of the Big Bang growth is time flowing leaving space as the remembrance or after thought. Once anything is in the Universe, nothing can leave the Universe and this also applies to time. The growth is time where time will always leave more dots in singularity as what space has already got because every new dot is introduced as time and time has this new dot standing motionless while it forms time and space spins around the centre dot. Even time has to remain in the Universe once it became part of the Universe, because except for going into singularity again, other than that route, time also has nowhere to go but to form space as time in remembrance. Using time, every element has a function distinctly different to all the other elements. This too is what time brings to space. Carbon stores heat while oxygen transports heat and nitrogen displaces heat in expanding heat into space. A fire doesn't burn oxygen because then all the oxygen on Earth would be gone by now. Oxygen carries the heat as smoke and once the oxygen comes into contact with nitrogen the heat transfers from oxygen to the nitrogen and the nitrogen reacts by displacing (exploding into space) the heat as space forms resulting in more space. Oxygen can't burn and nitrogen can't burn. It is the manner in which each element stands in regard to heat that allows a fire or the release and transfer of heat. This accumulation of heat in the cell gets into the cell via breathing in air that takes the heat by the way of oxygen carrying heat into the lungs and then through the blood to where the blood deposits the heat as a supply of sustenance. The heat will eventually form movement and all other life manipulating abilities, but also one form of cosmic qualities in gaining space. This adding of heat forms partly growth which uses a process I think of in terms of called controlled cancer and that is why we need to breathe. The breathing allows oxygen to take heat into the body and some of this heat is used to increase the atoms in the cell. Time does not increase the volume size but the heat bringing movement does the increase while time only increases size by forming a relevancy in the present relating to what was in the past.

The body keeps on growing but the size becomes genetically limited to a specific form. Then some cells die off and make room for other cells growing bigger with cosmic growth applying. The growth rate remains but is genetically capped at a limit. To compensate, some fibres grow skew and crooked and lumpy while the skin growth starts to form wrinkles as it becomes bigger than the size that is required to cover the adult body. Aging is a process of cosmic growth and can never be controlled by any human intervention. The brain grows bigger cells and to accommodate bigger cells the brain has to reduce the active number in use. Aging is a process of cosmic growth.

As the chicken occupies more space, its relation to time will influence the space-time value affecting the chicken. However, the reasoning is rather complicated and intense in explaining I shall leave it at that.

A cancerous cell is where the cell only uses the supply of heat to grow and become mad. Most of the heat taken in via the oxygen is used to apply motion to the body to fulfil the purpose of life, but some remain in the atom to increase the overall size of the cell. This principle we named aging and the process is cosmic. There is sweet nothing humans can do about it because as we can see with fossil bones, the process carries on in growth and the fossil becomes bigger as time proceeds, although the bones are dead for hundreds of millions of years. Time forms space and space is what remains of time as time went by. Whatever is in the Universe can never leave the Universe and that even includes time as a cosmic factor. But this goes beyond the conception Newtonians can gather, because it doesn't include or provide for mass to be calculated, and when they run out of calculations to play with, they then no sooner run out of thought. It is little wonder they have the notion that computers can replace humans because they are all computing power with as little thought and reason available (both being God-given life additions to the human mind only) as breathing would allow! But then again who am I to judge, for that is the atheist, being just another mindless animal.

That is the principle behind space-time. Space-time is not some rumoured idea that one may or may not find in a Black Hole when someone will have the courage to build some ridiculous space whirl and elevate down the mine shaft of the nearest compatible but dual Black Hole that is situated just outside our reach at present. Time carries on and to that there is no stop. That is a Godly order that can only end when infinity meets and joins eternity once more by eliminating all space that parts infinity and eternity. While there is this process of partition formed by heat expanding or cold reducing, the space that is in place between eternity and infinity will not allow for eternity to be united to infinity but will remain connected as a unit to infinity, space will be what time leaves behind as time soldiers on to once more meet infinity for reunification while infinity absorbs heat left over from eternity moving as time. Space is the memory of time that came and went as time moved on and has gone by. I am not elaborating on this at this junction because there is enough said about this matter in other books that I wrote. Space is the memory of time and time is the movement of everything in relation to anything in particular. Time moves through time where time then in the second act, forms the space that time moves through.

Aging or the process of deforming the human body is the same but on a much smaller scale than the systems whereby stars develop. Stars reduce in space at the same rate that space expands in time and the reason why stars reduce in size id because space expands. Stars shrink as much as space stands still and space expands as much as starts sustain size. Nothing is expanding and nothing is contracting and nothing is shrinking. It is relevancies changing as time forms space. The connection made by the governing singularity, albeit the main form of life constructing the body or the centre singularity in a star maintaining the controlling singularity, the time control takes relevancy, a format ($k = a^3 \div T^2$ and $k^{-1} = T^2 \div a^3$) as well as ($k^0 = a^3 \div T^2 k$) is struck whereby a **governing singularity** or the human forms the formula with which the **controlling singularity** exists in **relevance** (Π) to apply movement within every cell (Π^2),

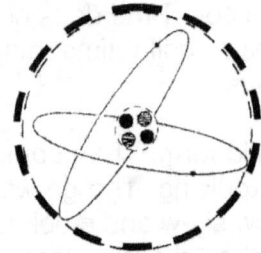
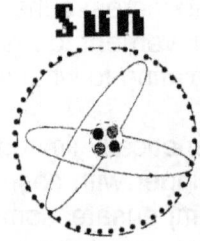

I wish to remind that if a **relevance** (Π) did not validate a **positional validity** (Π^3) securing a **governing singularity** (Π^0) in terms of movement formed by **the gravity** (Π^2) that produces the

controlling singularity (Π^2), a three dimensional status, then space (Π^3) would not be obtained and thereby the Universe would not be secured. However, in this ($k = a^3 \div T^2$ and $k^{-1} = T^2 \div a^3$) as well as ($k^0 = a^3 \div T^2 k$) is seated personalities that I am going to explain later in the book as the book develops and information progresses. Life uses time to create space within which it finds an ability to manipulate movement. Life developing the body in which life is housed is a micro-micro adaptation of stars developing and the time process in life is downscaled beyond what words can tell. As the star increases by cosmic development, so does the volumetric relevance of the star decrease in virtual size and as the body life holds increases in atomic size be cosmic development, so does the body decrease in its overall structural forming. This trend is of cosmic nature and man can never alter this process in any way ever.

Let us use a tale to try indicating time in future, present and past. Einstein's declaration of time, present, past and future as one is a lot of hogwash and such a statement is as far from reality, science, logic and truth as any untruth can be. There seems to be some confusion about what humans sees as history coming about from human interaction and what time is in cosmic terms. The cosmos rolls on growing regardless of life being present or being absent and life has no influence on the cosmos even if man wishes to claim such a giant feat.

While filling a carbon cosmic body, life serves as energy, where such energy is not being part of the cosmic creation because it can leave the cosmos and after leaving the cosmos it leaves what belongs to the cosmos behind in the cosmos. It controls what occupies space in time, and with that no one can disagree, because what life accumulates destructs after life abandons the use of the cosmic product. There is a point where life starts and a point where life ends, and these two points beforehand and afterwards has no cosmic ties. While filling the body with life, the life moving the body, is moving the body as energy flowing in the immediate. Time running is the immediate, the present, and holds the position of awareness while controlling the body functions goes on as unawareness. Then what is the immediate? The immediate is a point in a centre where whatever spins around that centre holds all other points, forming a centre of whatever in relation and in position to that centre in question. But time is what flows continuous from what was before to what is going to be and considering that perspective we have, that we think of as history. There is no history except for what we remember and what we care to highlight. History is in singularity and singularity is eternal. To the human mind history is what happened yester minute / yesterday / yester year. That is not history in cosmic terms, because human history is eventualities that either happened or did not happen as memory recalls. Time is what was that is that will be.

But the present cannot be without a past forming the present that then takes the future where the present will go. A body consists of atoms and every atom represents the body. Where the atom was in the previous instant is where the atom will be in the next instant it is going to be, in the very next instant following, which depends on time leaving space as a result of time flowing. That is history in cosmic terms but that is space filled by time flowing. My position where I now am depends on my position the previous instant and is going to be the very next instant. That is the cosmos holding time and my position which is not only is in relation to the centre I form, but is relative to every atomic centre point having a position in a specific allocated position somewhere in the Universe at that instantaneous one point reserved by every atom.

Time places any object in a specific arrangement at a point determined by every point at that point in time throughout the entire Universe. This time does because time flows by duplicating space from the past through the present and into the future. Every aspect in the Universe has such a point where the point has to be in terms of every other point in the entirety there is. Even the thoughts one has, and which we consider beyond time, is in time in space as well as taking up time. The thought is in singularity taking the future to become the past. While having the thought, the body occupies space and that thought is part of the cosmos and therefore has to have a place in which to be. However, without going into a thesis of debating, I must admit the place we refer to is once again another dimension devoid of cosmic light and therefore not in the cosmos we can see. That puts life in a dimension that goes beyond even singularity and that puts singularity in different dimensions we can only detect with intellect.

The following example used by me is done to explain time in the sense of human activity but be as it may in our importance it has no cosmic connection but is totally about the eventuality concerning life. It has no historical value and is not even a reality but only came about at the will of life forming energy movement. On close examination the entire situation is so human that in hindsight the evidence is so vague it only relies on human memory and human interpretation, which in cosmic terms are untraceable imaginary non-existing fantasy, but even the presence of singularity falls in such a category. Who can, by using what now is evidence available being in the present, can show evidence beyond doubt to prove beyond doubt that any part of ancient history is a reality and can serve as undeniable truth beyond question or doubt? History depends on interpretation and interpretation depends on intellect and intellect is part of singularity and singularity is part of the Universe but has no place in the Universe. How can any one prove beyond doubt there was a Trojan horse or that the Queen of Sheba did visit King Solomon? I can, on the other hand, prove that two thousand years ago the moon was at a specific distance from the Earth and position the Earth orbited at a specific predetermined angle in relation to the Sun. The cosmos places time as used in history as indicated by space – time, but human history is formed by events that does not actually exist but for being in the mind of the human, and on the condition that the human is wishing to believe the history. The history could have been or never was and in either case it has no bearing at all on cosmic development. Yet humans tend to give human history all prominence as if every aspect of the cosmos depends on the reliability of such recorded history. Let's inspect human history…

Thought is beyond the dimension light holds because thought cannot light up light. When thought establishes a current that flows in the spine to a nerve, the current is in the past of the immediate past onto the electric impulse that came as a result of thought. Thoughts are miles ahead in time of the electricity the thought generated and only afterwards released such electric jolt, which then started and commenced moving down the spine. The thought might be still occupied holding the same idea but then the idea is only a repetition of the thought that initiated the generating of the electron and the releasing of the electronic information down the nerves. But also the thought could be something totally different by the time the previous thought had an electron generated and sent off with information. The electron and the generating thereof happens in the space of time lapsed after the thought provoked the electron into forming. The electron is not the thought but is only a consequence of the thought through time using space.

There first has to be thought, to generate current and from such generating an electric impulse comes about. There is first thought. Then there is a generating of a current. Then finally there is a flow of such an impulse of current. The very, very, instant light becomes part of the picture, time has moved on, the person is with another thought although it might still be the same thought in repetition of the same process using the duplication of the process the thought went from the past going into the present and onto the future. This going on into the future brought in time lapse, the result of forming an electron impulse but forming the impulse came at a time frame after the initial thought came in place.

The light is current which is the same as the electric impulse, but when the thought provokes the impulse, the impulse is in a time frame behind the thought and that part we see through light, is the past space, carried back through space in time, putting that space occupied by the light, in a different position to the space occupied by time at the very, very instant it happened. Light or electric current or transmitting heat is not time because the thought that created the impulse moved on leaving the generated impulse as electricity using other time to establish a message between the nerve and the brain. Let's go and become very physical or in other words very practical.

Let's find the physics aspect that involves mass.

Kepler stated that the space (a^3) is formed by time in the instant (k) relating to time moving (T^2). So what did Newton miss in his oversight about all the aspects of Kepler's work?

The cosmos is time by gravity and so also is the Earth part of that cosmic reality with everything on the Earth. The fact that anything is on the Earth moving with the Earth unites whatever is on the Earth with the Earth as being part of the Earth. That is how the Earth awards mass.

According to Kepler, everything in the Universe is in the Universe because space gives time in motion dimensional qualities. The factor we think of as mass is the relevancy (k) that is attaching the object having the mass to the Earth where the Earth is providing the motion. The mass forms k extending from k^0 that is located in the centre of the Earth and the Earth rotating (T^2) around such a centre k^0 provides the movement time indicates which in case of the Earth (a^3) is the Earth (a^3) spinning (T^2) around the Earth's k axis k^0.

This is movement inherited as a cosmic reality, which we find in all circumstances and under all conditions throughout the entire Universe. In the Universe at large there is no mass but only relevant movement, thought of as gravity and confirmed by Kepler as $a^3 = T^2 k$ that becomes singularity holding space-time and in Kepler's mathematics it is stated as singularity forming space-time or is mathematically expressed as $k^0 = a^3 / T^2 k$.

Then on only one place in the entire Universe, life enters the scenario and as much as life is alien to the cosmos, movement of anything outside the cosmic movement is alien to the Universe. Life is the movement other than cosmic movement but this concept is only valid on Earth where it is the only place throughout the entirety of the Universe at large where life is a known separate ability to move and this phenomenon is only located once.

By moving the body, we find mass enters the frame, because by moving any object, the object becomes alien to the Earth or find the moving structure can be parted from the Earth to which mass unites the two. The Earth only moves what belongs to the Earth and which is part of the Earth. To move independently and in addition to the Earth is to alienate the object from being part of the Earth and in alienating from the Earth that object is moving in relation to the Earth independently and rendering the independence to become the factor moving independently from the Earth and only life possess such ability as to move where the reward in such alienating is independent will enforcing.

Only by the will of life allowing motion apart from cosmic motion, can mass as a factor play a part, moving through space, become granted a position in space. Mass and movement of life only connect to life and while anything is still with life, such a living object is not a part of the cosmos. A man sees something. The seeing is not the same as the thought, but is another thought that leads to the realisation of what the person is seeing. From the thought of realising what he sees, comes a thought as a response to what he has seen and the one thought becomes the past while the next thought becomes the present. To the human interpretation this is still one thing and it is all the same thing, but it is not. It all forms part of time moving forever and time could never stand still.

Newtonians are of the opinion they can have time standing still. I would love to see them achieve that accomplishment, because time in motion is God's active presence creating the Universe. Time moving is that by which the entirety is created and is maintained and sustained. The man seeing something is one of innumerable instances all flowing from the past through the present and into the future, and that is the Universe! It is the space forming the space we live in and in which time duplicates, in which time applies motion to ensure space-time. The fact that space is there is because time forms a centre holding a relevance from where a position moves between points, offering point one as the past, then comes the instant confirming singularity as it forms the present and where the movement is heading is the future and the second point of movement. Time moving through space is space-time $a^3 = T^2 k$.

a^3 **the space during which time this happened**

k is the instant in time where the event takes place

T^2 **is the moving of the time from past k_1 through present (the instant k^0) to the future k_2**

That is why time flowing is there in space, using the space it produced in the past as the past. Time going in repetition of the past, (space) forms the three dimensions (k_1, k^0, k_2) as time in motion T^2 to form the three dimensional Universe a^3 in time $a^3 = T^2 k$. Time has to move to sustain the form of the cosmos and it is there to give the cosmos the ability to change aspects.

A man realizes an idea at the one instant. The space-time he occupied during the time duration he had the thought, vanishes with light to the past, and the thought is stored by means of light, is somewhere in his brain. The storing of the thought came after the thought because it is the thought that is stored and is not the same thing. The thought in the current became history and then this process became currently stored as history in a cell as a memory of the past. Even the storing time, is no longer part of time in the immediate that the thought held, because light carried the message through space to another part of the brain. This means there are different events during the same time running parallel in one person. This happened while the person was already with another thought, using time for other purposes, which might be the same thought, but the time in space is another part of time in space. The position the man held in the space during the time of the thought is carried simultaneously by light, to the greater cosmos, away from the space he held in time during the thought.

Everything in the Universe has to be in a place, so where then is the place that the thought fit? The fact that there is a thought, presents the idea of space forming singularity but I am not delving into that because that I do in the **Open Letter About Gravity's Formula**. Life can only use what is available to use in the Universe, because life is unable to create anything. Creating is the prerogative of God alone. Thought is completely connected to life and where life is, thought has to be present. Even life existing in as little space as dual cells holding life as a form, is capable of thought because if there were no thought present, those forms of life would not recognise food or danger and would get extinct within one Darwin generation. If thoughts were not a contributing part of the cosmos, life would not have been able to manipulate the process, because again I repeat that life can only use what is already available in the Universe. Moving is the result of firstly thought resulting in movement and thought is the result of one contributing addition to the Universe in the form of life. It is thought that holds life and if thought is lost, the body life used goes cosmic in every way, notwithstanding how large or small such a body that served life might be.

Then the thought helps him conclude an idea. The moment of realization of the idea is a time period in which the idea was conceived, the instant forming the momentarily time duration that held the idea first, that period in the instant, already moved the idea into the past and by that time, the time in which the concluding of the idea came in place, was moved to the past where the then present pushed the time that was forming the idea into the past. Storing the idea by electricity in a brain cell as a memory happened at the speed of light that came after the time instant when the idea formed as a thought in the mind. The memory in thought and not the thought was carried by light to another part of the person's brain, which is his memory. (It is not the person in person; it is his memory, because the person and his memory stands apart and is separated by space in time.) In the simultaneous event, the space the person occupied during the time the person realized, the idea disappeared into space. Such movement is carried by light as slow as the speed of light. The space went to the past tense of the position the person held during the realizing of the idea and this took the idea into the past tense the person held seen from the point the person in present holds in terms of the position his body held in the duration of that time.

Both occurrences went to the past at the speed of light, which is ages slower that the frame tome moves in, which means the speed of light is not equal to time, because it takes time for light to move through space in time. Time is a much higher dimension than the speed of light occupies, because light carries the value or image of time gone by.

What I see in my present is the past of what I see. I can never see the present, because it takes time to move through space to bring a picture to me of space where I then have to conclude a thought about what I saw. That puts space in between my seeing and my realising what I saw, using the repetition of time and space is time delayed. Space is what was when time was there, but as time moved on, space was left behind as a legacy of time. If that is not the case, then where did all the space come from that increased since the time of the Big Bang? I can only see light as an image of the past bringing a message of what once was then coming to me as now in my present. It is not my present but light is my past.

If God did not create creation by thought, then life would not have the ability to establish a presence in space with time, because as God placed thought in the cosmos by creating Creation by God's presence as intelligence and thought, man finds the ability to repeat this process that is already a presence in the Universe and is placed there by the Creator God. Saying this fills me with sympathy for the single–mindedness and stupid idiocy atheists are suffering from.

Time is space changing and what changes more rapid than thought? Thought has to be in space to change space as time changes. This, I do realise is far ahead of what any atheist is able to understand, because they can cheat with mass to calculate and that is all. Stupidity found an identity and that is Newtonian atheists.

What our Newtonians fails to realise, is the moving or motion that is the most complex issue there can be. Nothing shifts, but everything dissolves only to be reformed in a new allocated position. What moves, breaks down into singularity and becomes reinstated again as time in a different relevance. This process does only involve one aspect of the Universe in the Universe in space-time using space to change by time moving space, and that is the moving of time through space. That is where Einstein's Universe draws flat. In the atom in the centre of all material that is forming space-time and from such a point in the centre holding singularity, space-time dissolves in the one allocated position and as time moves by changing and reapplying relevancy, time rebuilds the entire space-time that the atoms hold in the very next time frame. This happens to singularity maintaining space-time and does not extend into the entire Universe we think of as the entire Universe. In order to move from one position in time to another position in time the entire Universe has to disappear into singularity and be repositioned, because singularity in the instant or in infinity can't move. However, the entire Universe is the atom because every atom forms an entire Universe. We can gather time flowing by looking at life representing time flowing.

The thought comes first, forming a realising and that takes time. But to hold time, the though has to be in the Universe as part of the Universe, because if it does not hold time in space, it is not part of the Universe. The prelude to any deed is a thought that holds time and only then can movement come as a result from thought. The thought has to hold space, because the thought holds time and that is time in space. I know this entire idea and what this idea holds will prove to be above what the Newtonian atheists are able to realise and realising the Newtonian atheists' potential of comprehension about the Newtonian mind, that I dare say, is so simple to realise that even I can realise that much in spite of me being so uneducated. If all things created in the Universe were so simple as the Newtonians think it is, there would be a chance that something as simple minded as a Newtonian would be able to take charge of Creation, but it is not and therefore the Entity called God that is in charge of Creation is so big, the simple minded Newtonians are to dull witted to realise how involved Creation truly is. I am not insulting but I grasp the atheists' position. It has been long enough that the atheists held the picture that they are smart and therefore they can be atheists and I am a believer and therefore I am mentally incapable of grasping physics. I only wish to press home how utterly unscientific the atheists' mentality is and how little they truly grasp physics.

The importance of understanding the concept is most crucial, because this I show in **_An Open Letter On Gravity_** and moreover I show much more completely I show the importance of understanding the concept in **_An Open letter about Gravity's Formula_** where I use this evidence to trace the very point where the cosmos started and by that I show how eternity parted from infinity. Investigating this becomes the evidence showing us how the Universe formed.

Light is a number of spots following one another by duplicating. Light is massages coming from the past sent to the future where the past runs into to the future by means of duplicating the past as the future. This process repeats as one dot captures the next spot through applying motion.

Space is only space by motion of space and motion is the trend of time. That is the reason we have to speak about space-time because half of space is time and half of time is space. It is space a^3 that is equal to = the motion of space being time T^2 **k**. The photons are independent but follow one another denser or more widely spaced. These spots that are duplicating we call photons. If the spots are close in synchronising the spots can become so solid we can surgically remove body tissue with the light and cut metal. It is called lazar surgery.

By creating such a synchronised duplicating frequency between the photons following in the lazar photon duplication and the cells matching space duplication with time, the photons in a lazar can match the duplication of the cells as it happens and removing them. If the light duplication slows down in relation to our motion of duplicating we use to duplicate space by motion the photon spots may become invisible but still the spots will be identifiable as single units coming in a different frequencies. It is the manner that we hold a frequency to that frequency of the light and the space separating the spots. Our view on the density of the light depends on the frequency in motion that the light duplicate space as light puts time to space in ratio to the duplicating of our material in human flesh. This means we too have a duplicating frequency where we change space to motion to space. Our frequency might be so slow in ratio that some light becomes solid and have the ability to cut our tissue or our duplication may become so fast that the light will travel right through our seemingly solid tissue missing all the material or the light may generate so much duplication it will act as a solid structure that can (in a nuclear blast) remove all our skin from the body.

The duplication, which I am referring, to takes place at the point where the single dimension produce space by motion as it did the first instant. That is continuing since time began. The duplication of space is at a point man can only reach with his intellect and beast cannot reach such a point ever because of lack of intellect. That is why man can recognise religion and beasts cannot. At singularity every aspect of the cosmos is in a constant duplication of space by motion providing time. It is all in relation to the space being duplicated by the time used to follow one another by duplication to motion provided by time. The continuing of motion duplicating space of space from the one position to the next position to the next filling space is time. The space we have is equal to the motion providing the time to duplicate the space in motion by reproduction future space. In the same manner life is also in competition for claimed space and that makes up a very important part of life, which no one still discovered. Life is one action more than simply the duplication we find of space in time in the cosmos. Life applies motion other and in complimentary to the cosmic duplication of space by motion. Life can move apart from the Earth while the Earth spins. Life keeps the body in a semi fluid state but far from being fluid and in a solid state being far from solid (except for the bones) The body is not a precise solid, neither is it a exact liquid and it is far from being a gas. I would suggest we must begin to search for life's qualities at such a point. There are so many intellectuals that have so many opinions about global warming through fossil fuel burning and what harm and destruction may come to the Earth and life on the earth. Ask those with such a blazing opinion to define the basic principle of life and see what they say. Still they are of the opinion fossil fuels will devastate all life by heating the Earth atmosphere. They say this state of affairs is due to global warming from fossil gasses and pronounce doom and gloom while those scientists has no vague idea about nature. They may be good at mathematics but that does not make them wise, although that is not the way they see them. Ask those so clever what makes plants grow and they will tell you about water and they will tell you about nitrogen but that are all. A tree that collects thirty tons of matter in one life will remove something like a few grams of minerals from the earth. The tree removes a few grams of minerals from the soil but way thirty tons in al its collected material. How can the tree be thirty tons and only remove a few hundred grams of minerals from the soil if the tree growth was connected to the soil and what it may remove from the soil. The tree can only apply the carbon it removes from the atmosphere as plant material and release the unwanted oxygen. That is what plants do best. They remove carbon –monoxide from the atmosphere therefore they must replenish their future cells with some of the carbon they remove. This means that the more carbon we humans put in the atmosphere by releasing the carbon as fossil fuel waste the bigger the future plants in centuries to come will be. By finding a way to duplicate the carbon as the tree grows the leaves removes the carbon from the atmosphere and use it to produce the growth the tree will use as carbon material. The space the tree has is duplicated as the tree fills the space as it grows using carbon, which the tree accumulated in relation to the spin of the Earth as the Earth moves around its axis centre and about the suns axis. This is time. The motion of the Earth we have in degrees has to establish a frequency to the motion of the duplicating of the carbon that the plant use to duplicate the space it fills in the future. The duplicating and processing of the tree we call growth is directly in sequence to the spin of the earth. Growth is motion by duplicating space filling empty pace with carbon filled space is directly linked to the time to motion the Earth has is motion of spin in degrees in a circle coming from a straight line that forms a circle around the sun. Both is motion but both apply motion in relation to duplication of space in sequence to the sun. The Earth is filling the space by duplicating the space where as the tree is duplicating the space by filling the space with newly accumulated carbon from the atmosphere. The tree applies life above and beyond the duplicating of the Earth providing time. The growth rate depends on a ratio where the faster the Earth rotates in relation to the plant filling of cells with carbon is a growth cycle where the plant is not only duplicating the space it has but also duplicating new space acquired in going with the flow of time because it duplicates more than it had before while being alive. The duplicating of the plant material will provide the plant with time to grow. This also applies to life not belonging to plant material and it even applies to the fighting of diseases. The fossil fuel producing carbon – monoxide will bring all the harm to man because as carbon - monoxide proves to be food for plants it is most poisonous for man and plants remove the poison to replace that poison with food for the rest of us. We are not killing the Earth or plants…no…we are directly committing suicide so that the top one percent can be rich and the rest of us humans can be lazy. The sickness waiting for the future generations will be terrible as our bodies weaken from the poison we inhale to die slowly. We are murdering our future but we are doing little to damage plants or the Earth. Even life in a fight for survival is playing the duplicating game.

I collect a virus as the virus collects me as we pass each other on route to somewhere. Then the virus joins my space inside my body. The virus and me are fighting to occupy the future of the same spot. The virus holds a^3 but is fighting to secure $T^2 k$ and my cells are holding a^3 from their stance while they are fighting to secure the space in $T^2 k$ If the virus wins the virus will occupy that specific spot and that part I am holding will be missing. If I wish to win, I have to remove the motion the virus has that we refer to as life and if the virus wish to win the virus must end the motion that I have in my cells. By removing the motion life compute life by duplicating motion $a^3 = T^2 k$. That is the difference between life and the cosmos where life generates motion outside the motion of the cosmos. If the motion of life evacuates the cell the cell become just another cosmic object and will follow the duplication that the cosmos brings about. The second part of the motion is what the difference is between life and the cosmos. Through the motion life allow me to have I have the independence and the will to nominate a position where I shall duplicate the next motion. Life give the material life has the choice of duplicating and selecting a choice in locations available whereas the cosmic objects are in the moving clutches of the cosmos. It is either the virus ending my motion in will to provide motion or me ending the motion of the virus at free will that will bring along what we call death. When my motion stops permanently and is beyond restoring I am dead and the same goes for the virus. Life too is motion duplicating the motion of space but in choice of limited space and reluctant motion. In some cases the virus will beat the human tissue and replace the future position with a virus material while in other cases the human tissue will beat the virus tissue and claim the future spot the virus tissue holds. This becomes exactly a collision where there are heat generated we call fever and the destroying of tissue by denouncing the future space found by motion becomes heat, a by-product of motion destroying space. Ending the motion of life ends life and as long as life can establish motion life can defer death. That means when life relinquishes the ability to manipulate motion the abandoning of life in the cell takes place. Life is the producing of deliberate manipulated motion in the Universe. Let's us find the way space-time manage life.

With us humans being on Earth as long as we can remember we cannot take all blame of confusing cosmic energy with life. We are bound to get the cosmos and life all mixed up because we have developed a vision on life forming part of the cosmos. It is not our fault that life is all that we can recognise about the cosmos. Everywhere we look we see life. Even if the sunrise we think it rises in the blissfulness of life. The clouds we see are a fact of the cosmos but it is so highly significant we cannot associate the cloud without it being a part of life. If the solar system did not have the presence of life on one less important dot the solar system would be as uneventful as the rest of the cosmos. It is our human brainpower that gives the cosmos an event full history. Without human power nothing would occur so regularly it will flow from eternity to eternity without anybody noticing. It is life giving space the time we recognise but we must then recognise the time in space for what it is. In the cosmos the only separating factor that distinguishes particles is motion. Motion divides and identifies differences amongst material in space and material and space. In the centre of galactica there is little distinguishing because there is little motion parting the structured and giving such structures individuality. Cosmic life or cosmic energy is the result of heat finding a centre which then is promoted by heat forming a presence in such a centre that brings about individuality and an ability to part space by recognising relevancies applying as borders between structures competing for position. We humans have a natural inborn ability to confuse the cosmos and life in the cosmos. There are only a very few that will find some ability to recognise the cosmos for what it is…and then not even they really have such an ability. Those who landed on the moon and spent a few nights there might recognise the cosmos for what the cosmos represents. It is death. But still the death on the moon is as mild as it gets. After the moon it can only get wilder and much more deadly. That is the cosmos. The cosmos is death to life except this tiny dot we live on. The cosmos does not initiate or sperm life at a dime a dozen as the Mainstream science wishes to promote. The cosmos is about hostility towards life because life is alien to the cosmos. The cosmos is not about driving across some ocean and visit Mars. Mars will kill life as much as Mars is killing life by-products that life is sending to represent life and send back information. Life cannot just cross the space that is parting the Earth and Mars and conquer another planet. America was conquered and subsequently infested by many aliens from Europe but that is not what is waiting for those waiting to colonise Mars. We all have the opinion that the past will repeat because the past always repeated in the past. Those planning to infest mars are not repeating the past. They will die brutal deaths because of the relevancy of space-time. Every aspect about Mars will be as different as every aspect is about Jupiter and just because we find we can land a roving object on Mars does not make Mars any less hostile than Jupiter. The entire cosmos depends on relevancies forming as singularity that finds heat to support individuality and promoting new relevancies of which life cannot be part. Allow me now to investigate life as life is and cannot be part of the cosmos. Judge me by supporting or contradicting me but do not just merely dismiss me because if you do, it only proves that you cannot dismiss me.

The issue of establishing gravity goes beyond our most vivid dreams because life is the answer and the response to gravity. Without life (from a human perspective only) the cosmos loses all sensibility. If there is no life to establish and record established events which leads to the recording of time the recording of time disappear and with it the sense to record time. Only when it is important to life that the cosmos is $13,5 \times 10^9$ years since the Big Bang does time become an appreciated cosmic factor. In the absence of life the recording of time is senseless because who will be there to care. Remember that boastfulness and self-importance we inherit from the cosmos because that is what the cosmos is telling us by putting us in the centre of the cosmos. All light coming from everywhere puts us being every individual in the centre of the cosmos. Every point we choose to move too relocates such a centre for every individual. Every individual is in the centre of the Universe because light leads us to believe and accept just that. Life is the counter fact for gravity placing a secured central spot in the Universe. If heat was suddenly in the centre of the surroundings the heat will flow from such a centre out into the expansion forming light. By securing the heat of life in any given centre that centre secures the flow of heat towards that centre and not away from that centre. That becomes possible on one condition namely that life creates gravity from heat and not an expanding of space overheating, which is the case with the rest of the cosmos. It started with all being everything in one point holding all and everything in one spot. That mathematics teach us because what ever symbol or number becomes related to a power of zero (and not to zero as such) for instance a^0 or $10000^0 \times 10^0$ the only relevancy remaining is a point without sides and without any differentiation of established individuality.

Only by permitting heat to rise above all other points of singularity does singularity achieve individuality. It comes about as relevancy places conditions on the secured singularity and attachments on the responding singularity. Heating one point will bring about motion though the motion is only the expanding of space it will introduce the other reactions and the reactions will lead to a response development we named the Universe or Creation. Introducing heat brings about antigravity and by space occurring space becomes established with just that motion. The response is motion and motion is not being in two places in the Universe at the same time. That means time forms the second issue as time puts space in place. Every action brings about relevancy and there is truly no need to send off spacecraft beyond the ends of the Earth to find the proof to Einstein's theory on special relativity. All motion brings about special relativity as a responding factor. Gravity is the control of space extending boundaries or securing boundaries by committing gravity to such securing of space boundaries. The factor of motion brings about time by the double square and that four reduces again and afterwards by the root to divide by two. Expansion is part of contraction and that the Universe established as motion. The first motion committed the Universe onto a path that will lead from eternity bringing about eternity and ending in eternity. The proof of this is the Coanda effect where motion is the opportunity for new gravity to commit the boundaries of the special relevancies. But the Coanda effect is the initiation of life meaning it will end long before eternity arrives. If it came as a response from cosmic action it then only will end in eternity ending it. Sparking one point centralises such a point and from there gravity comes about to reprimand motion coming about from antigravity. That is "cosmic life" which is totally different from the life supporting our motion. Cosmic Life becomes part of the cosmos under specific conditions and ends with eternity. We having life have an artificial part of the cosmos, which is not part of the cosmos. We start life and end life without committing eternity as a part of this side of the cosmos. We can get off the bus before eternity arrives on this side being outside singularity. Life is not natural to the cosmos and while humans try to pretend that life is a natural response of the cosmos like gravity is, we can just as well go back to cave painting because we will never find a way leaving the dark caves eventually. Life ends with death and the cosmos ends with eternity.

Antiabortionists must not ask when life begins as they do in their quest to control the lives of others. It is no persons concern when life becomes a factor. For finding that answer we are too small. They must ask when can death enter as a factor since life is already established when the sperm swam to the egg. Life began when motion entered the equation of the cosmos and committed energy to occupied singularity. But if life starts at that point where life establishes motion and commits a carbon point holding singularity in motion the overwhelming number of sperm and of eggs becomes aborted in any case. Can the egg die and if it can then eggs get killed every time eggs are not fertilised. Then the sperm dies when not fertilising the egg. The answer is that the fetes is alive when it can die but when can the fetes die because when the fetes can die then is the fetes alive. It is death that places all and every relevancy attached to life on life. Understanding life in the cosmos and life being part of the cosmos makes us humans and not animals. All such fighting is only a measure to pretend to prove that every such individual supporting such a thought is the centre of the cosmos and such a centre does not belong to any other individual because all other individuals in the cosmos received a lesser centre than the one I enjoy. Even atheism falls in this category.

Man-In-Motion, Man-In-Mind, Man-In- Motive, Makes Man Blind

I wish to state here and now un-emphatically and categorically without any reservation of any kind mistaken double standard in talking there maybe. That I do not ever muddle with other persons religion. Religion to my thinking is as private matter as sex is. To dabble in other person's religion also goes as much against my personal religion as it is against the religion my Congregation upholds of which I am a member. It is not my place to convert whom ever or for whatever reason. I state here and now with no exception that never has this article or any other reference I may make have any purpose in converting any body to my way of view about the spiritual in any way. I believe it is not what you believe or not believe but how you live by your convictions and what ever your convictions may hold that is what does affect you personally in making you man or beast. Where I make remarks about atheism it must be seen as a challenge I put to intellectuals in their ability to think by reason and not feelings. My remarks about atheism are to show those practising in a religion those that believe they are intellectual try to put forward as to prove their intellect. By that measure I try to show how atheism is the reason of the thoughtless and what foolishness of exclusion it holds. If you are an intellectual, as all scientists believe they are intelligent also is a form of open mindedness and seeing what convictions others may have. Only the intellectual can see how other minds reason. Being intellectual is having clear thoughts and to have an open mind. All teaching of whatever nature has a positive and a negative connotation and the individual sets the standards. It is the way the presenter presents whereby the receiver receive the presented in a negative or a positive connotation. I am not intending or have any intentions in converting or changing any person's outlook on the spiritual side even in the slightest way imaginable. Me, living by my conviction to its full believe no bigger sin can there be than converting a person. Therefore no sin of man can be bigger than converting any person and even explaining that to others goes into my religious beliefs. My intimate religion I only share with my direct family. Converting others hold the most blasphemous behaviour man can divulge. Converting others to my religion is what more than any other fact holds the highest epitome of sublimation there is. Secondly and in line with the first is my belief is that I hold the view that the Bible has a base derived from ancient Egyptian teachings and I refer to that as I go along with my exchanging of thought in this article. To this day we are in and with all knowledge of splendour we associate with our times we live in, we can still not understand how the Egyptians erected the colossal structures and the manner in which they did it so many thousands of years ago. The Egyptians are miles ahead of us because the only advantage we have that is superior to them living back then is the manner how we are able to harvest energy and how we are able to distribute energy. If they were that advanced in all respects of the social standings it is not realistic to consider a civilisation being that advanced in one area exclusively and with no other wisdom or having other advances.

The hour in thinking has dawned where we humans must come to terms with the cosmos, with creation and with life. Mixing and matching was fine up to a point in the nineteen sixties where thinking about the cosmos and creation was a smart way to show superior intellect but being right or being wrong was only a case of honour and pride that could bolster one's ego or could hurt one's pride. Then came the age of serious space explorations and suddenly what we knew about space and outer space became serious stuff. Being correct or incorrect was a case of sending another person to honour or to the grave. Since the sixties life lost results from incorrect principles that people interpret incorrectly and matters of space and time got far more serious than it had ever been. Ever since man had a look at the night sky as a species it was what we saw and believed about the night sky that set us humans apart from the rest of Creation. We received intelligence when those we think of as our predecessors saw stars at night and then for the first time ever had a look at the night sky and by seeing the stars started a conversation. That first conversation was beginning of human civilisation when wisdom sprang from that what they saw. In another book as part of THE THESES I show briefly why I am of the opinion that man became human when he saw the funny silver dots in the night sky with a degree of admiration and recognition to the splendour of the unknown.

Now there is no longer only splendour in the unknown but the time arrived where man is on a quest to find the unknown and gallant as they ever may be, it is fool heartiness to send brave men and women on search into the darkness of the unknown and not know what are there waiting on them. What dangers will establish the outcome of their fate when, those as brave as they are, may find them on a route to travel through outer space without knowing what space consists of? It is not any longer plainly philosophising for the pride but it became an issue of life and death driving us to what we should find as evidence in distinction because where distinction should be became crucial to man and to machine and the survival of that. Man and machine survival is crucial to each other and life connects directly to a machine because man's life interlinks with machine as it did at no other time in development of the history of man. A great philosopher I may never be and I more than any other person realise this truth. A philosopher I will never be but on the other hand we are all thinkers. It is what we think about while we think that set us apart. Trying to think the correct thoughts might seem trivial and somewhat pretentious when agreeing about me never being able to be a philosopher but on the other side of the coin thinking controlled thoughts does not hurt me while I am thinking either. Please consider that even some advances may come from the weakest of thoughts when such thoughts have the intent to try to unravel a thread of wisdom while running along with a controlled thought. Wisdom can only come when running by measure of thoughts and it can come in no other way. With this I too wish to connect in sharing thoughts of woven patterns as I see them waving and for what they ever may be worth there may be some one somewhere that will pick up some advances when they consider my thinking.

My thinking is an accumulation of what I learned and what I learned was my personal foolishness that turned to my personal wisdom by paying in endurance and in personal pain. Most of my wisdom can be concluded in one realisation…I have learned to use as much as experience I can gain from other's pain because learning from their endurance save me my personal pain which I then might not endure. I have come to learn that it helped me to consider Life as being a dimension apart from cosmic material based dimensions where singularity is in a dimension that is holding a space capsule floating with time where we call such a capsule the Universe. We all are cover and moreover engulfed by space, we live in space and travel in time through space and no one knows what space is. Is space dimensions or is space a condition of the mind floating by paths through space…sometimes vividly and other times mentally but always presently…let us journey on a mind travel and find out. As I stand on Earth holding my first dimensional space-time displacement of our planet I can observe by using the second dimensional light source of the sun where my surroundings are made of three dimension atoms holding space-time in the forth dimension in time and space. I wish to move from point A to B and think in consideration about my planned action as my brain sends electric impulses to my muscles and that brings my muscles in mechanical motion. The electrons are three-dimensional travelling by means of the square to singularity from singularity and singularity is in the single dimension. Kepler taught us that when he taught us $a^3 = T^2 k$ It says that space is motion moving from a point to b point.

Arriving at point B I think to stop (not necessarily by thought or mental planning) and my brain stops sending electrical impulses to the muscle fibre concerned with the action in applying my body motion. At that point I come to a stop and my thoughts go to a rugby match played at Loftus, which is a South African provincial rugby team's head quarters. There is a rugby game on and from the commentators commentating I observe that it is where a game is in progress at that specific moment. As the proverb goes: I am there in spirit and my spirit being at Loftus are then some 400 km. south of my body where my body is on my farm in the Northern Transvaal. When hearing the commentator going on record about the progress of the match I do time travel with the exception that the duration in time it took my thoughts to travel is beyond human measure. Such a thought connect my thought to a thought I had a week before.

The very next instant my mind travels much farther back in distance and in time and space as I travel by mental motion to a rugby game played in Christ Church New Zealand a week prior to the day in question. In less than a heartbeat I went to the other side of the world. My thoughts took me not only half way around the world but out of the present time dimension. I relieved the party on the other side of the world and was in a dimension not available to any one else. Time travel surely seems to be a mind game. As I am standing in thought about the present forming in the presence of the past, I see my next-door Neighbour (to us in South Africa living in semi desert your next door neighbour lives normally 30 km. from you) coming towards me and my thoughts return not only from New Zealand back to where I stand on my farm in South Africa but also from a previous weak in the past changing and substituting the past for the present. The past became the present and I came all the way across the ocean travelling from New Zealand to the Transvaal in less than the wink of an eye. Now where I am in the very current time span I again return to my body where my body never left my farm or the present. My body was occupying the same spot it had before my travelling abroad came about and I was frozen in one spot all the time. it is my mind and my thoughts that return while I find my body moving towards my neighbour without my actual realising of this moving motion. My body and my mind are in two different spaces while they are in the same space. I must come back to this issue because if my body was representing me I have to be unable to leave my body since no object can be in two spaces in one instant. I use the second dimensional system in the wave to transmit sound with which I transfer the thoughts I think. The sound I transmit my thoughts with travels across the space using time by means of repositioning the three dimensional atoms between Old Neighbour and me in applying motion to the atoms. There is some confusion about dimensions because my thoughts are single dimension since I cannot see, feel or taste my thoughts, but my thoughts are transmitted by a vocal wave pushing atoms that are in the third dimension around where the pushing is part of the third dimension while all the while I use time which forms the fourth dimension to accomplish the fiasco we think of as understanding. The single dimension uses the second dimension to transmit the first dimension by using the third dimension and the changes in the third dimension is using the second dimension to re-arrange the third dimension in the time of the fourth dimension. All this complication is only bridging space between Old Neighbour and me and the space dividing us by the fourth dimension of space in the third dimension using time in the second dimension to transmit thought in the first dimension to convey what my mind harboured in the fifth dimension of my brain waves. All the dimensions all took part in sending a thought to another human and it crossed space in time to get to him being in the fourth dimension of space and time. He then uses the same system to convey a thought by message to me using his voice box and employing all the dimensions, which I just employed. Please note that it is a thought from the fifth dimension that I convey with the applying of organs in the forth dimension through ordering electrons in the third dimension to control matter in my body placed in the fourth dimension of space in time. While my words carry towards him, he drops down like a log as a result of not fighting something in the first dimension, which Newton saw as a force called gravity. Since my thoughts are obviously not in the control sphere of the first, second, third or fourth dimension that is all controlling space-time I have to presume that my thoughts have to be part of the fifth dimension. My thought bridges space-time within the first four dimensions at will which then has to be that thoughts are in another dimension. There is no zero dimension therefore we can only go up in dimensions and the next dimension is the fifth dimension. A thought from the fifth dimension prompts me to respond in the forth dimension by creating electrons in the third dimension while my body stands supported but restrained at the same time by gravity in the first dimension. The thought from the fifth dimension orders a response in using what ever is there to use in all the other dimensions available. All the other dimensions stand ready to be ordered by thought, which clearly forms part of the fifth dimension. My thoughts which is apart of a dimension higher than any dimension I find on Earth are ordering my atoms, which is in the third dimension to use the forth dimension of space in time to act by fighting gravity, which is asserting a force in the first dimension to respond and act on what I see in the light crossing the space where the light is sending a massage and such sending of a massage is holding a place in the second dimension which is restraining my motion in the forth dimension.

My response comes from some emotion that is reason and as such it is another dimension other than life in my limbs is. But life in my limbs is yet again another dimension from material not holding life but which is in the fourth dimension of space and time. If I was a gold fish or a virus I would not have the mental capacity of the reason to prompt me with thoughts as to realise that I am required to act in response to what I see. My having thought is a higher dimension as my limbs having motion to use as a response to what my thoughts command. The moving of my limbs can only be employed by my mind but it is not part of my mental reasoning or thought pattern. Yet my thoughts command reaction where such a message prompting my body to motion is directly conveyed form the fifth dimension to respond. It is aid that the message is the conveying of electricity but such simplicity can only be Newtonian by Nature. If I shock my arms with an electric jolt coming from an external source being conducted from an area which is outside my body my arm may respond but the responding will be useless as far as my control of such moving of my arm is. Lets get back to Old Neighbour and his problem of falling. Old Neighbour has a problem…he fell and after the falling he does not respond to awareness any more. I grab his arm because impulse in thought commands my arms to grab his wrist. I did not stand and reason for some time and then after debating with myself came to any conclusion. I reacted on thought where such thought was anything but deliberate while my responding on the thought was all out purposeful and most deliberate. I feel his pulse and find there is no signal coming from the heat as a beat. His breathing stopped and my next action of investigating his problem is to look into his eyes. There is suddenly a dullness in his eyes that was not present moments ago. Something went that was. I have thoughts in assessing his problem because his problem is that there is an absence of thought. My observation consists of thoughts relaying massages that is transmitted by my physical body in relation to my senses receiving and responding to electronic massage translations about Old Neighbour in the fact that he somehow relinquished all earthly responsibility and vacated all interest in his previously problems of an earthly nature. All burden suddenly befall his next of kin. Without lifting a finger (because he can no longer lift any finger which is the problem in the first place) he now saddled others with lifting his body and every one except Old Neighbour is now burdened with his last remains.

The heart shows has no beat, which is a sure indicator that there is a serious absence of a pulse moving blood around in his body. His lost of motion in terns of breathing shows that the longs lost all ability to provide oxygen for the transmitting of heat to enable the rest of the body the task of burning food. His eyes became stony marbles. Where I communicated with him moments ago, he has turned to stone because he shows the same ability to respond by hearing and speaking as a stone does and all senses has gone absent. My thoughts and breathing, heartbeat and hearing are still there. I can speak to him but it is his ears that have gone deaf. Even if I can squash air down his longs the attempt would prove futile because he is unable to use the air in his longs to convey any thoughts through his voice box. I can hit his chest with fury and support a heartbeat, and provide an artificial motion of blood where the blood will then be able to carry the oxygen but such motion can no longer create heat to keep him alive or with life. The air I force down his throat still has the ability to produce sound because when the air was in my longs I could produce sound with it. The air I can use to do my shouting to him creates and in that way the air has the potential of sound, but when the air goes to his longs there is no longer any of his ability left to establish a method whereby he can create sound from the air I push down into his longs. That ability even a newborn baby has, has gone away from him. His ears are still connected to his head and all the required tools equipping all previous aid that use to enable him to function normally are still there, all intact, but also it has gone forever. All the biological organs needed for hearing and making sound is still unscathed and remains in the right places where they are not damaged in the least, but the use thereof has gone. The electrons needed to translate whatever requirements enabling body function must still be in there somewhere, because I saw no discharge of any sorts flashing as electricity escaped from his body. Even by giving him electrons through an externally generated flow of electricity will not create any of the required but lost electron flow to generate life back in place. I may shock him till he hops around all over the place, but motion is denied for brain activity to function once more. His brain has gone empty, although it is full to the scull. That which I mentioned previously as being part of the fifth dimension is no longer present. Thoughts no longer command muscle and the muscle lost the ability to read the massage sent by the thought to the body whereby the body will respond with acting life. His thoughts are no longer with us or with his body. It is no longer Old Neighbour lying there, but it is his remains. That which is part of space-time is there. The body is there as the rocks are that is underneath the body. There is however no difference between what the rock is and what the body is. Even an atheist will tell you there is a difference in what is there on the ground and what were there present and part of the body only moments before he dropped to the ground. No heat or electricity can revive what he lost. It is a body without life. I challenge any atheist that thinks life is equal to electricity to charge his body with electricity and thereby revive the life that Old Neighbour has lost. When life has left the limbs, no electricity can replace life. Once the thoughts that Old Neighbour had has gone Old Neighbour left with the thoughts that went.

Minutes ago there was life in the body of Old Neighbour allowing Old Neighbour to talk, to think and reason, to discuss and to argue, be angry or glad, but that, which now is lying on the Earth has no more and no longer such ability. The source of energy giving life to the cadaver is no longer present. That which drove the body by thought is no longer driving the body because of the absence of thought. Jolt the body with whatever electricity you may deem it needs, such electricity cannot re-install thought to the grey matter. The grey matter is there in all the mass it had, but the grey matter prove to be completely useless in providing thought to enable to re-establish yet again control to the body as it did before the thoughts left the body. You may argue that it is energy that has gone missing and with that argument I have no quarrel since no one know what energy is in the first place. Science proved that energy cannot go lost but has to go from one form to another form. Energy can never destroy or vanish but has to replace form or attachment. The body of Old Neighbour is there, and it is holding all the organs and it is the organs that are there in their usual position that should still be performing. They are not performing. Old Neighbour has still got the required heat to perform because Old Neighbour has not gone but a few minutes ago and in the South African sun bodies do not go cold through lack of heat In Africa we are use to temperatures of forty degrees Celsius and more. The cadaver has all the essence to sustain heat and if life is heat life was unable to leave that quickly because even nights remain blistering hot in our part of the planet. If life was electricity, then I should be able to recharge him by connecting leads somewhere and call an ambulance. But supplying any form of current at any voltage rate will not bring back life once it has gone. You can heat him with a blowtorch while shocking him with a cow prodder (and does those things unleash electricity!) it will revive him as much as it will harm him or do him bad or good. He has become apathy in every sense.

His lifeless body will never carry his mind anywhere again because although the brains are still there holding all the mass it had when Old Neighbour was still with us, the brain is thoughtless and that has taken Old Neighbour away from us. He stopped breathing and such stopping is a serious consideration when one requires the body to function. The longs take in oxygen to carry heat into the body because as all atoms do, so the atoms forming the body also require heat to sustain cosmos growth. But in order to sustain life in the cosmos the carbon atom that is there to support life requires additional heat and by the process of breathing the oxygen alone takes heat into the body. In a process heat is taken in on an oversupply basis to perform the burning of food but the heat that is excessive and is more than that witch is required for use of the body's sustaining is dismissed. Two oxygen atoms links up and form a unit with one carbon atom. This is very significant because understanding the cosmos and the cosmic design of elements are extremely important to know what the medical process is within the body. By understanding the function of each element, we are able to read into the properties of each element to see what the function, the purpose and the duty is of every element that the element is there for to perform. The heat leaving the body is carried out by the carbon element that is supported by the two oxygen partners in order to keep the carbon element part of a gas. The oxygen no longer is burdened with the carrying of the heat and that is most significant because carbon carries heat completely different to the way that oxygen carries heat. In that manner the body repel as food waste as well as an oversupply of heat. That is another way how the body manage to rid itself of waste. But that is only part of the function of heat. In order to support life, the cell carrying life in the body uses additional heat to perform the motion of space- time that has the ability to manipulate space-time. The heat the body uses to apply motion and therefore perform a function we call life. Requires an additional supplement of heat as an energy taken in that enters the body in a process which we call breathing. Using the heat to move, is what makes the tissue grow and that is what causes aging. To fight off diseases the body also require extensive amount of heat and wear such heat becomes excessive the body may lose the function of life. To be alive requires the function of heat intake in order to enable the body to expel excessive food, which became a waste product to the body. Old Neighbour stopped breathing therefore all the factors above has no more purpose in that which still is considered as his body. Clearly such considering is incorrect and yet man insists to respect the dump of waste that was his body as still being his body but clearly does not function as his body any longer. Our dearly has departed although his physical remains stayed with us to rot if we do not take care of the cadaver and the sooner the better for everybody involved. We that are part of the living now have to move Old Neighbour because he no longer has such abilities. Minutes ago he still had the abilities to move about in a manner as he pleased but from him went energy which we consider as thought or life. It must be energy that he lost because all other necessities required by him in his for filling of his duties as a living being, such duties he still has (that is if you consider his body as Him). On the other hand the abilities might still be there but it seems his needs shifted as he lost the ability to control the ability to function. The minute Old Neighbour lost control of his body other life started controlling his body and those now in control of his body are not friendly to us being in and still part of life. His body does not perform in the manner it use to when he was still in control of the body by thought of the mind and since the thought no longer charge the body to function in performing body motion the body we attach to Old Neighbour cannot be Old Neighbour because Old Neighbour's body is there in the third dimension and by not moving independent it is part of the fourth dimension in space and time. The forth dimension of allowing the third dimension to flow with time as such flowing of time still is securing all his abilities to function as a human but those abilities has disappeared with his thoughts which also has gone vacant. All the effort he may muster will not allow him the ability to perform a wink notwithstanding the presence his eyelids still have as being part of his body.

The only visible something that Old Neighbour lost that makes him less of a human being than he was this morning when he woke from a nights sleep is the loss in energy which enable him the power of motion. The thought, which provoked motion by order of the fifth dimension, is no longer able to command such motion. His body with all the parts still hold dimensions in the first the second the third and the fourth dimension, but clearly it is the fifth dimension that has gone absent. The cadaver is still part of every dimension excluding the dimension of life and life then has to be a dimension above and beyond that of the fourth dimension in space and time. Thought to enable him to cross the oceans and land in New Zealand by thought is no longer a viable option to Old Neighbour as it was before. The cadaver is at present what we refer to as being lifeless and dead. The generator or power source or dynamo that drives the body by commanding motion is absent. That which serves to drive life or that which we attach such an idea named as what ever you may consider it to be that abilities and that dynamics providing energy in sustaining motion of the control of the human body has gone away never to come back. Those in science and those in atheism are of the opinion that life is just another cosmos tool, but there is no manner in the cosmos by which we can re-install life once it has gone.

Whatever any person may try to re-install life back into any cell, in is futile since the machine that gave drive to motion is no longer able to provide motion. All the wonders that the human body possess in motorised function is no longer in motorised function mode although it should be if it was only a matter of replacing the lost energy by providing an electrical shock or some fuel of some sorts. If life was an energy we found everywhere in the cosmos, there has to be energy everywhere to re-install life. Life is the only function in the Universe that may die. I know Newtonians cherish the idea that stars can die but that idea is as Newtonian as the magic of gravity is Newtonian. Newtonians whish to put fuel into stars as they did with coal stoves and when the coal in the stove or in the star ran out, death came to the fire in the star and death came to the burning of the star. On Earth people progressed past burning coal in stove and got electricity to provide stoves with usable heat. Not so in cosmology where all stars still burn by using material that will in the end finally burn to ashes and the star will die. In cosmology those promoting cosmology never did once look for a way to change a star to at least work on the principle of an electric stove. In the cosmos only life and that which has life can burn fuel to find drive. The rest of the cosmos accumulate heat around singularity to find drive. Nevertheless no fuel can get that motor that use to represent Old Neighbour's carbon based body running again therefore the energy that Old Neighbour lost is not a replaceable kind as in the case of ordinary heat from fossil fuels, food or electricity. The part that gives the orders to run the part we call the machine of human motion has gone for good. The part that did the running and took the orders are still there in all splendour but that just isn't enough to keep the body that the atheists say is Old Neighbour alive or back to life. Surprisingly the problem of energy and life becomes far apart when logic replaces Newtonian atheism and illogic. Newtonian atheism demands that whatever drives the body is only energy. Well so far they were pretty unsuccessful in replenishing that energy to substitute life!

I can shove a ton of coal down his thought and it will do him or you whom are still being part of the living no good at all. Coal is a source of energy but since coal is not digestible coal does not form a usable energy of consumption…where is the "all energy is the same" now? Roast him with electricity and see how far that will convince him to return to life. Push a gallon of pure glucose into his veins while you go about pumping his longs with air where he will start to inflate and see what his reaction will be on returning to life with all the receiving of energy he is getting. Newtonians all share the opinion that energy all over is the same. Energy is not merely energy and once again Newtonians got every thing very wrong. With Newtonian's incorrectness all the sheepish atheists go about and utter an echo that one can hear for miles around, but all the echo is only an echo after all in sheepish muttering with no ability to substantiating individual testing of thought correctness about and amongst the lot of them atheists. Let us test the statement and find the truth… Those who are thought to know best about science says that when energy is not used it becomes latent or so does science proclaim in any case. Let us take life as life is promoting motion and performing where life then is one of the many sources of energy and test the energy statement from that. Those that know best says that energy cannot be lost but transforms from one state it is in to another state that it is in. It is said that by rolling a rock up some hill one transforms the energy the rock had at the bottom to another form of energy, which the rock then has on top of the hill. That is oh so smartly thought of but for one minor technicality… They miss the part where life is absent and yet you find rocks quietly rolling uphill as energy drives rocks uphill. The fact that the rock goes up hill has little to do with the cosmos or with nature but it has everything to do with life having the ability to manipulate space-time and rolling the rock up a hill. That effort life is going into is just that type of manipulating that makes life what life is, the manipulating ability and controller of space-time by motion other and above that of the cosmos.

One cannot ever consider an instance where a rock rolled up a hill and its committing to accomplish such rolling by unleashing the rock's free will of setting a motion into time in space as part of changing space-time. The rock cannot muster the manipulating requirements to get into the situation where the rock is having the same latent energy it got without the help of life to role because to do some rolling up any hill all by itself the rock needed the same support of life as Old Neighbour had but that has gone absent from Old Neighbour as much as it was never present in the rock. Old Neighbour had to be very different from being what the rock is to be able to role the rock anywhere and in particular to role it uphill in the first case. When inspected closer life is the energy keeping the human body running as a motor and by distinguishing life the motor stops. In order to change the energy of the rock from latent to potential is due to the effort of life intervening and changing or amending the status quo. Something about life in the body of Old Neighbour's went latent and but by that which went latent forming energy it did not vanish. It seems the referring to energy is somehow intertwining as much as committing life by some involvement in the process of using or distributing the results of the energy distribution. Little distribution of energy can take place without involving life directly and one such a case is the weather but we will get to that later on. Life was part of the fourth dimension by commanding the four dimensions from a higher vantage point from and at the fifth dimensional commanding post. Now without commands coming from the fifth commanding post. Old Neighbour is as dead as the rock is and has now the same ability as the rock has where in the past and with life Old Neighbour's use to be able to role the rock up a mountain. The rock previously never rolled up the mountain without life supporting and enabling such rolling but now Old Neighbour's has the same insufficiency as the rock has to role up a mountain. Old Neighbour had the ability to role the rock anywhere it chose to do it anywhere and that ability was part of Old Neighbour's up to the moment Old Neighbour went latent. Life shared time in the body and space with the body thus it was part of the matter of the body that carried life all around. That ability of independent motion that life had before life renounced the second it vacated the body it no longer uses. Without doubt is the fact that life was the indisputable source of energy driving the body that life occupied for years by the person we know as Old Neighbour every where (on Earth) through out the vastness of the third dimension that the Earth has to offer. This motion came about and was separate from the motion life inflicted but formed the basis of motion while the Earth was taking everything in its sphere of influence around by employing the fourth dimension of space travelling by time. The Earth and the body share space-time by the third dimension while travelling in the forth dimension and that may or may not include Old Neighbour but it does include the body Old Neighbour use to use. Then Old Neighbour went missing from the body that did not go missing and even the atheists have to agree to this part. But having Old Neighbour ending his sharing partnership with his body does not end the body being present in the fourth dimension. The body no longer takes orders from the thoughts of Old Neighbour's and therefore what ever Old Neighbour was cannot end just because of the fact that the body and Old Neighbour ended a lifelong partnership. If it was energy that was driving Old Neighbour, then the energy cannot go missing...it is either that or it is millions of teachers in science classes are telling small children a pack of lies! Teachers all over the world are teaching children about how energy never goes missing but only change form by transforming. Where the space-time sharing between a carbon body and the thoughts that use to drive and control such a carbon body then ends, life cannot end just because the partnership went sour. The two factors no longer share space-time but the two factors cannot disappear because of the functions of the body ending. The body is still there but without the ability of independent motion whereas the motion may have ended controlling the body but being energy could not abruptly end being... If energy was as invisible as Newtonians lay claim to, it cannot end and with life manifesting as energy while in command of the human body it formed a duty as energy which proves that as an energy it may transform or reform but cannot go into a mode where life could simply vanish. Remember how that energy, which drove the body had the ability to leave the body and journey with my leaving my body by thought in order to take a trip without the body to New Zealand and revisit a rugby match that was played in the past but also in New Zealand? The body was not tied down to the time restrain it was in and neither did distance have a grip on the life in the body. It could drive the body while visiting places outside the body. The leaving of the body did not affect the energy driving the body while life was inside the body when it gave the body a function of movement, which the body had and that gave the body an ability that no other cluster of atoms enjoyed sharing one unit of space in the precise measure in motion of time and the precise quantity and form in one construction as that what Old Neighbour had enjoyed being what Old Neighbour and his body was in the Universe. It gave the body of Old

Neighbour the means to displace space-time not only by gravity and motion as all other structures have but it gave the body a means of changing the space the body occupies in the time duration that life controls and directs the body that life puts claim too while life occupies that body. It gives the object carrying life more independence than any other structure through out the Universe can claim. No mountain can move a little in the morning to avoid the blistering sun and shift to another place at night to escape a blistering cold wind. Being the mountain does give the mountain independence from all others in the Universe because the atoms forming the mountain puts claim to a specified space in a specified duration of time. Being the atom within the mountain gives the atom an independence from the rest of the Universe, which puts the atom down as forming an entire Universe within another Universe. That is my Universe…there is always some Universe within some other Universe… The independence that the atoms has forms the mountain in such a manner that the mountain is a unit and enjoys independence by form all other Universe units formed by atoms that is joining forces in motion by motion. Spinning in a unit the atoms forming motion as a unit is the way the Universe allows the independence from which there may be Universes allowed forming other units independent and bonded to other Universes. The human body including all life on Earth can shift by free motion from the position the body presides over to alter change or convert the position that the body has by thought that is establishing motion in the forth dimension as to suit the needs and requirements of the fifth dimension using and commanding space the body with life enjoy in time duration. This means that what ever body carries the ability life, life then bestow on such bodies an independence greater than all other independence and such independence is very exceptional as nothing ells known to man in the cosmos can achieve such motion by pure will power exerted by thought. It puts that which carries life in commanding space-time far apart from that being in space-time but disadvantaged to the point of not carrying life. A plant may not be able to run to a better position but when in competition for sunlight it can try to outgrow its neighbouring plants and claim a larger share of the available heat that the sun has to offer. That effort to accomplish motion as to secure a better position is completely out of the domain of any rock. A plant can grow its seed in such a way as to ensure distribution and gain advantage over the spread of its ability to control as much species space in the general space it holds on Earth of what space the Earth has to distribute in territory. There is no chance that a puddle of mud can run after water to keep wet. There is no chance of the puddle of water slowly crawling in the direction of the reclining stream. No rock can go for a quick swim to cool and then return to its position afterwards. Life can manipulate the space-time it holds by changing the unoccupied space-time surrounding the space-time under its control to its advantage in the sense of bettering its chances on survival as well as its species chances on finding relocation in a more suitable environment.

That is the overall advantage that life holds and that prove that life is not merely a type of energy referred to in general terms that does some work in relation to the growth coming from time that develops the Universe. Life is what no other form in the cosmos can be with only one disadvantage going in life direction. Life is very much temporary whereas the rest of every aspect in the cosmos is shifting from one eternity through development of singularity at a rate of one complex cycle after the other of eternities. Eventually singularity unification will once more form one commune in the final form where singularity started from again being in some unknown but frozen form in eternity. In contrast to that shift of eternity the cosmos endures to gain in progress, life finds progress much faster since life has the privilege where life can die. Dying is not stopping to exist because that only nothing can accomplish and the only thing not present in the Universe is "nothing". Dying must be going to where singularity rules and where death is taking life to a position from where what is in control of cosmic development between positions in eternity is extending to an eternity from where the Universe is being controlled. The controlling measure of space-time is coming from singularity, which is by dimension not part of the Universe space-time is in. Since life shows the evidence through life's relation with light that relation can only be if life was part of the centre of the Universe. The centre of the Universe holds singularity and from singularity life is a focus point in the centre of the Universe. Life in the centre singularity is then also matching singularity in being in the centre of the Universe, and with the eternal qualities we only find present in singularity, life therefore has to retract to singularity by dying. The eternity from where life comes and where life has to move to is the same eternity one will locate in the eternity where we find singularity. But life is much more than just singularity, because life controls from singularity as much as life is in control of singularity and as much as life control singularity. Singularity cannot move, yet the singularity life takes charge of, life moves around by manipulation. Life is in motion of space-time being motion of space. Yet life is not in control of time. Try and measure the time a mountain holds space and compare that to the time given too any one in any form that life holds and measure the duration granted to life measuring from birth to death being in the time life occupies space on this planet. Then after getting an unbelievable answer a person can appreciate that life is the energy and without life the structure we call the body holding life becomes the equal to what a mountain is from the onset of the lava flow to where gravity brings the final demise. Life makes the body it control special and where the cosmic structure which holds life, loses life the cosmic structure we think of as a body, just becomes one other cosmic construction with no special qualities that sets such a body apart. Life is the manipulation of space-time and the higher the degree of advance is, the more life can manipulate space-time. Connected to that is the duration that is granted to the higher developed forms of life.

In the case of humans life extends beyond the human body holding life. Life can extend by being a development of the manipulating qualities life is given. Humans started to build machines, which has only the purpose to serve man in his manipulation efforts and in those machines, are built some qualities that is only given to man. An aircraft flying may be as dead as the next mountain is, but through the aircraft, man as the ultimate form of life can manipulate space-time far outside the reach of lesser species. I do not wish to start comparing life as being advance or more advance so I leave my argument at that as far as life development goes for now. Nevertheless man has not developed wiser but only more clever. Man has found a way to harvest energy way beyond other human or animal power. That has also given man the ability to destroy and kill off his own kind in destruction where that tenacity in killing had the ability to kill all of mankind if man had that ability available a few hundred centuries ago. With mans cleverness he got more destructive in his killing lust. Let's leave corrupt politics to corrupt politicians and return once more to Old Neighbour and his problem of having no more problems.

In the very beginning I stated that through the way the mind travels, it has to hold a higher position than what the body holds because I showed how easily I could travel around, even as far as half way around the globe in no time at all. Sure I was not there in person, but my thoughts conveyed some understanding of what was happening on other places outside my range of vision. Our arrogance we that holds life show about our importance in the Universe is not totally unfounded because we see our position we have as being smack in the centre of the Universe. We take so much light for granted never thinking for one second how impossible our relation with light truly is. This totally extraordinary relation we have with light must be one of the reasons why we humans put our position we have in the Universe in such a pivotal place. There is some substantiation in the fact that all of us in person with the ability to use our hind legs to walk on Earth, we all have the inborn idea that the Universe was created especially for us, us being those holding life while we are on Earth. We all realise that such an idea is absolutely bizarre. It is so ridiculous I wish to compare what I said with the importance we give to an ant. It would be the same as if the ant in the park thinks a thousand people are paid to maintain Central Park in New York with one purpose and that is to please that one ant…and yet that is happening with the light and us. Every person is standing in the Universe and all the light through out the Universe is directly flowing to the very point where that person is standing. This being in the centre of the Universe and finding light flowing precisely to where we are standing is what happens to all of us. The place where I stand or any other individual for that matter is standing is positioned in such a manner that every beam of light travelling from all over the Universe is directly flowing to that very specific spot where that individual is standing. From all the corners of the Universe one line of light is especially directed to that location where I or from another persons point of view, that person is standing. The light departed from every location following in that one direction and that is leading the light all the way to the spot where I am filling that spot. All the light in the Universe is coming to me. Straight to me because where I am standing I am filling the one spot on Earth. That is also the centre of the Universe. If where I am standing is not the centre of the Universe, then why would all the light flow all the time it takes to flow to come to that centre spot? If anyone does not believe what I am saying or does not appreciate my saying that then go outside and see the vastness from where the light is coming from. It is coming from all over. It coming from areas so large not even Einstein can calculate the size of the Universe at that far location and all the light is rushing towards me specifically. There is not one ray of light that is going to miss me by fluke or accident. The left where it is coming from as the light through all it's travelling had one purpose all the while it was travelling and that is to meet me at the point I am. Every beam has my name on it and it is coming for my eyes. Can any one imagine if a person was standing in a location in a specific city centre and found all the persons in that city was running towards him where he is occupying that city centre point, how frightening such a person must feel. Yet it is happening to every one from where ever the vastness of space is situated and light may be located, that location is sending the light off to meet me where I stand. The light is coming across space to that very specific point the viewer is standing. The only reason why we would form a pivotal role in our position we hold in the Universe is because what we are, occupies space-time from within singularity. Since singularity is the centre of the Universe and we are in the centre of the Universe we are in singularity, and from singularity comes the control and establishing of space-time.

That is just one small part of the fantastic affair. Some of the light left the stations they come from some 12×10^9 years ago to meet little old me in this spot I am filling. The light has been travelling 12×10^9 at the speed of light, which I might add is much before my birth. All the while it was travelling it was crossing space and time, rushing all the way to meet me at this point. No one ever questions how was it possible for the light to know I was going to stand at this point and wait for the light to arrive. I have to ask such a question since only I am standing here and only I am receiving the light at this position and all though I was on the moon, the light would still trace me at the location I am. How did the light know I was going to take centre stage at that moment and fill the specific centre of the entire Universe? I have to be in the centre of the universe because all the light is travelling to this spot filling the centre of the Universe. The light takes two million years only from the closest next galactica to meet me here and after all the time it is meeting me in the centre of the Universe. How important can I ever be? Light is coming across time measured in millions and billons of years through space measured in millions of trillions of kilometres, ignoring all other places it could go too and came to meet me in the centre of the universe.

To the light on route time means nothing and space even less. Light it seems cannot be more motivated to reach me at this point where I am filling the centre of the Universe at this moment. Not one ray is by accident missing me except by my choice prevailing. It flows through the Universe in time and in space in the hope I whom is filling this spot where too all light is coming, the light has to be having the hope I am graceful enough to notice the light arriving when it arrives. If I were not in the mood to acknowledge the light arriving after all the time it took to travel and meet me at the centre of the Universe which I am filling, the light would have done all the travelling just to be disappointed by my not meeting the light. An effort spanning billions of years and an effort stretching trillions of mega kilometres will all be in vain just because I neglected to meet the light. How important can any person be to find light coming that far to meet me standing where I am standing. From everywhere the light is coming my way but that miracle is not passing me by although it is passing me by. Without noticing it I am noticing it because from my attitude and the attitude every other person on Earth has, I am very aware, as is every one else aware of his, hers and my position where I am filling the centre of the Earth. I am feeling even more important as to acknowledge the total importance I have in filling the centre spot the Universe is offering me. The light is tracing me specifically at the location I am occupying just to please me and serve me with all the information about the history of the Universe. I can accept and acknowledge the effort the light is enduring or I can dismiss and ignore the lights efforts. The choice is mine and the entire Universe has to take my sedition and leave it at that. I suppose that will allow me some arrogance and encourage me to think this all covered by the Universe with only me in mind was specially created with only me in mind and if I wish to draw a map about the Universe I have all the right in the world to place me in the centre of the Universe where I so clearly belong and from where the all of the everything is meeting me to secure my importance. After all, all the light travelling the cosmos is doing it!

Even if I shift to another position anywhere in the Universe I will find that the light will change direction and trace me in my new location. Even if my new location is on the other side of the Earth, the light will still get me at that location. The light flows to me from where ever and to top that it is also flowing to all other persons. The only precondition is that the object filling such a centre has to have life in order to realise that object fills the centre of the Universe. If it does not have life it will still find itself in the centre of the Universe but it would lack the realising of that fact. That means it is not the Earth that is that important but it is where the location is that is filled That point the observer is using to view from, is what is so incredibly important since that is the location of singularity. Singularity is the point in location where the Universe was born. If it was only the sun that the light was streaming from which is choosing me as the centre of the Universe it then cannot be that very exclusive since the sun is my prime source of light and I am the sun's prime donor of light. The sun is close and the light is plenty. But it is coming from all over an everywhere there is a possibility that light can come from.

The mind sets a norm that the body can follow or not follow but the body never sets a norm that the mind cannot follow. All cells forming my body and even those holding life has an electron a neutron and a proton and very deep within the very deep within next to the truly unknown is a structure that holds position in relation to singularity. When a cell holds life it is different from a sell not holding life although when the cell is not holding life it still constitutes of the same composition it had when holding life. Yet when life departed everything changed when something changed because something is different from what was present when life was around to fill the cell with motion.

A life-carrying cell not carrying life while being a part of a body that is other wised filled with life has gangrene a most deadly disease that kills as none other. A cell absorbing heat normally is showing growth whereas a cell in abnormal heat intake is cancerous and again is deadly. I can go on and on about this but it is apparent that as soon as life looses control over heat and heat distribution to carry out motion that connects life to the living cell the stabilising factor or things holding life in the cell goes abnormally wrong and such conditions can, may and will lead to life vacating the cosmic position it is occupying and the vacating is permanent while the occupying the body is short lived. To understand the way I wish to direct the argument please allow me to indicate how I see the normal as we will find in the cosmos in life carrying and non-life carrying matter. The cosmos began when the principle placing the line, the half circle and the triangle equal to the only available value there was at **_the time_**.

It is an obscure thought but was the Universe created to be useful to life or was life created to make the Universe useful. Let's face it the Universe will not end when life disappear from the Universe but what will make the Universe worth while if there is no life to appreciate the usefulness of the Universe. When life disappears from the Universe and the Universe return to being an eternity driven cosmic engine, what will appreciate the engine driven by heart. If no one is there to appreciate the measure by what greatness we may admire the Universe. When life disappear who will then wonder why creation established the equality between the straight-line, half circle and the "what comes after that which comes before". It all will seem to be somewhat wasted when the Universe runs it s course without any appreciation of admirers to admire the spectacular but the spectacular will never disappear because of the loss of the admiration. With the admirers present to bestow admiration we must investigate what we appreciate because by our appreciation will we learn what there is to know. It all started with one spot so small there were no sides to the spot. Everything that is here and is present as well as all that was in the past and has overgrown its presence being part of our presence as well as everything that will grow into the present when we deteriorate into the past was present in that one spot. That spot was singularity because that spot still is singularity.

The line $=180^0$ ➡   The half circle ⌒ $=180^0$ The triangle △ $=180^0$
Since almost before serious recorded history dating became scientific principle mathematicians knew that the straight line holds 180^0 degrees matching the half circle as well as the triangle. But never have I read any definition about this phenomenon and how it comes about or what may cause such odd connection. Heat occupying space has the cube that can apply r, as a straight line bringing about the cube with all its other names that may find attachment to specific form but nevertheless still remains only a six-sided cube with angles changing in some cases. This was when the spot became the dot and the dot had possibilities of forming a Universe which may (if all runs well) to fit life into some dot that was till part of a very tiny dot in a dot that would someday be called the Milky Way. From such a small spot came a dot and the spot still forms part of the dot but the dot grew larger than the spot.

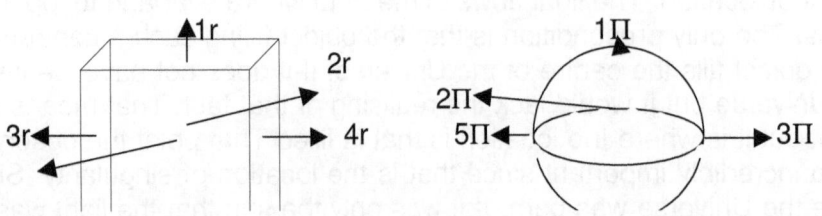

In the sphere there are no radius but only the extending of Π from the centre Π in six opposing directions relating to one another by the square but remaining Π because of the unity the matter holds in relating to space. It is not possible to draw a precise line that would form a precise ring and not cut some atoms in parts. Because there will always be an atom disallowing the precise positioning of the circle the circle continues on a solid basis holding Π as a positional reference and not r. In every sphere there then are the seven Π relating in precise dimensional and positional equality forming equilibrium to the centre Π as well as to one another by 90^0 and 180^0 implicating the dimensional positioning. Therefore the sphere holds 7 ^{Π} and the cube holds 6 r^2

Where space comes into contact with the sphere the cube loses one of the six dimensions it has to the more dominating seven dimension of the sphere whereby the seven dimension in equilibrium will dominate the six dimension loosely connected bringing about that the cube then has 5 sides to the seven of the cube. This means that in the cube the "bottom falls out" and without a "bottom" to support objects they fall to earth. Remember that a body "floats" in space, but at one specific point it starts to "fall" to the earth. That is gravity and it is a dimension change much more than any force.

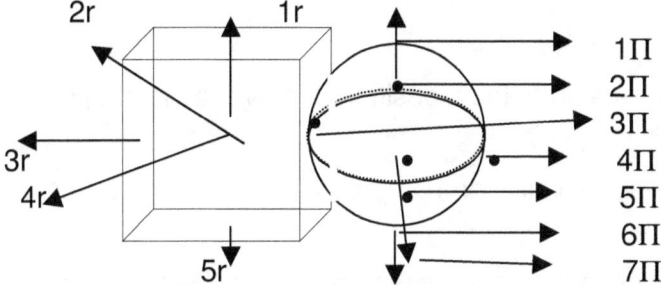

From such a point every other point will be opposing any other point not pointing in the direction to which the first point is pointing, whereby it extends the direction it holds. No matter what the point is or where the point leads, such a point holding a specific direction will be unique in the direction it is rotating because at that or any other specific point wherever, it will be directing not in the direction it spins but in the direction flowing from the centre point outwards.

All atoms are a minute form of a coming black Hole and viewed in the structure composition it is clear why I say this. On the outside there is heat trying to get inside the atom where the heat is needed. On the inside of the atom there is a need for heat and the inside is in constant regulation of the heat flow as to keep stability. In understanding the dynamics of physics we must understand the cosmos where the process begins and where the process ends.

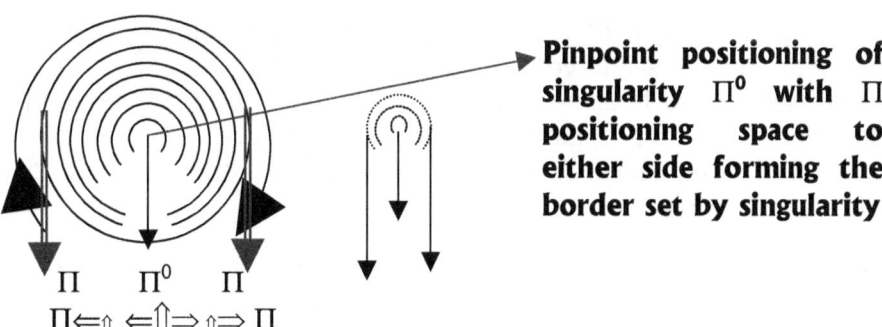

Pinpoint positioning of singularity $Π^0$ with Π positioning space to either side forming the border set by singularity

The atom holds a very unique position in that it links three dimensions to a forth dimension and this part is where I came to understand Einstein's thinking but a with Newton I could not accept Einstein's thinking. Only bringing in religion could I get further about the formulation of singularity because in that I found what connected the Universe whereas Einstein left space as space and tried to link time to space as an additional factor.

That would be the same as not linking life to the body or exclude life from the body while trying to argue about life being part of the cosmos as Newtonians seem to do. This occurs in all atoms through out the cosmos with no exception on the rule. But life-carrying atoms in carbon $_6$ commits life as an additional supplement to the atom as life can become absent from the atom leaving still in the normal range of a cosmic structure. In the past number of pages I brought reason to those of reason that there are more to the body supplemented by the presence of life than merely carbon fibre. It runs much deeper than physics can intrepid. As far as pure physics go, nothing changes when life goes absent and yet everything alters when life abandons the atom.

I saw a very neatly outlined connection that the atom has in its position in the Universe as it was the evidence of the smallest all connecting matter tying what is matter into a small container.

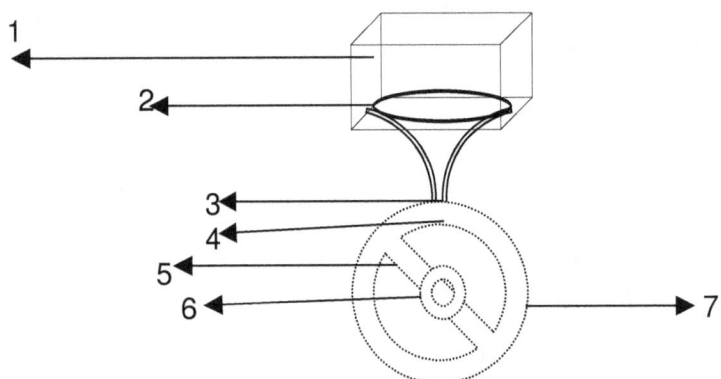

1 The square value of space-time in the fourth dimension holds a positional relevance to singularity by 10 points or places.

2 The space – time in the square of space loses the value of 10 by entering the atomic relevancy formula and become 3 sides to the cube.

3 The space – time holding space as the square loses the value of 3 by replacing 3 with Π thus going on down the line of the atomic relevancy formula and become Π sides to the circle as $\Pi^2\Pi$.

4 The atomic relevancy of space-time displacement changes once more to Π^2 where space goes flat in time.

5 The atomic relevancy of space-time displacement changes once more from the neutron square of Π^2 to the proton's double square $\Pi^2+\Pi^2$ as space unoccupied disappear and time forms the square by the square. The relevancy flowing from this figuration is so very important when archaeology presents facts and with this the archaeologists (may dare to say) became the blabbering fools they should not be as they are presenting serious science as a qualifying joke with the funnies they come up with in setting every one with thinking minds laughing.

I shall return with this argument in a later time when more facts relating to the argument are exchanged between us. For now we still have a cadaver on our hands to dispose of and quickly we must because of dire consequences that may follow if not done in urgency. Rotting corpses bring untold diseases as the great influenza epidemic of 1919 can support in evidence. Then why the danger about the corps all of a sudden if life within the corps have such minor importance as the Newtonian atheist wishes us to believe.

The molecular structure is still the same and so are the chemical composition within in the body and the mind. If it is chemicals making up man then, the man should be there because none of the chemicals went AWOL. If life is about the electricity that runs down the spine, our distinguished atheists should replenish life with some minor application of current as a means of stimulation. All the ingredients are present yet the manipulative nature of life making life so very exclusive to all other cosmic ingredients does no longer function.

Everything the fourth dimension can provide is still present within the body, yet that substance beyond the forth dimension, that ability we cannot detect but with a lot of intellectual thought, that ingredient Newtonians deny them of recognising as very special exclusive to life is not there any more. The ability manifesting as the energy or energy supplying has relinquished its role within body and mind. If it is only a matter of electrons generating pulse consistent to flowing from sector to sector in the providing of artificial current from an external source should supplement life's conducting of the work of the body. But shock the body as much as you may, the body functions disappeared with the disappearance of life from the body. With the exit of life the vital electron distribution seized and the cadaver became just another particle containing what all substance other than life contains. The cadaver became just one more structure in the cosmos with the same ability as a rock or a mountain. The electrons conveying the massage may be replaceable but the sender and receiver and the decoding of the massage has no longer any function within the body. We may send some electrons into the body, charge the body with oxygen via a machine doing the pumping of the air to precise rhythm as did life, we may stimulate the heart by nerve pulses artificially supplied. We may contract muscle by supply of electrons. But replacing life can never be one of our accomplishments once life has left.

All muscles including the heart and longs can be stimulated artificially and in doing it may prolong the writing of the death certificate, but having the person stand up straight once more or pronounce some wish to be for filled or just a simple effort as winking eyebrows when asked to do so is far beyond the abilities of the lifeless structure occupying space-time in the fourth dimension with the aid of all other dimensions excluding the fifth dimension. Previously the body seemed to manipulate space-time with the utmost ease, walking wherever the mind chose to walk through unoccupied space-time, only adhering to the restraint compiled by the other dimensions and inflicted on life to startle the manipulating abilities. Now such abilities disappeared, vanished with life to some place we see as death. But life as energy has permanency that can never become denied and disappears. It cannot become washed away, wished away found to have disappeared for it is energy, a something of eternal power in being.

As energy has linking to eternity it has to come from somewhere as much as go somewhere it came from after it left. If it went latent even then it has to be stored in some place of gathering energy of life's nature. That storage facility is not part of the fourth dimension as the rest of the body is, so it will no longer be in range of the detectable, yet it must be somewhere with the linking energy holds to eternity. We also can tell that lesser forms of life will destruct the composition of the cadaver feeding and preying on it till what is left has no longer use in sustaining other forms of life as food. We learn from the esteemed and well respected Brainy Bunch the food we eat provide the energy we use. Sure that is very to the point and easy to swallow. But what uses the food in maintaining the energy. This is my problem with the SUPER-EDUCATED-WISE-OF-THE-WISE as they will forever give information that has no substantiation but only scratches the surface and leads nowhere. How does food give the energy and what uses the food for energy. If it is the protein filled body of the human flesh, then what makes the intake of food become meaningless after life departs. Should the food be energy then just feed the man and revive him. Food is all he needs to regain life and if such an effort fail to revive the dead then one should seek to find deeper meaning to what is obviously not obvious.

If life only connects to the fourth dimension of space in time through energy supply such as food may supply then the body is there to be nourished back to life. Also in the opening of this argument I showed how life travels time with no limits to boundaries in time. I did admit that such travelling only applied to the mind and not the body but since it is the mind that has gone absent and not the brain, then the travelling time by the mind must still be in affect as it was the mind that did the travelling and not the body. The body did the travelling of the time constraint but it is the same body that cannot even permit travelling to its grave because of the absent of the mind.

Time did not restrain the mind and if it did not restrain the mind time must have little control over the mind. Since the mind is no longer present in the body and time is still doing all the restraining on the body we may conclude with some extelegence that wherever the mind goes in storage time will hold no constraint over it. All the atoms that were in use by compounding a human body would one day again form some flesh of another being. It will ultimately not be human and it will obviously never form a body as one group of solid fleshy matter and it is logic that the compliment will become divided when forming a future body amongst millions of bodies all containing substances previously located in millions of different bodies to form millions of different life forms but it will become use full once more in the future.

With such a remark I do not say that the atoms scattered after destruction of the body of say a dog will form a dog in future. Such a presumption is madness. There will be dogs in future claiming some of the matter and there will be other life forms finding use for the ingredients that once constituted Old Neighbour but the principle is that the atoms that were in use will again find use because of the atoms eternal connection to time.

The carbon fibre is on Earth placed for the use to carry and support life and as it did in the past, so it will do in the future. It is part of the eternal qualities of the atom to maintain space-time for the foreseeable future and the foreseeable future I suppose is the duration of time that the Earth has to sustain life. To us humans such a concept of time on Earth holds all the factors we connect to eternity because to us the Earth and eternity is almost alike, but in thinking that never should one forget that in the realms of the cosmos the Earth is but a flash in the pan, a wink of the eye and it is gone. But not straying that far into the future we are still measuring the chances of Old Neighbour becoming Old Neighbour once more for he was quite a likable chap and some people will miss him (I suppose).

What will the chances be of resurrecting Old Neighbour to his former self? Well left to the simplicity of the arguments held by the astonishingly brilliant Newtonian atheists we must consider it better than one hundred percent and according to their superb argumentative powers it is as good as done with the aid of a pump, an electric generator and a shovel to push food down his throat. But beware because when gauging by their record of previous successes notwithstanding the simplistic manner they go about denouncing the complexity of life, my prediction is also my advice: if you are a betting man do not bet on a positive outcome because you are about to lose money in such a bet! Your chances in winning will be as good as that of our atheists' wonderful arguments being correct.

Well now Old Neighbour is going to push daisies, or is he? Who is who and what does the daisy pushing? We know very well it is Old Neighbour that is going to push daisies because he is not with us and the "not being with us" part means gone away. Should one force he argument that his body, the compliment and assortment of DNA sells arranged to a specific order matching a pattern profile that belonged to Old Neighbour exclusively is Old Neighbour, or at least that is what our Newtonian-Bright-Boys insist on being the case. You, well any person, makes up a compliment of DNA sells and according to your sell arrangement whereby you become you and by pre-selecting sells and arranging them to a specific order where they form one totality and arrangement by assortment giving any person the prospect of life. Our distinguished atheist loses all other related arguments past this point in order to conclude what they believe to be correct. Considered in the utmost simplicity, yes, that is correct and as that alone it leaves no doubt.

Through this a rat cannot be a horse and a dog cannot be a lion

It is so very simple to understand when explained with such excruciating simplicity that even us living on the other side of the Universe where the Lame-Brains belong are can accept without arguments because we are so scared of putting the least of effort into the simplest form of thinking giving the Brainy Bunch the scope of miles around to come up with the most idiotic answers they can dream up and we the Lame-Brains are too willing to accept as long as we are excluded from any form of thinking. Therefore we allow them so gracefully and with all dignity applied to both sides of the intellectual divide, to bullshit us to a stand still and make us feel great full that we were so privileged in accepting they're demising and diminishing mentality bestowed onto us. I say this from a stance where I am part of the idiots ranks and stand amongst my fellow mindless admiring those of the fortunate and privileged with they're wealth of thinking power because they achieved so many a splendid degree and are therefore the rich in thinking making me just one other poor beggar in thinking-power. The human being as with all beings having life connected to the body structure they occupy which are the compliment of arranged sells and such an arrangement exclude my being a horse and it excludes the horses chances of being an ant.

With things that simple and sells going nowhere as they did not go anywhere in the dying of Old Neighbour why are they not functioning? What made them go on a permanent strike? Why can our Brainy atheists not once more persuade or force those sells on strike into accepting responsibility for their work responsibility because all of the world needs Old Neighbour around and the medical profession did not yet receive they're rightful chance to drain his money like a broken dam wall under the banner of keeping him alive for his family. Well at least until his medical aid runs out and his bank account has gone bust. With that simplicity being the case of life the atheist can at least replenish the life to the sells until everybody in line from the chemical manufacturers down to the cleaners washing the hospital floor had they're chance of becoming Old Neighbour's inheritors and not his wife and children. With Old Neighbour circumventing the money draining system it becomes totally unfair and what is more is why did the system spend so many billions in creating a net where they made Old Neighbour so scared of death and disease he will gladly part with all his money as long as the system gets the chance to help him cheat death (should you not believe me look at the cancer and other advertisements and think for yourself who is paying for the brain washing). Why not only tell those with cancer to do the fighting? Why charge everybody up to come out with they're six shooters a blazing in spraying lead. Who is paying for such advertisements and who receives the benefits of such advertisements all done under the banner of securing a longer life for every body.

I am a diabetic and a smoker that does no exercise of any kind but to get out of bed in the morning. I was medically ordered on so many occasions to quit my smoking, and I not sooner did that then they started feeding me anti depressing pills and anti anxiety pills and sleeping pills and stimulants to fight the sleeping-during-the-day-attacks and the.... The list goes on almost indefinite. Once I pick up my smoking habit again I suddenly do not need one of their pills to keep me "normal". While my smoke may kill me the exhaust fumes of the cars in use which pours the most deadly of gasses into the atmosphere being carbon monoxide is not maybe but definitely not only killing me but also nature in every aspect. Carbon dioxide is a natural element on Earth while carbon monoxide is a chemical acid eating or more accurately said devouring even the likes of statues chiselled from granite rock as well as things manufacture in iron to rust. That aspect no one ever comes to mention BUT SMOKING is the killer destroying life by the billions! The doctors are reluctant to allow the tobacco industry to kill you because that will deny them the chance of killing you chemically and making the profit themselves either through driving their luxurious cars or stuff they prescribe and you can only purchase through chemists. So the doctors scare the daylights out of you about death (which you will never escape in any case) to feed you pills (so chemically poisonous they can only sell on prescription as they are sure killers and most dangerous) and the system is creating another slave by making another fool so brainwashed he truly believes he will eventually cheat death! And Old Neighbour had the audacity to escape the loose of the system and die still with money in his bank account! Such a dead is outrageous and cannot be tolerated. Believe me if the medical profession got to Old Neighbour before I did, in his dying effort they would have kept him alive for another few hundred thousand reasons, reasons you keep in a bank vault and pester his wife and children with guilt so that they part with the money so willingly they will even pay anybody to advise them to part with the money. (If that is not why you pay the doctors treating a man that is ninety nine percent dead already then why are you paying him in any case). You the reader may not see it but this is all resulting from atheism and a system promoting atheism and is an all out war world wide making every breathing person on Earth a slave to milk until death does its part. Convincing people about the simplicity of life will encourage them to fork out money to be kept alive so that the slave will gladly allow more milking.

Slavery so I am told and so I do believe from the bottom of my heart is wrong. But the slaves did not have it so bad in the days of the Greek and Roman Empires. They were much better off than us being the slaves of the current World Order. Slaves under the Roman law were fed clothed and accommodated on the Master's account. The law was that the owner of a slave had to feed him and provide accommodation for his slave. Then the slave had the right to ten percent of the income the owner generated from the services of such a slave while the slave had the chance (if he could) to buy is freedom Slaves in the current World Empire of the Hoggenheimers an Mammonites enjoy the pleasantness of a just system where the system does away with the need to bay slaves, the slaves join the system or die. Furthermore they make the slaves pay from their wages for food logging transport and clothes while the Hoggenheimers do not even pay them ten percent of what the Mammonites earn from their services. Under modern law, modern slaves are worst off than slaves two thousand years ago! And to top this Old Neighbour had the audacity to escape the slavery without even paying his last bid for his freedom. How criminal can a man become in such a manner of escaping what was rightfully his dues to pay. With all the simplicity about life and the promoting of escaping death why can the atheist not bring Old Neighbour back to do his last part and fill the already overflowing money caskets of the Hoggenheimers and Mammonites.

There is this wife of one certain pop star a member of a very well known group in the sixties and one of the four members in this very well known group. This wife of the famous pop star made millions on promoting the abandoning of the use of animal meat as food. She told about her and her husband having lamb chops one afternoon while some other lambs were grazing nearby. As she saw the lambs with her mouth stuffed with their friends she then and there got thinking about cruelty and the humane aspect about eating lamb in the presence of lamb nearby. She was devouring the flesh of sheep that was killed for the purpose of feeding the human population and that gave her the idea to make millions on that thought and selling humanity in the process.

For some sake of sanity let us scrutinise the situation and for once go just a shade deeper than just being prognostic in our conclusions. The lamb has carbon$_{12}$ as a mixture of forming the composition that we named protein. What will be that different from eating grain and eating flesh? Both holds life and both holds death after life. The grain is an infant that did not yet start life whereas the sheep is an adult whose life was cut short during life. Both faced death before they received the honour of completing their sole purpose on Earth and that will be to feed man. She went on a campaign promoting vegetarian dishes that did not even contain fat as protein but included the biggest variety of plants imaginable. While on the tour of promoting the eating of plants (and selling her book to millions of other fools that run on emotions they do not understand, cannot control and where such emotions totally outsmart their thinking capacity) she stopped far short of explaining why she would consider plants lesser life than what sheep are. Can the reason be that the sheep think nothing of devouring the grain and she allows the sheep to do the thinking on her behalf? Is it because grain does not run around when "chased to become grained" for food. Or could it be that the price of the book and her selling power of the content of her book allowed her to sucker some idiots (and I believe the number of idiots caught in the scam runs into millions) tinted her perceptivity so very slightly in favour of the consuming of plants that cannot make any sound or request any human emotion by running and shouting in protest trying to escape the butchers knife or in the case of plants the sickle.

In the case where we consume fruit as food the fruit we eat is food still alive in the same manner as does lions starting to eat a buffalo that is still standing on all four legs. If someone somewhere came about the promoting of eating animals while they are still alive I would surely go on the same protesting crusade as she did in her bit to fight the food supply in the form of meat. We now are faced with the same cynical questions our friend Old Neighbour left us with. Is it his corpse lying there or is it he lying there. If it is he then I have to admit that we are eating lamb. But if it is his corpse then we are not eating lamb but merely the remains of what was lamb once. I am not wasting any space on arguments about killing to eat because kill to eat we do because we have to do it. There are no other options open to us but to kill or to become killed through starvation.

The bottom line underwriting everything said about what form of food we should or should not eat is the human capability of becoming completely self- absorbed in sublimation. We think we know exactly how God created all around because we know exactly God did not create that which is all around. Therefore it is our claim to right that we may take the place of God and decide what should count where and what is food and what is not food, but for god sake keep it simple otherwise we will not understand why we may think ourselves as gods. As long as science portrait matters simple excluding the not very popular complications of thinking every thing thought through decisively we may find that being god can be a very pleasant way of living and un-complicated. If we do not complicate everything we may even think of ourselves as very clever gods without the excruciating effort of being clever gods. Just go about and visualise our brilliance in reason and tell ourselves how kind-hearted and humane we are without any deep philosophising about truth and matters of complexity.

If you are in support of the humane aspect then consider that the deed of eating fruit will be far worse when eating the unborn and defenceless or robbing the unborn defenceless seedling of nourishment so dearly and lovingly accumulated through severe hardship and unquestionable devotion in loving labour by a caring mother than a developed specimen of any specie. Remember that when eating the unborn fruit or the food meant to feed the unborn seed will be denying life the chance to be and that is very unfair! At least the meat eaters gave the sheep the feeling of being sheepish before removing the feeling permanently but in the case of fruit eaters the fruit never had a chance of feeling fruity. I should add that to my mind humanists are the worst practising sublimation because atheists deny the fact of God but humanists are in criticising of Gods way in creating the balance we know as the echo chain. Humanists are constantly trying to show all that are willing to listen to their senseless rambling how much better a job they have in mind for all life on Earth than that which God established up to now through giving man reason to think with a mind and not an emotion and forgetting that the methods applied got civilization in such a tested and tried state as those methods did but still they whish to change it because they think they know so much better.

If our pop-star-wife did not have the pop star fame and all the pop press in support and with the wealth of food supply around how far would she come with the cheap mentality and the thoughtless advocating of the shameless theatrics to support her promotion of self enriching by selling books. When any nation is in total starvation as the Germans were just months after W.W 2 I wonder how many hungry men and women with children crying starved to almost death would applaud her madness as greatness. She got through because there were abundant and not because she had sensibility in her quest to make money.

She could manipulate others while the others were swamped in good times and rolling in the fat of fortune fed to burst while gloating about how their humane hearts bleed for the helpless sheep all over the world knowing very well none of them ever had to skip one meal because of want. They never had to live through one night of agony where their children were crying because the children were too hungry to fall asleep. When thinking about such conditions their gloating in self-praise is quite sickening. From me and mine to you and yours I am telling you this shocker: the total destruction of mankind may only be as far away as the swing of a telescope, and the announcement of a funny little dot that seems to grow as it is heading our way but more about this later on. She is merely one of millions making senselessly money without thought of dangers larking

This I say because nature tells the truth about man and the way mankind evolved. All predators on the hunt have eyes pointing foreword to find the maximum advantage in three-dimensional sight. By focussing in hundred present accuracy the predator can pin point the kill and act swiftly and abruptly minimising the chances of the hunted from escaping such an attack. On the other hand when looking at animals that is mainly vegetarian we find their eyes on the side of the head to secure maximum vigilance and response to such an attack. When looking at the human face we find the eyes even more in the centre of the head than in the case of an eagle, famous for his hunting skills and such a small but obvious clue demolishes the entire bleeding hearts cry for passion.

All animals dependent on meat for food sustaining have eyes pointing to the front the very place humans find their eyes to be. The road our humane idiots genes followed took them through a ancestral path with a long range of meat eaters that brought the gene carrier to what he or she is in the modern age, but being smart they make themselves the fools they are. If we humans were fruit eaters only and had no natural inclination for meat then our eyes would be next to our ears instead of being rite above our noses in the centre of our faces. Those placing meat eating in so many disputes should then also change their eyes position to the side of the head and denounce their ancestral trace of meat eating.

Man has a vision allowing 180^0 sight where as animals born to be the prey has a sight range of 360^0 and none of the humane intellectuals ever came that far in reasoning. With such direct and undeniable evidence about our eating of meat, how on Earth can those shouting no in support of meat eating show their faces around as intellectual beings. This also goes to some religions denouncing the eating of meat but as long as they keep their religion to themselves without trying to convert me to such rubbish they can believe what they want and exclude me. I say this because on occasions I got into debates with such people that wanted to push their religious ideas down my thought about some Indian god living in India and you send him money with a prayer where he then fixes your problems rite across the ocean providing you do not eat meat because of his say so.

All species on Earth are what their history made them. They are moreover the road they followed down to where the specie currently is than what they are at present because when circumstances change the genes with idle qualities will arouse the complexity of the specie and old habits that saved the specie from extinction in the past can come to the front and again save the members from extinction. The Sudanese can survive by eating leaves from trees until the rains arrive to bring about new harvests (although the rains never return permanently). On the other hand the impala cannot start eating lions to keep alive until new vegetation grows again. But even the harmless impala is not that harmless to grass, as grass has to grow meters every year in order to sustain the impala's nourishing needs and at the same time secure the survival of the grass as a specimen of life on earth.

Man too, if need be, can survive on grass and that puts man on top of the evolution ladder and not their misguided impulses in correcting the ways we developed. It is great to play god when God gives in abundance. It is great to play god when God brought your specie this far. But try and play god when God closed the clouds bringing rain and hunger with facing starvation. Then the mind fills only with thoughts silencing the hunger pains and the obsession comes as the hungry wish to fill the stomach with food without filling the mind with cheap sentimentality. How brave will the Super humane then be I wonder. Being humane is closing life to a very single minded approach and in this the massage of the atheists simplistic views about life ring out loud.

All this may be fair but there is another side as all things in creation stand in relevancy In my quest to find answers one question I could never find an answer for is why do the world not import the Sudanese to Britain America and Australia instead of exporting the food they donate to Sudan. Sudan has become a country that will never again support such massive numbers of people and the food will forever be needed. The growing desert claimed the country and it cannot sustain human populations. Declare Sudan uninhabitable and take the people to the countries donating the food. It will be much cheaper to feed them in the countries I have mentioned and at the same time it will please the bleeding hearts, give the Mammonites more slaves and the Mammonists more slaves to drive while not hurting the unemployed one bit for jobless they are because jobless they wish to be. Change the relevancy in the equation and take the people for once to the food and not the food on a yearly basis to the people.

Before every Anglo American starts demanding my immediate and successful castration without precondition let me add why I say what I say. By feeding the population the bleeding hearts are getting their wish but in it they are sadistic and devilish cruel. Before any aid can be requested a disaster must be in progress. Being a disaster in progress means millions are suffering. There has to be an enormous lack of food supply to wake the caller. Babies go hungry mothers weep fathers run off because they wish to find food and disappear in the process. Suffering runs deep as it runs wide and no aid can prevent that as no precautionary measures will ever be good enough. By helping once you are spreading the suffering to last longer and with more pain next time around and we all know there is a next time around because of climate changes going on. Feeling good about your self because of proving once again your good nature, your blessing heart and empathy by the giving aid helps no one because of the coming of the next time. The simple truth is that those in power and those with influence give nothing as much as care for the helped victims. The philanthropist collect money on behalf of the Hoggenheimers from the bleeding hearts while the philanthropist encourage the bleeding hearts to donate in giving for the simple reason the philanthropist share in the spoils of the unselfish act. The Mammonites bay the food as cheap they can in names of companies they own with as little money possible from stocks the donating parties would trash in any case because of poor quality, then bay the food from their private companies with huge profits going to the private company because the selling party is also baying on behalf of the relief organisations with the money the bleeding hearts donated not because out of true sympathy, but the bleeding heart wish to kill the guilt they feel as they know they have it splendid and therefore they need to prove to all but mostly to themselves they're godlike generosity by donation.

The Hoggenheimers take their cut with excessive profits by distributing the bleeding hearts', which the philanthropist collected so unselfishly as proving it by taking their fare share of the profits going around giving the money to the Mammonites, which are baying on behalf of the bleeding hearts from their firms as they sell the stocks they previously bought for next to nothing with excessive profits. At this point the Hoggenheimers bring in the Mammonists to do some slave driving as the spoils has to be sent across the world. In this heart braking act of generosity some more unbelievable profits go the way of the Hoggenheimers and the Mammonites because the firms involved just so happens to belong to a shared venture between the Hoggenheimers and the Mammonites and by some more overwhelming generosity they share crumbs with the Mammonists doing the slave driving.

Now you tell me who is unhappy while all this good heartedness goes around and is there any blame to be where the rich becomes richer as that is no one's fault. If the bleeding hearts were serious about their conviction in generosity they would not bay some guilt relief. If the other parties were serious about their convictions they too would try to find a permanent solution but then there will be less profits to gain. The bleeding hearts are quite satisfied that big planes are used to transport and distribute the food but they know very well that that is the most expensive means of transport and someone somewhere is changing very unselfishly a dime spent to a dollar wasted. In this way the relevancy is getting the rich richer, by giving the guilty guilt relieve and helping the luckless to another round of heart ship in hunger.

Change the relevancy around if the act is in pure kindness and brother love. Take the luckless out of the equation of desperation in cycles by removing them from the problem. In that there are some more relevancies involved. If the bleeding heart were serious they would never mind bringing the luckless to share in their abundance. The other part of the option is to let nature take its toll rectify in natures way and be done with it but then the profit issue stands to lose millions of reasons why neither option is an option. The relevancy will lead to a cheaper solution although more expensive the first time around. Everything is about relevancies. On the one side of the relevancy is the Earth became unsustainable to carry a human burden in that part of the world and on the other side of the relevancy is, the western countries have food to donate in tons through baying and selling agents, (and I shall gladly eat my farm if the politicians were not sharing in the bounty of tax money donated in generosity).

The one side of the relevancy is the Sudanese will never be self supporting because on the other side of the relevancy is in the long run a desert means drought and water will never again be abundant. The only solution to the equation in solving the problem is by changing all aspect around in the relevancy and through that finds a permanent solution to an unsolvable problem that will forever remain unresolved until the relevancy changes to finding an answer instead of avoiding a solution. If the cosmos can tell us one thing it is that changing the relevancy brings about solutions. By creating the Big Bang it solved a problem of overcrowding as we have in Sudan and by creating space as we should in Sudan the cosmos separated matter from space as it is still doing with the Hubble constant proving that space is on the increase. But if space is on the increase and all is about relevancies something else must be in decline on the other side of the relevancy to find equilibrium between the problem and the solution.

The cosmos brought in space on the one side and matter on the other side and between matter and the factor of space growing must be some sort of problem solving. If we wish to find the answers to the cosmic mysteries it should be the most obvious starting point because there is one side of the relevancy known to man and then looking on the opposite side of the relevancy must be the solution. Where one thing is growing something else must then be in declining and in that comes the answer of the relevancy that I share with the introduction of my theory on matter holding space in time. Most prominent in all relevancies there are must be the atom, the one little container giving matter character and different uses in the Universe and by adding or removing one small part it changes in character as Doctor Jackal and Mister Hyde never could. Every one knows what is in the container but what is the container in? If someone ever gave that thought the light of day I have missed it.

Animals use eyes but we humans have more too see with when using what we have by using our minds to see with. We should see the Universe with the light of understanding shining in our minds. When one look at the night sky one see darkness with little specks on light. Why would anybody see darkness because darkness has no light. Yet we see the darkness. The darkness should be invisible if we are seeing light because the one contradict the other. If the night sky was black then black is what we should see but then again black is the absence of colour and colour is the visibility of light.

We see the darkness because it has light it is withholding from us and while withholding the light from us we see the withholding as darkness and the darkness we do not see because we are not suppose to see darkness. That makes the darkness we see not darkness to be but it is in fact light we can see as darkness. On the other hand there is the brightly lit dots we can see because the light shining as dots are darkness as the darkness are stars giving us the light they are not withholding. As they are not withholding the light but pouring it into the vast container of light the stars then become the darkness we cannot see because they give us they're light and by giving us they're light they then have no light to have. That means by giving us they're light they withhold they're darkness from us and that makes the stars filled with darkness. That means what we see as light is light and what we see as darkness is also light instead of what we ought to see as darkness because that we cannot not see is the darkness.

Therefore when God gave the command "Let there be light" it was the command "Let the Universe begin" because the Universe we see is light we do not see and the stars we see is the darkness we do not see. When one is looking at the darkness as an animal you will be seeing the darkness because the mind you use is that of the animal. Then, yes you may be an atheist because all animals are atheists. I have never herd of one bleeding heart or philanthropist of whatever kind convert animals to any religion there is available. If you are an atheist and you see the night sky as darkness that would mean that is what you see as an animal; a darkness that if you had the sense of a man you ought to know that it is impossible to see darkness therefore it must be light. On the other side of the relevancy you also ought to know that by seeing the light the star is giving away the light and when giving away the light it has to hold all the darkness it is claiming for own use because only by claiming back the darkness can it give the Universe the light as it does otherwise it would give its darkness by withholding the light. The darkness is singularity uncommitted to specifics, spinning at the speed of light never pointing in one direction long enough to shine as light but shining long enough not to be darkness. But being light we can see it but because it is in random spin, spinning at the same speed we use electrons to convey massages when translating information we cannot see it. The light then becomes darkness because it is extracting all the light through the one singularity line uncommitted not energising it because of the absence of a replacing source converting new light. In contrast to that is the light we see because of reasons forthcoming from not being able to see darkness we can see it as energised uncommitted singularity with the aid of a sustaining in singularity from a committed form replenishing the uncommitted singularity to maintain direction.

I do not see how one can be an atheist and put claim to being a human while observing what there is out there in the way animals observe by only and purely relying on the eyes without incorporating the mind. I cannot see how any human claiming to be human cannot see past the barriers restraining the animals from being human. With minds it is so clear what the Word of God says, but to be human and not see what humans should see is a dangerous reflection on the mind you use. God did not say, "Let light be visible" or "See the light", He said "Let there be light" and that is what there is. If humans then see darkness where they know that one cannot see darkness the darkness are within their minds and therefore they are atheists not withstanding that they may or may not claim faith as part of their thinking. If you cannot read the Bible through human vision and as a consequence not understand the Bible don't blame the Bible for your inabilities but blame yourself and your inabilities. It is not the Bible you cannot read it is you that cannot read the Bible. Place the relevancy where it and as it belongs.

We are human therefore we have light in our minds and ought to make use of that! This very afternoon as was writing this part I took a break and lo and behold, one of my sons came to me with a problem of a religious nature. I shall not go into detail about his problem but I asked him to define religion and what life is. To strike some sense between his problem and the size he sees it in I asked him to tell me in his view about the contents of the Bible according to the Bible and the dominie (Afrikaans for preacher man) how would they define life because some parts in discussion about his problem was the discussion involving tackling the issue and thereby the issue turned to how far can you go in solving matters and leave the rest to preying and doing prayer. I am of the opinion and will die by that opinion that prayer only serves a purpose in thanks when you yourself completed the task without preying for some force to help you complete the task at hand. Life is the manipulation of your surrounding and that means you do things yourself if you want things done and you do not prey for things to be done on your behalf by God. That is the definition of life. It is the manipulation of space-time and involves neither magic nor divine prayer but you go about changing your surrounding to match your needs. What all preaching never advocate is that we are in the seventh day of creation where that specifically states that God went to rest and from that I draw the conclusion man can and man must do everything by himself because God clearly says He has gone to rest. We do things on merit by ourselves or not at all. That is the energy we think of as life. The fact that we have the ability to self-sustaining and not being fixed to the universal position space-time landed us in gives us life. With life in hand you manipulate what ever you can as you replace positions to suit the required changing of objects where changes are needed. Then your acquired needs changed them to be to your taste and there is no other way out. Life is about changing your surrounding for the better of yourself or others and to improve all around you. The ability to manipulate space-time is the energy I have and Old Neighbour lost. Still it is energy. It is neither food nor electricity but it is a more advanced form of energy than the energy mentioned. Another part of life is tacking the responsibility for change your manipulation may bring about and the effect such change may have for other beings sharing space-time with you. Never confuse the needs of others with needs of your own and project such needs about yourself as beneficial to others without consulting others. This is very typical human behaviour.

With the Newtonian confusion raging man has mixed matters bringing about a highly unsatisfactory climate where we try to pin cosmic value and pre-conditions on life and place very stringent condition suitable only for life onto the cosmos. That leaves science in disarray and confusion. Heat sustaining life as pre-condition Xepted science projects to stars and where stars fade we allow them to die as if blessed with life's changing and renewing. Stars certainly do not have emotions and when they erupt it is not in anger. The chemicals stars need to maintain singularity is very poisonous to man and the matter making life sustainable will have no chance of surviving even as a flash in the star. We think of a star being hot in the manner we translate life's pre-condition to what is hot. It is to the letter the same way that we take outer space as being unsustainably cold where it is quite the opposite applying.

While looking at the Earth we think of the cosmos. We reflect what we conceive as conditions to match life being normal to the cosmos. Planets have to be plenty full because even we have one in hand and eight others in the back yard as spare should we make this one we have untenable to life. And should we run out of planets to ruin there then should be others nearby carrying life on one in nine, as is the case with us. We try to find life everywhere because life has such abundance on Earth in everything we see. We even reflect our vision of time to mach time in the cosmos giving the start of creation an Earth bound time range never thinking that the Universe is growing and not dying. In the same manner we think of the Universe as a living organism while the Universe constitute every aspect we relate to death. In fact, the Universe is the ultimate death. In the Universe everything will only be once and never again whereas with life there was as much as there will be and even more will come than what was. That is the last thing one will find in the cosmos. If time ran out for whatever time will not replace or bring back what ever. Even the way we portray the earth's surface we wish to reflect to space using the same methods we use on earth. One mile will be one mile wherever you wish to take the mile. After all one mile is one thousand seven hundred and sixty yards (if my memory serves me correctly because this is still part of my culture when I was at school and South Africa used the British yard stick). Not once comes the thought that man cannot step one yard in space. Still one yard will be one yard wherever the yard may follow man. Man has acquired the inability to divorce life and the cosmos for some reason we can presume as cultural. Unfortunately we go in accordance to what we see and that is more cultural than culture it self. We see a shining light and presume it is a star in the same manner as we see a large dark antelope with horns on it head exactly in the same way as that of a buffalo and presume that what we see is a buffalo. In the case of the buffalo the past thought us such observations are correct and hence we grew accustomed to the culture of believing our eyes.

Never do science take charge of thought and divide flesh from energy in the manner I have done during this the writing of this article. Outside the view we have we can locate a something that is there but needs some vision in extelegence to locate. It is a small part of life that has an attachment to the physical but an overwhelming comes attach to something indescribable to define. In other books of mine I try my best to prove that our view of trailing outside the sphere of the sun is a myth and even travelling to another planet is not the same "as going abroad". There are so many dimensional barriers attached to what we can see without our locating or even knowing of such existing barriers because they remain unobservable barriers. There are so much more than what ever may meet the eye. In part 7 of the Theses I touch on the subject about the age of the Earth and how short sighted (once again) the Newtonian view are on this matter. The Earth is in truth not 4.5×10^9 years old but the core was part of the cosmos during the birth, the very first moment of the cosmic birth. Many processes came to change and shape the Earth to what we enjoy today, but the inner-core-value came from the first parting of the singularity Alfa.

How life started as such I do not wish to speculate on, but logic tells that what ever was at day one of singularity Alfa, nothing since was added or removed and that puts the carbon carrying life at the very start as well. It would be reasonable to suspect that all cosmic structures holds the carbon but not all structures can present a satisfying environment to sustain and protect the singularity of the carbon in order to bring it to a point of holding divinity secured.

One opinion that I strongly hold is that Chandrasekhar is as misinformed about his carbon-a-plenty theory as he was about his crushing stars in weight. Carbon cannot come from the cosmos and go through the Π limit unscathed to infest the earth. That is as Newtonian as all other bullshit can be. Life in carbon was a part of the Earth as it was part of the sun, but it had its being burnt to blisters and could on that account not develops on the sun.

What ever the Earth went through was also a survival test for species on earth. What ever the sun threw at the Earth the form of life that was dominant then, had to make do or die. The fact that life made do is testimony to life's survival skills. Life will last, no matter what man may throw at it. It is man that places man in jeopardy. Man is the prize of life's achievement that I do believe. Man is the accomplishment all other species carried the burden of. Life is built into man and all qualities of life manifested in man.

That makes man the youngest and the least protected. That makes man the weakest link in surviving. I have my sincere doubt about modern civilised man's ability to survive even the onslaught of a brake down in civilisation. One harsh winter and not one in a thousand would be able to see the next summer rains bring relief. Picture a big city without electricity for one month and think who would survive even such a limited test. One hundred years ago such a remark would have made me as silly looking as the claim I make about gravity. But man has gone down the tube, at the end of the ladder although to man's thinking he is at the top of where he ever was before.

We are launching a chemical war at all pests we do not seek. We kill and destroy them without thought. Bacteria, fungi and, viruses have been at tests far grater than man can produce and survived to tell the story to the next generation. It is written in their life code for the next generation to read and fight. When a species are at it greatest danger of not surviving an onslaught on its very existence a factor much dormant in normal conditions kick in. That factor rewrites the coded massage and the following generation find armoured protection. Man is weakening with all the chemical aid we see as medicine protecting us while we put the most dangerous forms of life on a survival course we cannot afford the luxury of. The day will come when there is no stopping these killing-surviving machines and we, man will stand defenceless while they go about killing and maiming on sight. Every little headache is a call for aspirin. Every cough is a call for anti biotic. One day we will find the disease and ourselves defenceless well and truly developed. Man will die and the count will become more than man can destroy human bodies. That will leave corpses for more viruses to grow and plan more attacks. Payday has to arrive we must see that coming and not be as arrogant in our self believe.

The fashion of the century is to place all, as equal and life holding space in a dog is equal to life holding space in an ant. That can never be for the single reason that all life in the body of a human cannot be equal. Any person can go without a limb notwithstanding the sacrifice they endure in whatever function. Losing an arm or a leg does not risk life at all on the condition that it is removed before it may infect disease to other organs in the body. Losing a liver is serious but machines may provide such an organ function replacement and life goes on, fairly difficult but without eminent danger of death to the rest of the body. The same argument can be said about the heart longs kidneys and such. The function organs play in maintaining the body is crucial but not vital. Losing such an organ does not mean death by necessity and can become even to some of minor significance. When losing the head or part of the brain things turn to a lot more serious nature.

I have witnessed friends of mine that were motorbike maniacs like I am, falling off their bikes and receiving head injury. After the recovery those persons changed in a manner where they became alien to themselves. They became another person no one new before and none can recognise. Such an injury is very serious and lethal, more to the persons that love him than lethal to himself. The persons that love him has lost a love one and gained a stranger they do not care for. Even in one body all life does not stand equal let alone from specie to specie. Losing my arm is not the same as losing my life because I can still live (more unpleasant but that is not the argument) with such a loss. The conclusion of logic is that the arm is not the "me" I lose, as did Old Neighbour when he went missing leaving his remains behind. Some of my body is life in issue for use to be discarded when no longer required for service but other part is much more closely connected to me as life.

This brought about the atheists campaign that life comes as part of a wholesale package wrapped in a carbon container and all philosophy centred around this argument went missing when some connection was proved between electricity and motorized motion of body muscle and fibre. This was the dawn of electricity and the wish-wash that went around with miraculous curing by only sending impulses of electric devises that could cure all and almost bring death back to life. Some devises remained proving through time their worth but in general it was a lot of quack and most disappeared where they came from.

Then came the theory that life was only electricity flowing from the brain to where ever body motion required the flow and all other philosophy went silent. It is not hard to imagine why because physics place electricity as a force with the same presumption (though they will die before admitting it) that a force has a control in similar fashion to a ghost or some unknown free spirit running around to every one's amazement. That mentality sticks like glue and much of that influenced scientific arguments to be in apathy to the philosophical and since 1945 when the physics got hold of the German nuclear bomb and let it loose on Japan it is mathematics ruling logic to the point of madness. No one since then had any inclination to touch this aspect again since all were satisfied that everything was flawless. Flawless indeed but at the heart of mathematics and in the very start of physics lured a flaw that became more apparent every year and the flaw eluded every one to date. It even diminishes all sensible argumentative possibilities to a stand still.

Losing a limb might not kill and it might not change any personally but it is loss to life. If some one acts promptly and in time doctors commonly have the ability to connect the lost limb and with some minor complication the limb may even restore to normal application. Would such prompt action work in the case of Old Neighbour being officially dead for say twenty minutes. The answer may be yes and more likely no because it depends on the brain damage that occurred in time laps where the brain fibre were starved of blood and more important oxygen. As was the case with some of my biker friends brain damage can and more likely will result in a mild to drastic personality change and in some cases dangerous insight attacks may occur.

Changes of such a nature are very serious and symptomatic of injury to the brain. In the brain damaged victim likes and dislikes behaviour pattern and mood swings will change the personality of the individual. The changes may result from a blocking of the flow of blood and it may result from a nerve area that lost function culpabilities but life still remains present. From physics point of view I am of the opinion that it is a natural phenomenon gone very bad and such changes in personality takes place with or without injury. The Romans believed that when a person breaks a mirror he is doomed for seven years because the broken mirror damaged his sole. This we modern people know is just another folk law tale but with some angle of truth. Of course the mirror part is the untruth but there is quite some truth behind the personality changes with an interval fluctuation of seven years. I would not go as far as putting a stop watch to the date in seven years but in a more or less manner we all show some changes in personality and a man of fifty will not find the company of a few teenagers to be friendship bonding and neither will the teenagers like a fifty year cold going gallivanting with girls very pleasant. Of course once again there are many exceptions to the rule and as with all else in the cosmos there are relevancies changing circumstances that may occur. What is without doubt is that the link between life playing a part and the fibre connection playing a part and it will be as silly to claim the carbon has no influence on the life energy as it would be to deny that there is another energy present above and beyond the fibre. With this I wish to introduce my Theory on the Seven Dimensions and I put it to you as I originally started with without changing some of it to fit my present day views.

1.4   THE SEVEN HEAVENS

Although from the name one may have the idea the article is exclusively attached to the spiritual as much as it is about religion and has nothing to do with physics. When a friend of mine saw my article in one of my scribbling pads (this was years ago before any idea of writing a book ever entered my head) he was astonished by my claim that it was pure and unadulterated physics. This was my advance from nowhere into physics. Justifiably you may say as my friend did so many years ago that the seven heavens have no bearing on physics but by saying that the biggest mistake comes into the open. I admit whole-heartedly I did not realise the importance it had back when I wrote down the loose ideas but in retrospect that was my initiating although not my first ideas. Every aspect of every aspect connects in some way leaving only nothing unconnected. It should be somewhat obvious by now that I see "nothing" having no claim in any form of nothing as part of mathematics or physics and to my view that is the main difference between arithmetic and mathematics. In arithmetic there are an allowance for a number or a marker such as zero or nil whereas in mathematics no such number can be found because no such pointer can claim any position from the origin.

Even when I wrote the thoughts down that many years ago I did not yet dispute zero as a number, but I have to admit I had some difficulty with the value of nothing. For instance what was more nothing and what was less nothing when there was two of nothing facing each other. In all of mathematics there has to be growth as much as there has to be decline from wherever any marker may be. In the article I show that the line the half circle and the triangle have on common factor in as much as all being 180^0

A straight line cannot start at zero and still be a straight line because zero extending to wherever brings about a full zero. A straight line starts at the point where the pen point meets paper. That point may be any distance from infinity to a measurable dot, but it cannot be zero.

180^0 X 2 = 360^0

Any straight line is also half a square be cause the line forming the square cannot start at zero for the reasons I just mentioned. That is singularity pointing an eternal direction from a point of infinity and that is the basis of the cosmos as much as that is the basis of mathematics. To escape from nothing one has to become something and by doing that one could not have been in nothing in the first place. If one holds a point in nothing one cannot become something because of the nothing value.

 To back this argument that no line can ever start at zero is to ask the simple question: what will the length of the shortest possible line be. It must be a line where the starting point is so close to the ending point the distance parting the two is incalculable yet there is the line therefore the end and the start is apart still sharing the same spot.

The difference in the circle and the square is the direction the indicator follows and a square cannot spin, as a circle cannot be motionless The factor of Π indicate eternal motion and NOT zero motion. There is a massive difference in that concept. If no line can have a zero point to start with where will the circle get the zero to indicate motion! This principle is the most basic mathematic rule The method applied when calculating a wave is by finding an average in the triangle continuing from the straight line to the pitch of the wave and then the decline will form a duplicate presenting the other side.

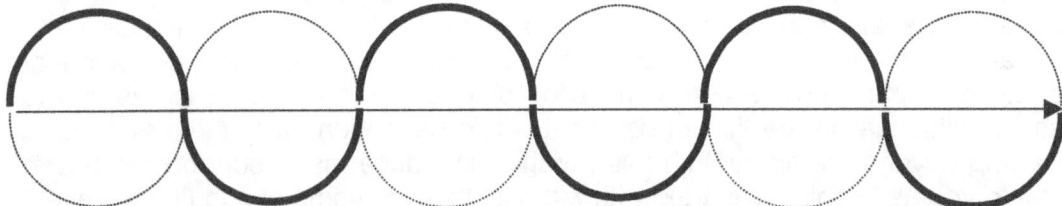

When the end of the rotation arrives the end rotation also announce the beginning of another rotation and not nullifying of the previous rotation because the rotation will have a line showing the effort it made and as it forms a wave, the wave will be there forever. The pitch may decline to a straight line, but the line remains. The wave confirms rotating directions followed by the circle as it spins. By stating that a wheel has a relevancy of zero by completion of a rotation such a claim denies the wave its rite of existing. The wave going flat, as it becomes a straight line also has an indication to singularity.

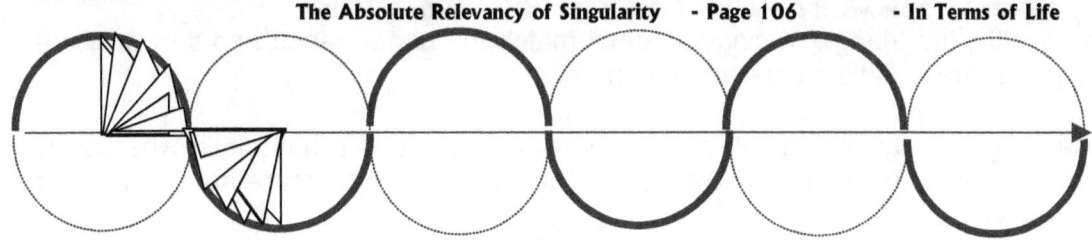

Being a circle means the thing must be round and spinning. In that case, let us take an example well known to all, the spinning top. The top spins on the thinnest of points, and still maintains a balance. By being a calculating value to match the work done in the rotating half circle the triangle depicts the flow of the straight line.

The straight-line holds a duplicate value of 180^0 to the half circle as well as the triangle all being part of singularity as much as being positions from singularity. That alone has to confirm the connection existing in the dimensional aspect. The dynamics behind the two principles is much, much more complicated than what the illustrations as shown above would suggest. However by using such basic of illustrations the simplicity might be tending somewhat to come across as misleadingly simple, but taken down to the core of factors behind the principles that forms the most basic of the principles, the illustrations prove rather effective in explaining the crude idea. However, please do not be fooled by such simplicity, in the very detail analysis it is as complex as can come. From the star holding a dominant point or most valued point in singularity it affirm all five other structure each holding singularity individually.

The Universe link in so many ways we will not begin to realise the manner within the next thousand years. Electricity is one part of the link, but there are other links we may never come to know about because there is always another part of the cosmos above and below our perception and abilities that will elude us.

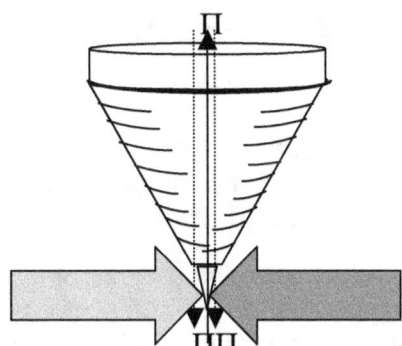

The network of individual singularity not only provide spinning through governing singularity in the sphere but also provide spinning in the geodesic through out the cosmos linking all matter to matter in a network no one will ever come to understand in full. In the sphere the foursquare triangle holds space in time maintaining singularity of different assortments. In view of the matter-to-matter Roche factor where the factor consists forming relation between particles occupying densified space-time of where ($\Pi / 2 \: X \: \Pi / 2$) relating to the foursquare triangle the value of gravity Π^2 comes in position as $\Pi^2 / 4 \: X \: 4 = \Pi^2$.

A STRAIGHT LINE , TRIANGLE AND HALF A CIRCLE WILL ALWAYS HAVE EQUALITY IN DIMENSIONAL CAPACITY PROVIDING EQUILBRIUM BEING 180^0 BECAUSE EACH ONE SHARES A COMMON DINOMINATOR IN SINGULARITY. As the straight line averts a zero it holds another straight line in place to set about such an averting where the two lines will always carry a relevancy in elation to progress (the triangle) and a common denominator in the start from singularity.

At this point the equality of the straight-line dimension to the triangle and the half circle holds prominence as a straight line, a half circle and a triangle is dimensionally equal. The common denominator will bolster all factors to an equivalent ratio,

When singularity by the straight line increases the singularity by the triangle will also bolster giving equal potency in singularity by the half circle. As the singularity of the major component revives the lesser singularity to equality, the triangle in singularity will match the performance and so would the half circle respond in precise ratio setting equilibrium in order.

With the normal extending of singularity it will always form the triangle in a half circle whereby Π relates to the cube by 5 points to either side of the line singularity forms. Thus there are 10 standing related to seven and visa versa.

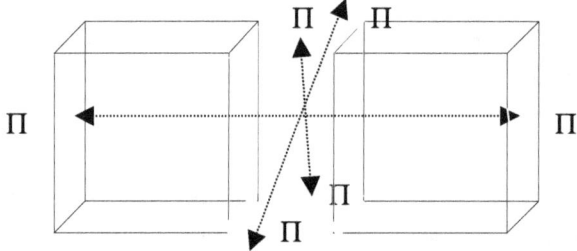

From this the lesser partner will fill by the extent of the larger partner and as soon as equilibrium sets in the growth will duplex in both accounts, normally to the fatality f the lesser partner as the lesser partner will not be capitulating under the straight of the duo. The Titius Bode configuration in accordance to orbiting formation holds a slightly different explanation to the explanation that applies to cosmic structure surrounded by space. It is moreover the individual singularity in maintaining the major singularity, which sustains the governing singularity providing equilibrium in space-time.

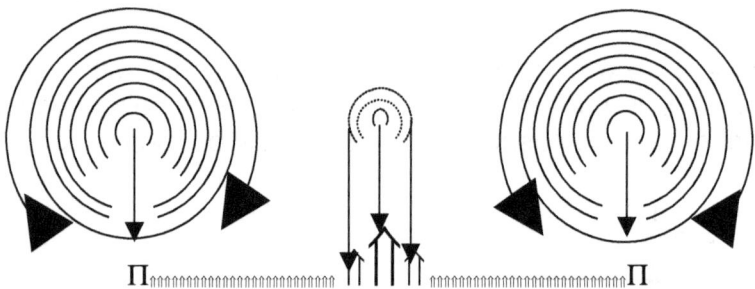

Not only does atomic individual singularity maintain self preservation, but in doing that it also sustain a governing singularity holding structural composition and form within a cluster of matter for example a star. Between stars there are a mutual or bonding singularity between atoms and stars.

The sectors provide individual singularity a means in sustaining governing singularity by which provision comes through maintaining governing singularity the required spin in maintaining cooling. If this process did not apply, there would be no connecting individual singularity to major singularity. The sectors provide individual singularity a means in sustaining governing singularity by which provision comes through maintaining governing singularity the required spin in maintaining cooling. If this process did not apply, there would be no connecting individual singularity to major singularity In this maintaining of cross referencing of singularity providing spin to the governing singularity many factors of singularity all form a close knit network inseparable one unity but also strictly individual to a point of destructing.

Singularity has three part and five points with Π as matter being sixth and space (r) as light the seventh.

First let us consider the dimensions accepted by science and later argue those forms that are excluded. The so-called heavens have another name, which are dimensions. These dimensions could be regarded as planes or spheres. In looking for the dimensions that form the universe, one must look for sides, which has a combined value, but exists in total isolation from one another. Any object that can be visualized has to contain at least three sides with six obvious different spheres. These spheres do have single applications in the universe. Allow me to explain.

The space between the spheres divide in half, but because of the extending of Π and not applying r as ordinary mathematics will suggest where Π replaces r the singularity extending from $Π^0$ will be half of Π in the square of $Π = (Π/2)^2 = 2.4674$. In this lies the dynamics why planets have a positional (be it rather a dimensional) relation of 7 / 10 The second Roche limit is within the sphere as $(\Pi^2/2) = 4.9348$.

According to me, as a believer of the Holy Bible, there are seven heavens. The Holy Creator sits in the seventh heaven. His Word tells me that. That means if his Word says there are seven heaves that means there are seven heavens, that means there must be six undetected heavens. Before those atheists start shouting their lungs out, first go about answering the question put by me in a previous article. Answer the question where do life, being energy, go after it has left the body after death. Let us presume that the question is still without an answer. If the Creator lives in the seventh heaven, then He must have created the other six. By acknowledging the creation, I echo what millions of intellectual people believe. The so-called Christians, Judaism and Islamic faiths all accept the first five parts of the same religious Bible. Therefore, I feel free to speak on behalf of millions of people who consider themselves religious and these people come from all lifestyles and all over the world.

THE SINGLE OR FIRST DIMENSION (GRAVITY)

In the single dimension, one finds gravity or a pulling force that every cosmic body has.

If any person does not believe In gravity, then try to jump high of far. One would find that there is a force, which pulls your body back in the direction of the centre of the earth. This force is calculated to be 9,81 N m/s² and is almost precise the same value right over the world. This force can only be overcome if an object is hurled into space at a speed of 11,20 km/s² and an angle of 90° with the earth. That brings about that any object on Earth is moving at a speed of 11,17 km/s² at any given moment.

As far as my knowledge goes, this force is perpendicular to the earth, with a distortion of 7% due to the inclination of the earth. A person in China will be pulled in an opposite direction to a person in America, because China and America are approximately in opposite directions to one another on the face of the globe. This means a person in America moves towards the Earth at 11,17 km/s² which is directly opposite in direction to the 11,17 km/s² which the person in China is drawn to America. Taken these viewpoints into consideration, gravity is a single directional force moving only in one direction that depends on the body position to the earth. Where does this force stop? Nobody knows, because the crust of the Earth is being drawn towards the centre of the earth. Even the see and the crust underneath it moves statically towards the centre of the Earth at an even pace. If one looks at this force, one gets the impression that there is only one direction towards the centre of the Earth without any given point where it will stop. On the other side there is no given any definite starting point. Without a star or an end, only a direction remains that envelops the whole idea.

I hope I was clear enough about the idea I tried to explain. The whole idea of gravity consists of only one single directional movement in one direction seen from all directions. There is no starting point, no point where it ends, only the definite direction that the body moves towards the centre of the Earth relative to a point where all directions are measured. Another strong candidate of this dimension is magnetism. The polarity of the atom in the iron core is only towards one direction. But I shall explain a little further down the road, why magnetism do not comply with the single dimension force when I have brought some more facts and arguments in explaining what forces come into play.

THE SECOND DIMENSION : THE WAVE

In the context of the wave, for example water, the circle of the wave flows from one given point outwards. That implies that there is a definite direction. However, there is a second dimension in the wavelength. The moment that the breadth of the circle is determined, the wave has already altered it. That means there are only two values to be considered realistically without freezing time and that is the direction and the frequency. Sound and light are two such values. The moment the wave is frozen in time, which means it is standing still, it is no longer a wave, but an unnatural structure man made for his own benefit. It does not occur in nature because a wave can never stand still.

Light, sound and waves are two-dimensional values. The dimension consists of a point of origin and a frequency. There is no distance, because the wave beams out in all directions simultaneously. The Doppler effect in sound has an influence but that is because the point of origin keeps altering due to the movement of the source. The speed of sound has a definite ratio to the density of the medium it moves through and that applies to the time value of the space-time occupied by matter. In a later part, I shall explain this in detail.

Due to the changes in space time occupation of the matter the frequency alters and tarnish directly in ration to the space and the time the matter occupies. As the frequency of the space deteriorates in time, the wave becomes less in value but still remains. Therefore, all the sounds the dinosaurs made are still with us but the space-time occupation that the sound waves had, rendered it undetectable to us. Proof of my argument is found in the heat that still can be detected with the Big Bang event. Although the space-time value is only round about 3 °K, the waves can still be measured, because these waves form part of space-time. The light wave is transmitted from its source in a sphere and will transfer itself through space-time until such time that a few particles hits a solid object. The wave consists of billions of light particles that move by wave density from the transmitter. The photon moves at a rate of ± 300 000 km/so. That means that the photon displaces space-time by negatively at the above-mentioned rate. This negative displacement of space-time by photons is called a light beam.

The light rays follow wave upon wave in endless motion through space in time. As soon as one wave of photons hit solid matter, only a very small number of the total wave is stopped. These meaningless stopping causes a shadow to form, but the large number of photons available would fill the gap formed by the loss of photons. That means that a shadow as large as the earth, will cast darkness for a very small distance / time span that means a tiny part in space-time.

The reason why the heaven does not light up is that the beam is scattered as it spreads out in a balloon formation and therefore the number of photons lessens in density as it becomes "duller". The more space-time it has to displace the thinner the layer of photons would become and the less intense the beam would seem. It is not so much the space (distance) that should be regarded but the time it spends in space has the ultimate segregation of the light beam. In space and time, the beam would become so minute that it would alternately become unobserved by the human eye.

All that we know, all the knowledge we have accumulated through time is based on the light that sends information to us. The size, the distance, the structure, the density and its position are all determined from the light that sends information to us. The light that reaches our planet determines all the size, the distance, the structure, the density and its position. All our facts are based on this small amount of evidence. Light is so widely connected with insight and knowledge that we assume that we have seen the light when a picture of a person is projected with a light bulb next to the person's head. Light is seen as the same as knowledge. Light is actually a very poor source of information. Magicians and con artists use the information of light reflected by mirrors, lenses and colours to mesmerize our wits.

Let us consider light as a source of information in daily use by humans on this planet. Light is a very poor medium of information, due to all kinds of illusions, which it causes. Take for instance how mirrors and lenses can disturb a person's image. Why then can we categorically and without doubt, be assured that the information that we collect are the truth. Still, those in power disregard any phenomenon they find as the truth. A very practical example to prove my point can be put as follows. Light shines on a green leaf. The leaf is reflecting a certain variation of light and the rest is absorbed. Let us picture the leaf olive green. The olive green leaf accepts all the colours in the spectrum, except that olive green. The olive green are not accepted by the leaf but are rejected. The part of the light spectrum that seems to be unacceptable to the leaf is the very colour we associate it with. If the leaf accepts all the colours except green, it must consist of all the colours except olive green. Another example we use every day, is "How bright the moon shines", when in fact, everybody knows that the moon does not shine!

This might seem trivial and irrelevant, but remember, because of such trivial terminology, people were burnt on stakes. When Georigiano Bruno tried to persuade the church it was not the sun rising and setting, this became his fate. Man's attachment to his visual sense is actually a little funny and a lot deplorable. Science can calculate the density of a neutron star, they can determine the heat value in such a star, but when this star becomes more dens, to such an extent that the density causes the neutron star's atoms to disintegrate, these very same scientists declare that:

1. the star has vanished;
2. it has gone through to another universe
3. it does not exist any more
4. because it is dark, it must be cold
5. it is lost in the creation, never to be recovered again
6. with this, all other laws of science are disregarded that says an atom consists of frozen energy;
7. the gravity fields can still be mesmerized therefore the matter it consists of must still be there;
8. All that matter can still be placed at one certain, predetermined place, and therefore it is position is a fact as far as the Universe is concerned.

When taking the above mentally into account, one cannot but wonder, how far did man actually progressed from those darken middle age mentalities.

With these very obvious facts in the fore ground of our minds, we still reassure all the doubtful thoughts of our ancestry, which regarded everything the unknown comprised of as magic and mysterious.

We still cannot appreciate anything that falls outside the boundaries of the visual senses that the instruments can detect and determine. How small we are and how inflated we regard us to be.

I have in the past been asked the question "Why are all these other stars and galaxies necessary , if we as humans can't use it?" The human arrogance has no limits and the common position man occupies in the micro as well as macro space, is completely distorted by our sense of self-appreciation and self-importance.

THE THIRD DIMENSION (THE ATOM AND MAGNETISM)

In the article that deals with the meaning about "nothing" the structure and lay out of the atom is explained to some extent. The atom is the smallest and the largest single unit that the Universe comprises of. In the most basic form, the atom exists of one single electron that orbits to energy levels around a nucleus in a single electron lay out. This electron to nucleus balance is the precise force that keeps the Universe in perpetual notion bound by time. In a later stage, I shall explain this in more detail. The energy levels that the electron positions itself to the nucleus are valued in quantum leaps. However, even this complies with a three dimensional length, width and depth. This means there are three definite measurements, although they seem to be microscopically.

Taken one step further, the atom is made up of frozen energy. Any matter that moves at the speed of light is pure energy ($E = MC^2$) according to Prof. A. Einstein. That means this atom cannot be changed in shape or size without enormous energy loss or gain and this would lead to extremely serious consequences. The Japanese at Hiroshima and Nagasaki can declare what extreme consequences are hidden underneath the structural disfiguration of the atom structures. Therefore, can the people in the Tunguska river valley give evidence to the outcome of atoms that gain in mass.

In this process, man has released the worst kind of destruction available. The energy released is of such vast dimensions those 50 years after the explosion the shadows of the victims are still edged out on the background in the cement and bricks. The intensity of the light that was released in a billionth of a second has almost forever changed the face of the material it shone on. Spare a thought for the people who received that light onto their bodies as their shadows remain as testament to those in power who damned them forever.

This process was not caused by atoms that was demolished, but merely by changing some of the element value from one atom to another. This leads to a spontaneous thought: "Why has no American ever been brought to justice for horrendous acts of war crimes?" Needless to say this act was the mother of all war inflicted war crimes by killing and maiming hundreds of thousands woman and children and civilians. These bombs made those in power, who ordered the release of these bombs, the biggest sadists and mass murderers ever known to man and when compared to evil-minded monsters like Nero, Nero with all his menace, suddenly becomes a silly and naughty boy.

Even decades after the release of these highly toxic energy sources, they still kill and maim the innocent. However, this only ties in with another admirable fact of the 20th century. Almost all nations on Earth have produced war criminals and warmongers, except the English speaking nations on earth. This group is blessed with the innocence to such an extreme that not one has ever been charged with one single act of a war crime. If none has been charged, none could be found guilty and then not one can be guilty of war misconduct. Let us go back to the atom's structure. Because the atom moves about at the speed of light, the structure is timeless. All matter that is timeless will last forever or eternally.

This brings about the atom's structure to be forever and timeless forming the third dimension. In a later part, the reader will find that I disagree totally with professor Einstein's assumption about the fact that the speed of light has the same value as time itself. In order not to start confusion this early in the book, we shall accept dr. Einstein's theory as for now. The message I am trying to bring across in this article is the comprehension of what the third dimension contains. It is made up of three dimensions without having time as a factor.

MAGNETISM: THE MISCONCEPTION OF THE ATOM
As I already pointed out, the atom is pure energy. Seeing that the electrons rotates about the nucleus at the speed of light, and the nucleus vibrates (?) at the speed of light, that brings about a confined cell made up of pure energy.

Seen in the whole picture, the fact that the atom is driven at the speed of light and is in total balance and harmony, its permanence lies in the fact that it exists confined to a structure, but not to time itself.

There are length, width and height. The tree dimensions it composes of will render the structure eternal life, or that is how it is regarded by science. The property of the atom might change from star to star, but the structure remains with its three dimensional qualities varying in size, but it remains the same. You, the reader may ask: "What is the common factor between the atom and magnetism?" Magnetism on the other hand flows between two points without stop in a closed circuit, not influenced by time. There is a direction (length), a circuit (height) and a start / finish point. This forms a closed ring formation, which has the same qualities as the atom. Time plays no role in this energy displacement, because this movement is coupled to the speed of light.
To prove its existence, lies in the fact that the human civilization is mostly driven by electricity.

The energy determines the magnetization, but the circuit remains permanent although some times in a latent form. Because of this, it actually consists of all the ingredients to qualify as a third dimensional force. The ferromagnetic field proves that space-time is being displaced as it is being done in the case of the atom structure, but the displacing of the space-time is in a continuous and constant closed circuit.

The poles that attract each other displace space-time in the same direction and those that repulse each other, displace space-time in opposite directions.
In bodies as insignificant as the Earth, the difference between gravity and magnetism will be enormous in comparison with structures the size of the Sun. The magnetism in the sun is comparatively much stronger than the earth, but the difference between gravity and magnetism would be less. In a structure, the size of a White Dwarf the electromagnetism might only be twice the force of gravity and accordingly as the star becomes bigger, the force would become equal.

In structures that compose of the mass of a neutron star, magnetism would be dominated by gravity to such an extent that the force of gravity would not allow any magnetism to exist. In the so-called black hole, there can be no such a thing as magnetism because the electron does not exist any more.

Electromagnetism is the "short circuit" in the flow of positive space-time displacement and can only exist in a structure that compiles of an iron-based core.

THE FOURTH DIMENSION: THE TIME FACTOR OF TIME IN SPACE
The previous three dimensions all dealt with the space factor in time in space, disregarding the time factor. Let us consider an everyday household item in normal use like a table to explain the value of time. The table is made of wood. The wood is comprised of timeless atoms. (That is, if the reader

accepts the previous argument about the atom that lies outside the boundaries of time). The form and the shape that the wood is in, is not timeless. It started as a seed that was enlarged by cell multiplication as it germinated to become a tree. The process of germination took a certain time and the growing of the tree took another period in time. That means the tree occupied more and more space in a given time period in the space-time it shared with the Earth in the form of a tree.

Afterwards the tree was chopped off. This felling of the tree comprised of a certain given space in a certain given time, as did the falling of the tree. Both these periods consisted of different space in different times that are coupled to the space-time the Earth occupied. For a certain period, the tree remained in an upright position occupying a little more space as time moved on. Then at a predetermined point in time, the tree was felled. Every blow by die axe displaced a certain piece of wood to a different place in space and time. That meant that every splinter of wood that broke from the tree was given its own place to occupy space in time. The position of the wood has been altered and even if one try as hard as they may, every piece of wood has received its own space in time to occupy and could never be regarded as a tree again. It occupies a complete different position in the space it shares with the Earth in time.

Afterwards this tree is stripped of its branches and leaves. The stripping takes a certain period and position in space and time. Each branch occupies a different position in space and time and is forever dispositional from its original position in space and time, because every part that is not part of the tree anymore, is in its own position in space and time.

Then this, the tree was taken from the plantation to the mill and moved through a different space in each fraction of time as it was transported. Every millisecond held a different position in space in relation to the next time fraction. At the mill, it was sawed into planks and the structure's position was altered even more in space and time. The planks were bought by a carpenter and were given a completely new form as a table in space and time.

Although this particular table I am writing on now, at this present moment, has occupied space in time since 1921 as a table, the wood occupied space in time much longer than the table has in its present state. This wood can never be made a tree again, because it is space and time has been altered indefinitely. So, how does this fit into the big picture of the Universe as all things in the Universe are connected and related?

I was born on a given second, minute and hour and I shall die on a certain second, minute and hour. From the day of my birth until the day of my death, I shall constantly alter my position in space and time and will never be able to occupy the very same spot of space-time because space and time will not allow it. According to my visual observation, I can presume to occupy the same space in time, but the geodesic outlay of the Universe is altered as every millisecond goes by. That means I can never remain in the same space in time, because my body's position alters by the rotation of the earth, the sun and the universe.

When my time is up, my space and time is altered to such a position where as I am no longer in control of it. From then on, I will never be able to determine my own position in space and time again. I am in a state called death. In this state my body will be broken up by microbes into gas and heat, every second time moves on, my body's occupation of space, and time would diminish. This will carry on in space and time until only the elements my body comprises of, remains. The atoms I am made of, have previously been used to form plants, trees, animals and even humans. After my death, it will never ever form another combined unit to match my exact replica. Therefore, I have departed in more senses than one.

That means my position as Peet Schutte, in space, and in time, is suspended, unconditionally, forever.

This means that for certain duration of time a certain combination of atoms is forced together to form the elements that are dedicated to me. This dedication is temporary, which means that I was for a certain designated space in certain duration of time-sharing space-time with the earth.

Although this is a known fact to every person on earth, it is astonishing to see how every person yearns to maintain a youthful and vigorous maternal structure. This structure is condemned to destruct the minute a person is born, yet everybody guards his attachment to that structure with a jealous observation. The cosmetic industry cashes billions of dollars a year and all that income is based on this fear of ageing, which ultimately leads to death of the person and destruction of the body. The ongoing process in the envelope of the fourth dimension connects all the previous three dimensions, we regard as time. This means that time itself is not created by man, but is part of the physical Universe and is only recognized by man, the very same way man has recognized the existence of the other three dimensions.

THE FIFTH DIMENSION: UNOBSERVABLE - THAT MEANS TIMELESS TO MAN'S SENSES

In the article "Life after death," I touched on the subject that life is a form of energy, and therefore is indestructible cannot be destroyed. If any reader cannot accept this argument, then please prove the opposite and let me know. Therefore, until the opposite is proven, I shall regard this argument to be correct. Because the generator, which I regard to be me, I of the electrons, is not the" me" that will be placed in a box and will never be able to share space time (because the worms are going to feast on me) I am above time and space and I shall not be in that coffin. Only the "me" that am made of flesh and bones and was considered "me" and which I had temporarily control over, will deteriorate in that coffin. If the generator of the electrons that is somewhere connected to the brain, and from where I control the me, which is the muscle bone and tissue held in place by the body I consider to be me, where do I, the generator of electrons go after I was disconnected from the me of flesh and bone me. I know where I, the energy-less dead body is going. I, the part I consider me, is to be thrown into a wooden box, dumped into a hole which is dug in the ground and where the worms are going to enjoy a feast of a meal.

This feast of a meal by the worms cannot be me, because I am not part of that decomposing structure. I was part of that decomposing structure, until the very second, I lost control of that decomposing structure. Thoughts, that travels faster than the speed of light is the actual part that is I. I can elaborate on this line of argument, but those that do not wish to be convinced will remain so because they prefer to remain unconvinced, not because they know they are right, but just the opposite. The I which is regarded as me, has received the ability, for a short while at least, to manipulate space-time, whether it was in the form of my body, or other matter I came into contact with, or unoccupied space-time as I moved about, occupying unoccupied space-time in a random fashion at free will.

The part that is I, and which is not part of the corpse any longer, can think of one thing and then think of something completely else. The very following second I can change my surroundings as soon as I change my thoughts by creating new thoughts. These thoughts are not connected to time or space. It moves arbitrary and involuntary through time and space, from present to past to future. Sometimes these thoughts are so strong that a person loses track of reality.

The terminology we use to describe this, is daydreaming. When I was teaching as a pedagogue in class, this condition was my biggest enemy. While I was conducting my class, the students would sit there wide eyed, listening, but at the same time they were miles away from school living a daydream that had no connection to matters of schooling. When they were asked to reply what was just said in class, they were flabbergasted and completely unaware of their surroundings.

I think I may presume with some certainty that the reader would follow the two parts of the same person I meant. I agree that an argument can be made that these thoughts are part of brain cells that are stored by nerve tissue, but the information in those cells are created by emotion. These thoughts can be depressed by emotion or be prominent on the foreground because of emotion. Man is made up of a stream of emotion that flows continuously through these emotion fields and the flow of emotions is that that puts meaning into a person's life. One has to consider emotion to be one of the most pure sources of energy.

Energy creates thought, creates ideas of a spiritual as well as physical nature, it establishes a flow of electrons that drives the human mind and body, it controls all muscle groups in the body like heartbeat and when the body is in danger, it produces chemicals which enables the body to react far better than it normally does. That means these other organs are also under the control of emotion. The emotion drives the body to produce chemical substances, which enables the body to perform at levels far above its known abilities. This emotional control defines heroes from cowards, sportsman from the ordinary and even philosophers from the masses. That means the emotional part is the part of me that cannot be destroyed and that is the part that generates emotion. All this comes down to one value that life possesses, and that is the manipulation and control of space-time.

When this emotion driver, that generates electrons, leaves the body, a person is considered dead. When one takes the moment that life leaves the body, nothing physical leaves the body. There is no lightning like electric conduction, there is no immediate spontaneous combustion, but there is no emotion either. As I already pointed out, the only reason why a person is considered dead is that he is considered lifeless, meaning without energy.

A lot of energy has to be deplaned somewhere. It is no longer part of the physical world and being energy it cannot just vanish into nothing. What is factual, is that life as energy, is no longer attached or bound to the fourth dimension of time. At this point, I have to prevent my ego and self-importance not to get the better of me and put myself on a pedestal equal to our Creator.

It would be much better if I were grateful and thankful for the time I was allowed to use the atoms loaned to me for my own personal use. The loan period was of such a short duration in space and time, that the extent of the period of loan becomes oblivious in space and time and space time.

THE SIXTH DIMENSION: THE LIGHT AND THE TRUTH
I must confess that in this my faith and religion plays a large role, as one of my beliefs is that the Messiah has already come.

Maybe the Jews, Moslems, Hindus and other religious groupings would find their own explanation according to their faith. My Messiah said, "In the house of my Father there are many mansions.", which I interpret that these groups should be left to find their own salvation. I may not condemn them or denounce them or try and convert them, because each should have its own mansion. However, personally I accept my Messiah and He declared that: "I am the Light and the Truth and no one can enter the house of My Father but through "Me."

If my Messiah points out to be a light, it stands to reason that there must be darkness. That will be a dark fifth dimension and light sixth dimension. Seeing that the majority of English speaking persons do not share my beliefs and religion, and I do not believe in converting any person to my faith, I shall leave this matter at this point.

It is not that I am ashamed of the religious beliefs that I follow; to the contrary, I believe that only Israelites may be converted to my religion. I think I made enough argument to prove the existence of a fifth dimension to which life has to go after death. I presume, because I am not familiar with the contents of other religions, that their religion will allow them a passage by what means their religion chooses, out of the dark fifth dimension.

This dimension, the sixth can only is entered by human life. Let us then see what human life is all about. A human does not have to be a human because the creature is compiled of human D.N.A. D.N.A. cannot form a human. The gorilla is 97,8% human and the orang-utan is 98,2% human. However, there are obviously big differences between these species and a human being. No cultured person on Earth will consider one of the ape species to be human. That means a human is more human by culture than by physical appearance. The additional 1,8% that a human have, is not even enough to explain the physical differences that exists between humans and apes. However, we know that the main difference between the species is the human's ability to reason, think, argue and control their emotions and instincts.

The more a human explore and scrutinize his feelings, himself, his surroundings and his universe, the more such a person would qualify to be human. After all, that is how other species evolved away from the animal and that should categorize us to be part of the sixth dimension.

THE SEVENTH DIMENSION: THE SUPREME ALMIGHTY
Because so few English-speaking persons consider themselves Israelites, they do not fall under the law of the Israelites. In such a case, I would consider myself blasphemous to share such knowledge with those that do not regard themselves to be law-abiding in all ways. Any person that does want to know more about this matter should read the book of Henog. I will say this much, that those that

read Henog would find out why the book of Henog has been left out of the Holy Bible by the Roman Catholic Church and the other churches. The Supreme Being lives in the Seventh Heaven as Creator of all. Because I regard my fellow Boere as brothers, I did explain to a very small extent the seventh heaven in the original Afrikaans version. But all Christians and only Christians read this...all churches are part of the Anti-Christ being the Body –Of- The- Anti-Christ. Do not look for the coming of the Anti-Christ for he is among us. They crucified Christ for His throwing over the money tables and throwing out the money offerings (a lucrative business in any society) ridding the Temple of money some two thousand years back, and today all Christian religions fight one another to feed that which Christ threw out...the money tables...and best of all is they feel righteous in doing so! All denominations are a part of the Anti-Christ and BEING THE ANTI CHRIST. Christ threw out the money tables because He said you cannot serve two Masters...you cannot love God and Mammon for one you shall love and one you shall hate. You cannot serve Mammon and God. If Christ came back today and once again threw out the money tables all Churches and worshiping priests will once more shout for His crucifixion as they did two thousand years ago. As the Pharisees were the Anti – Christ back when… so is all Christian churches and denominators, Priests, Pastors, Reverends, Bishops, name them what you like, small and large...they're all taking part in the crucifixion every time they ask for money "In the Name Of The Lord" and if He came to destabilize they're Money machinery today, they will hang him tomorrow morning at they're earliest convenience and even on the same cross (if they can find it). Every preacher is more into collecting than pouring out the Word. It is a trade off that the preacher will bring the Gospel in exchange for collecting offerings. They heel, bless, pray, condemn and condone on behalf of the name Mammon. I challenge every purist of heart to show me one preacher of the Gospel that sends donations back with the message that such donations is condoning the Crucifixion and as preacher will not except having the blood of Christ on his conscience. If you cannot show me one, I can show the body of the Anti-Christ for they will kill again if some one should try to diminish the lucrative trading done in the Name of the Lord.

9.6 THE TRINITY… and you.

The only renewable energy source in the cosmos is life, and even life submits to time. But that only confirms the reusability of life as energy and not the "replacibility" of life as energy. At birth, a baby is born with life valuating its position to the time applying life to the proton. In the manner, only life has the ability to accomplish, to divert from cosmic set rules but also in a very limited part of the cosmos because life is part of a small bubble and not a large Universe. Life's manipulation of space-time by occupation, through growth, movement, repositioning densified space-time, and even for the past fifty odd years repositioning unoccupied space-time, and the biggest accomplishment of all; transforming time in space to space-time.

This rendered a false security in life to such an extent life believes it can manipulate time. Weather the con artists in the medical profession truly believe they possess such ability to extend life to a limitless duration being such fools, or weather they know they are only bleeding the social slaves to death in cold calculation extorting their money by exploiting their fear of death, that I do not know. Either way, the outcome remains the same, as they pump the social slaves full of fear of death, they do compensate for their criminality by promising the mindless stupefied and brainwashed slaves an escape from death. They gain access to federal funding which then allow them legal means of bribing politicians into furthering their quest in an effort of finding a way in defying death. It will only eventually benefit the Mammonites and Hoggenheimers, those holding the money and the power should a miracle occur in finding such a remedy. The Mammonites and Hoggenheimers will se to it that the expense of such treatment will secure that only the best provided for will afford such treatment excluding all the brainless that paid for such research in the first instance. If they will allow some suckers to share it will only be so that the Mammonites and Hoggenheimers can then keep the social slaves in production longer to bleed them longer. In the meanwhile, they put the profits to the research in their coffers and by using the profits, they steel from the slaves, as they have done since class and power first became part of thee social structure. This time they apply such social discrepancies to beat their fight against time. This is the second biggest fraud on earth, coming in a close second to the biggest fraud in our social structure; the Industry Of Insurers against the loss or destruction of personal life, limb and property. They the big insurance firms, they take the cake for criminal conspiracy and even slightly beat other fraudsters such as bankers or politicians and other vested criminals without morals or character. A person in prison once told me inside prisons you couldn't find one criminal because inside are the stupid and the brainless and outside the real criminals controlling the all. Let us get back to cosmology and leave social studies for those that run around and bullshit one and all.

At birth, life uses the proton growth to its advantage by securing a better grip on the proton and benefiting from the proton's ever enlargement and the multiplying of the growing in size of the body. Life, at that stage, has the upper hand to the proton. As life and the proton grow, the benefits are to a mutual advantage, up to a point in time, where the advantage shifts slowly to favour the time component in the proton. That is the point where time starts dominating life. At first, in the baby phase, life is in full charge of the protons time intake or gravity produced heat. Then the status is mutual, but slowly time in the proton is taking charge as it is gaining the advantage, as life can no longer compete with the growing advantage that the proton receives from the heat supply gravity is force-feeding the proton. This advantage is ever present, always dominating because gravity is about dominating even life and the favour changes position after the midpoint of life is reached.

Where the time component becomes out of alignment with the rest of the time factor in the body (cancer) there is a chance of remedy. That, however, is not extending life it is merely correcting a flaw in the body. To claim the ability of extending life in whatever way is fraud. When a person swindles money from another person with promises of huge investment profit, it is punishable with removing such a trickster from society. When the Mammonites, Mammonists, and Hoggenheimers promise unduly and unsubstantiated life extension, in order to gain unbearable access to illegal profit margins, it is quite a different matter and very much excusable. Life cannot extend past its expiring date; time shall never allow it, no matter what they feed you, or how they force-feed you.

The fraud in science started with Newtonian misrepresentation of truths and runs at present through all of science like a cancer's growth, a killer disease that in the end will destroy all mankind with the poisons they sell in agriculture and household items. From all of that action those involved, they do all of that only to gain profit in money. They are prepared to destroy man's future by going against nature and the Earth, to gain a short-lived advantage to money and its control. Putting this into perspective I thank my Creator for making me a small, insignificant and mentally impaired creature, just the way I am. If that is power, wisdom and greatness, how thank full I am to be power less, stupid and small.

WITH THIS, I CONCLUDE LIFE AS A COSMIC FACTOR OUTSIDE THE COSMIC BALANCE. In the Afrikaans book, I devoted one chapter to life, but it includes some religious views I hold.

As my one and my exclusive aim with the English book is to prove the Biblical views correct, and disprove Newton's Nonsense, personal religion has no place in it. It can only lead to religious arguments, which I wish to avoid at all cost. I do not wish to convert any one, as much as I do not wish to be converted to other religious views. It is outside my power to religiously save any sole, as much as it is outside my power to condemn any sole, so why then, bother.

As all things in the universe are related and never can be a separate issue, even life, as we know it, stands in the same way, where two halves of the same but opposing sides form one being. Two halves that oppose each other are stuck together in the middle form all life.

DIVINITY OR infinity

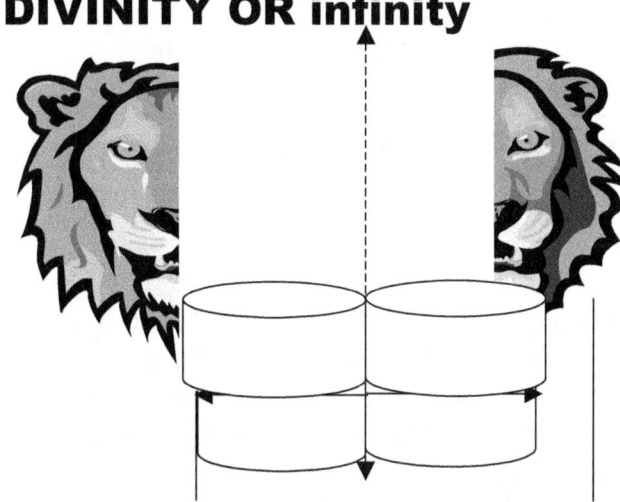

The view I express in this part of the book I challenge any one to disprove. The same challenge of course extends to the rest of the book but since this part may expose a slight view that may be seen a stepping into religion it is not meant to be representing religion in any form or in any way. It is pure physics.

$\Pi^1 \quad \Pi^2 \; \Pi^0 \; | \; \Pi^2 \quad \Pi^1$

In this chapter the motive is conveying the idea why there was a reason for life not to have been part of the Big Bang when the Big Bang was in process and any and all nuclear explosions is returning that space to a period where that space was in time eternal because the nuclear reaction took the concentration of heat back to a space just about space less with gravity in that spot enduring heat to a limit heat was part of space and time in eternity.

Singularity forms part of matter and the evidence even presents itself in life. Life also duplicate in the same manner as the proton but life is more than just duplicating Space occupied by carbon carrying life shows signs above ignoring it or just brushing the evidence aside that singularity holds a key value to life. Only if life is amidst or even deeper than the proton can the protons spinning around singularity, hold such a strong influence on life controlling the element of occupation as each individual proton occupies matter in space

NOTING THE RESEMBLANS IS SO OBVIOUS IT STICKS OUT LIKE A SORE THUMB. Life IS THE SECOND ENERGY NEXT TO HEAT REQUIRING HEAT BUT IS NOT HEAT AT ALL. The physical aspect in the manipulation of heat be it the body holding life or the heat the body apply sustaining proton growth is a huge component carrying all the ingredients of making it vital, still it is not life. I have not the right to claim the universe came about to the sole purpose to benefit life but this much I know and prove that the solar system came about exclusively to the benefit of life. The odds of another structure forming as the solar system did are even by cosmic standards remote. I wish to think that life's presence is a great contributing factor to the universe but such a claim constitutes the arrogance humans carry as a natural implication of being human.

I do not wish to start duplicating THE SEVEN DAYS OF CREATION but I need you extract some of its theme to present the warning I think is due to the war mongers amongst us. I see many youngsters wearing T-shirts with a slogan like: "NUCE THE BASTERDS" or something to that effect. That is the worst wish to wish upon any form of life and none worse can ever be closer than the true damnation of the life within the carbon. No life in carbon or any other element was present when the Command came to fore fill mint.

Divinity lies in singularity and I am not taking on arguments about that because any one can take and accept or chuck and condemn. That is the right of being person holding conscious and that right stands free to any person. Let us have a look where the double proton came to duplicate because that duplicating presents the fact of life from divinity and from singularity.

Π^0 | Singularity holds a value eternity as Π^0. Only stupidity and obstinacy can dismiss such a claim. Any arguer must then prove otherwise because singularity presents eternity running down the fragments of singularity to eternity where unifying will again hold eternity away from infinity. The edges even small as they are dividing singularity is the double proton so small and there still is no r to present a circle.

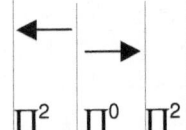

I wish to quickly and very shortly explain THE PHYSICAL PROTON BECAUSE THE PROTON HOLDING LIFE IS FAR TO GREAT A RESEMBLENCE TO NOT SEE SIMELARETY.

Having edges where Π^0 duplicate to present the edges singularity lost the value of Π^0 to the value of Π^1 with the same value singularity had being Π^1 to the one side and Π^1 to the other side, the cosmos received the eternal value of the first dimension outside eternity. It was the square of Π^1 being Π^{1+1}. That was the first dimension outside singularity Π^0 where singularity has a value of Π^1 in the form of $\Pi^{1+1=2}$. The first claim to space had a value of Π^2. This applied to both sides of the claim to space outside singularity, and the double proton became the dominant factor on matter.

In life, life has the occupying position applying $4\Pi^3$ the value carrying time directly to singularity where divinity is the life addition to matter. If not then what is the difference between rock and man and why can man move rock at will but rock cannot lift man without the aid of life. In the time value and R^4 in the space-value of thought.

THIS EFFORT IS TO TRY AND EXPLAIN THE UN-EXPLAINABLE. IN THE PICTURE THE WAR LORD FILLS DIMENSIONS 1 TO 4

Then through man's insanity, the process that happened during the creation period reverse in a turn about, admitting, on a small, small, small scale. However, life was never mend to be in this process. Star destruction and the Big Bang is one and the same thing where matter converts to heat. A star is the process where heat converts to matter, in as much as directly to the spin inside singularity.

5th Space with out time (R^4)

4th R^3 space to time (space-time) being Π^3

3rd R^2 time space or $3^2\Pi$

2nd R space in time $3\Pi^2$

1st T^2 space less time ($\Pi^2+\Pi^2$)

0 T^2 time in singularity Π^3

? T^2 time outside the cosmos $4\;\Pi^3$

We can see what the star discarded, that which the star found unacceptable, in the same manner as the discarded products in the rubbish dump the rubbish dump we visit with our eyes. The inside, which is precious to the star, remained in safekeeping and absolutely protected. The growth the star accomplished it holds safely contained in the inner nucleus. That precious part we can only reach with what the Creator supplied to the human, distinguishing it from other species, that which I call extelegence.

Finding the reason behind fusion follows a similar path, as does the sound barrier, because as stated in previous chapters, putting pressure and heat will lead to an outburst; not to fusion The links with the unoccupied space-time to occupied space-time ratio plays an all crucial role in the process of matter joining in one nucleus. At one point the water vapour stays stable in the air, suspended, and then something change enabling the vapour to form water and fall.

Species change not only their preferences, but re-apply their whole way of existing, and beside the fact that they might feel adventurous, there has to be a far better reason for the changing. Life becomes tenacious when in a position of "adapt or die". Some very significant factor in its surroundings forced the change onto them and in a case where a reptile has to learn to fly, or die, the logic conclusion is that the surface it used to walk became unbearable. The cause leading to such drastic changes in life style would be far greater than the unwelcome presence of a couple of meteorites.

The reason I bring all the deliberation about is to show the cosmos came from protons. The accumulation of protons was the universe. The big bang became all about forming protons, NOT SPACE.

What the nuclear bomb showed was that matter was only heat in a frozen state. This brings me to the point I wished to make and motivated all the blabbering on the way as we could arrive at this point. If heat is matter and matter is heat, what on earth is antimatter? Xepted science has many books of extreme highly academic insight written on the \subject of matter and antimatter, but as in the case of gravity the force, bringing along all its baby brother and sister force to keep it company. Giving it a name is all that science requires establishing one more useless fact. What is anti heat? It must be cold. Ever heard of cold destroying cold or heat destroying heat.

The truth is that I am not sure whether the inventor of the game "packman" got his queue from Xepted science or whether Xepted science was playing too much packman, but the similarity is somewhat disturbing close. Half the matter ate up half the matter and only a small matter remained. MATTER IS ENERGY, ENERGY THAT CANNOT DISAPPEAR, but Xepted science has it that the one piggy ate up "all of the other piggy" and remained half a piggy. That is the findings of persons not regarding the Bible accurate enough to take seriously! Energy; coming along as "anti-energy"; destroying energy; without leaving energy in another form. Boy does our Brainy Bunch need a trip down to the school classrooms because along the way they lost the way children learn about energy. Then again, such small matters are hardly what the Brainy Bunch waste energy on. They invent cosmic packman where matter dissolved matter only to leave small traces of matter, and of course, small traces of antimatter.

No such nonsense. Newtonians and atheists should read the Bible about creation instead of inventing their own dismal failure. The order was "let there be light…and there was light, Light there was not in the beginning, and by the end there was light. Ask yourself what will bring light and the answer is as clear as the bible. We know that the heat was enormous totally off the scale. Again it proves, not so, because the heat came from a frozen state going hotter. At first it was too cold for light to be. From our perspective the heat was incredible, but from the cosmic stance the heat became incredible.

You go on and ask any mechanic like I am about SPIN AND ANTI SPIN.

Adding any, more (which I can prove, as I did just now with the above) will contemplate religion, and I have no role in explaining Divinity. If one is born blind, another cannot explain the colour dark green, to such a person. If one is born religiously blind, you cannot explain the light to such an individual, and to what purpose will, arguments lead.

Every time the lesser time value relates to the higher time value, it assumes a position in the order of space-time. From our perspective, in any of the inner-Core-values mentioned above, space is long gone, but space is relative, as time values, space. The chain of time placing just mentioned forms part of every atom, in each proton from the hydrogen proton to Uranium 92, and higher.

Every time-value in each dimension already is part of the proton, waiting it's, turn to utilize time, as the proton progress in time. Life has the Devine position that it connects to both borders of the universe, in the dimension from $\$T^{2T2} = 4\Pi^3$ right through all sectors of the cosmos.

Divinity's R^4. This position life has to the cosmos, gives life it's unique " god-like" role in controlling, and manipulating space-time all the way through to space less time. The only sector life cannot manipulate is time $\$T^{2T2} = 4\Pi^3$, and Xepted science fail to realize this

$4\Pi^3$ is the position to life in time.

The position (R^4) where consciousness, thoughts and dreams in the mind, in time.

The time position $\$T^2 = 2(\Pi^3+\Pi^3)$ to contemplate on actions, reactions and decisions created in the dimension of the time less space (R^2) dimension.

The position $\$T = (\Pi^2 + \Pi^2)$ where life's connection (nerve centre) to densified space-time is in time.

The position $\$T = (3\Pi^2)$ where life's command connection to densified space-time is in time.

S$ = C/(C-V) X C/(C-V) is the space-time surrounding life.

When the warlords will fully and deliberately sets off a nuclear reaction, life in time with an earth value of $4\Pi^2 = 39.48$ becomes separated from all other dimensions except time in the eternal time value of $\$T^{2T2} = 4\Pi^3 = 124$. The flow of time development reverses through life's intervention and manipulation. As this new space less time creates a position on the border of the cosmos, the same value to time less space has to establish on the other border of the cosmos.

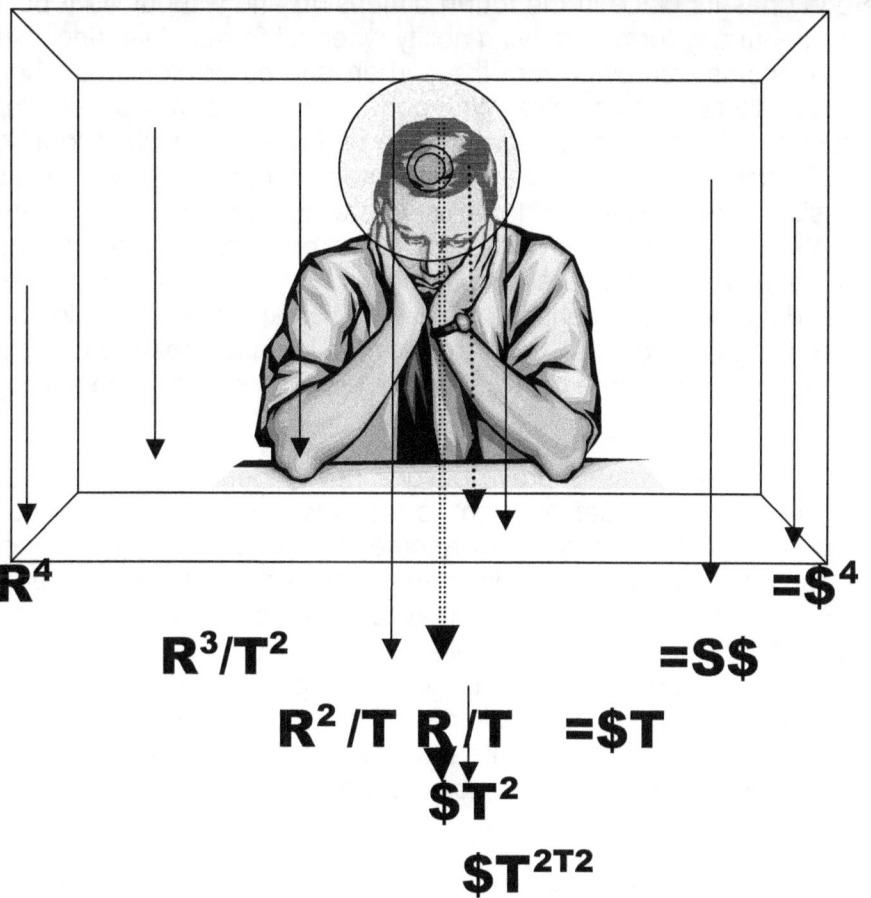

**The Destruction, creation
In addition, replacing of existing space-time, time-space**

Life then becomes locked, from the time less space, through all the other dimensions, it had the ability to manipulate, into the only sector it has no control over. In this way Life seals from all factors of space; for the rest of one eternity, until the cosmic time development has developed past the dimension to the 124 time-value.

Weather even Xepted science will ever accept this fact, which has nothing to do with religion, or faith, but is as much part of cosmology as fusion is, is an open question.

The warlords, Hoggenheimers and Mammonites, backing the warlord's efforts, I realize, will never accept this fact. The one thing that should be remembered; as they now are prepared even at this point, to send the earth and life to hell, with their chemical poisons, they will continue doing so. As long as they feel secured in some way of protecting their lives with bomb shelters, and their wealth by whatever means available, they will be prepared to cast every living aspect other than themselves in to eternal hell, to benefit and protect their own.

It is up to science, those members with conscience and vision, whom they cannot bay at a price, to instigate awareness to this fact. Politicians cannot be trusted for they spend half their working weeks begging the Hoggenheimers and Mammonites to bay them off with "contributions".

It is up to the honest academics that do not also rely on "contributions" for "research" which in the end will only and exclusively financially benefit the Hoggenheimers once again, to spread the awareness to john and Jane Dow.

Even the most influential academics though, too are in the grips of their MONEY POWER. I cannot see how the world will escape the nuclear fate.

WHAT WILL, BE WILL BE.

The challenge is to find a manner that life repeats this tendency to connect in four stages plus one extending. In singularity there are five positions extending to space-time with five positions and therefore life connecting to singularity should indicate five positions. As far as the aspects concerning the physics side goes I could prove in some way that my initial way of reasoning brought fruit to bear, but what about the dimensions past four and even past five. I think I may presume that in some way I did manage to show that the body has its place in the fourth dimension and without life it becomes just one more cosmic structure without any form of moving ability when all forms of life (including the bacteria that will decompose it to atoms) removes from the carbon and other elements where the elements then are exclusively and only cosmic particles. With good reason one may believe that the ability in conducting the manipulation of the space-time within the body has diminished to all extent and none is any longer present. To find a means of putting mathematical formulas to use in applying proof to indicate the fifth dimension is beyond me and that is where human extelegence plays the part. By the same token that atheists can say they wish for more proof about life and the fifth dimension I can demand the explanation to prove otherwise and ask an explanation about the energy presence in life-holding bodies and the energy absence in lifeless bodies and where as I doubt I may force atheists to sound understanding they too must admit that something does seem out of place in they're arguments about the energy being of a pure physical nature and only stubbornness will win by the days end.

So to them and those, as I might not find a way to prove beyond doubt them and those also must admit I have sown some doubt in their minds. Getting them to admit with some gallantry about the doubt factor, well that I must gallantly admit is a horse of another colour. With the fifth dimension seemingly impossible to prove mathematically the sixth will be much more difficult to prove and I shall not even attempt such an act. Fortunately man is not mathematics but much more complicated than mathematical equations can ever bring forth. Only good old fashion arguing with a dash of logic sprinkled when and where necessary will pave the way to understanding man and beast. To find out what man regard as good or bad and what beast evidently regard as good or bad must be different as everything else about man and beast are different and we must scrutinize beyond where mathematics and physics can prove. Even the most convinced Newtonian should see that there is a point such as that.

Being human every reader must have an opinion about good and bad and what is evil and what is not. I can scarcely imagine a lioness feeling bad about a kill while her cubs are filling their bellies with mouth-watering flesh of an antelope kill. Neither have I detected sorrow and anger as a new male kills the previous litter to establish his new domain. Neither the female nor the male bears sorrow after the deed of him destroying her litter sired by a previous dominant male although the lioness will protect her cubs while they are alive almost to the point where she may put her own life on the line. That how ever has nothing to do with rite or wrong good or evil and after the lion male did his killing of her cubs she shows no remorse or blame for that matter as she follows him back to the rest of the waiting pride. This is not exclusive to lions or even to predators but has a wide range of animals following the very same living style. Horses kill without thought because when a stallion wants to find a mare that he knows may follow him but for the foal by her side he will kill the unwanted foal and not be bothered by her reaction, being in the knowledge she will follow him after the foal is dead in any event. While the foal is still alive she may do some protecting of the foal but she will normally not go as far as the lioness in preventing the death of the foal. Baboons, monkeys and a variety of animals have this approach to life. During such an attack by the new dominant male baboon the females will fight off the onslaught either by grouping together or by fighting him off in ones and twos but the usual is that the new male is big young and strong and even a group of females are no match in a fight. However after all the noise and the shouting blood sweat and the rage of adrenalin has died down, the babies are dead and no female attacks the male after the fact in heart felt sorrow for the loss they feel such remorse over. No, it is clear that the deed was done and life goes on.

In humans such behaviour does take place and every time we read about such a deed committed against a harmless infant even the biggest humanist find a moment where he or she wishes the death penalty on the criminal for acting in such brutality. Why will humans shout for blood in punishment while other species take it in stride?

To find a solution I do what I always do. I turn to the Bible for an answer. (You atheists deny yourself a wonderful encyclopaedia of information and when reading the last book in this Theses you will come to understand my saying so.) In The Theory which is the seventh part of The Theses I explain to very detail the events of the first six days of creation as recorded by The Authentic Biblical Author but even after that some more explaining arrive when it is correctly translated. In the book called the Bible there are two trees described with distinction and I am of no opinion to whether they were in wooden fibre or just symbols to explain the complicated issues to persons with even lesser education back then than I have at present. It is an undeniable fact that man has an inclination about what is good and what is evil. Man would not kill an infant because he cannot find the infant's mother and the infant is crying of hunger. In the animal world any adult of the specie will walk past such an infant in distress with no feelings of care what so ever even if the specie does not show a normal tendency to destroy such an uncared suckling. Where one mail may kill another mail in a fight to establish dominance the group does not cry for justice as they loath such a deed. When a superior member of a pack relieve a lesser member of food or eating rank they do not hold congress in judgement to provide sanction for the lesser member with accompanying reprimanding about such incorrect behaviour by members of the group. Such is the caring of man that any human notwithstanding whatever urgency drives him at that particular moment that human with maternal instincts or not will stop to care when coming across a deserted human infant hungry and all alone. Why will man show such behaviour as normal through out all races on Earth without any cultural distinction in any way and this may be the only distinction that races share because some eat their dead and some burry them with pity filled emotion and… Oh, I can go on writing a book on this topic alone but that will be useless because every one should know what I mean. With this shared by all where and what does mention this distinguishing behaviour of man's ability in judging between good and bad for the very first time as a landmark to man. Accept the Bible or not, it remains the oldest Book available on matters reported by man since no one knows when because Moses may have assembled the research on information but the information as such dates to times predating even what Moses' research may indicate. In addition it may be correctly presumed that many or most of the facts he recorded he was taught as a prince in the house of pharaoh and that then may explain some detail about matters he actually knew nothing of. His being adopted by pharaoh's sister must be a plan with some significance and must result in some meaning other than to give Moses a childhood of luxury alone.

One of the trees the Bible mention carries a name specifically as life and from that I draw a conclusion that may reflect that life chose to go the way of having a variety as sells with complexity in the evidence we now gather from DNA strands whereas a choice of life would indicate a sell of simplicity in structure as we find with lowly developed insects and other specimen of life. I have seen how a corn-crake of a specific kind only found in the desert and semi desert regions where I live can start a pest becoming so out of control that when run over on the road they form layers of millimetres thick trampled and squashed by cars to the extent the tar on the road is no longer visible. It truly is a pest of Biblical proportions but fortunately to unleash this pest it has to rain in November in one specific week. If it does not rain at that specific date, and I do not mean approximately but precisely they don't show at all. That makes their return very sporadic and it only occurred about five times where two of the five times became a pest like none can repeat in the more than twenty years I farmed on that farm. The pattern also follow a distinction where at first one or two may show very sporadic in places. Then mating starts and the pest develop where by it truly come to a climax at the end of February and dies down in May. From a few eggs they develop millions on millions and I do not exaggerate in the least when I refer to this phenomena as a plague of Biblical proportions. Poison does not kill them and when one gets hold of another one the bigger one just start feeding on the smaller one. At the end of the meal where the bigger one devoured the smaller one in totality (and I mean boots and all) the specimen will shed its skin immediately, eat it or not eat it and walk off. The one is a precise duplicate of the other with no distinction amongst them of whatever nature. Seeing one is seeing the lot. They share genes in precise replica with no differentiation of any kind what so ever whereby the one may have even the tiniest of difference in any form. They come from a line where life was still very basic and the mother specie of the very original has not changed in any way through millions and possibly billions of years. That I say on the grounds that to my judgement that specie is of a very basic nature and has developed with one single aim in life and that is to survive. They eat everything from the most poisonous plants to fruit to meat and bones of animals lying dead in the veld to dry hide and even one another. Fortunately too they only occur in the most severe droughts and when developing into the pest, which I describe, there is little to nothing for even grasshoppers to feed. Seeing the specie for what it is it made me realise that life somewhere after them made a choice to form complexity and variety or remain as they have and form a universal gene where the original mother is still present in all her offspring even after so many billions(?) of years having the opportunity to progress from where they were. They made the choice to remain the same where as the line that man developed by the original parents may have made the choice to evolve through complexity.

With all the explaining I do not wish to prove that man had a nibble or never had a nibble from that specific tree but I only wish to indicate that there are a variety of interpretations and clues around and when sanctioned they may deliver a vastness of possibilities. There is one other tree of distinction mentioned and also a mention of some eating by the female at first and later on by the male. This tree also was named and it was the tree of good and evil but according to the Afrikaans Bible the mentioning of the name says specifically and I quote: *Boom van kennis, kennis van goed en kennis van kwaad"*. Directly translated it reads as follows "tree of knowledge, knowledge about good and knowledge about evil" and that is where my argument starts in my attempt to indicate the possibility of the sixth dimension belonging exclusively to man.

In the detailed analyses the specifics concentrate on the "knowledge" and then distinguishing between "knowledge of a good nature" and "knowledge of a bad nature". Please note there are three mentions of knowledge one only about knowledge then about the knowledge of the good and thirdly about the knowledge of the bad, but most important separating the three by distinction.

After clearing this part we may return to the animal world of some being wise and not very wise. Animals by nature and by genes acquire a base for knowledge to carry the specie through dangers and more important even, to the survival of the future of the surviving gene pool. Surviving as far as the animals go is the good and the evil and there, at that point all other definitions stop. All intelligence the specie holds and all the intelligence the specie acquired contains the one underlining element being individual and specie surviving. In saying that I do certainly not say that rules amongst members of a group of animals is non-existent. There are certain criteria the individuals have to meet to establish rank in the tribe. At the same time such rules do not centre around emotions of ethics but they are practical well placed and directed to ensuring stability and it seems the higher evolved the specie became the more sophistication there are amongst rules constraining some members to the advantage of others.

The Matriarch in the Elephant herd is Boss and that is in capital letters! No bull will dare to push members and lesser infants around and she takes much less shit from young elephant males than the females. She is the rule and the law and every one abide by that. Should a male wish to afflict his attentions on some elephant cow the Matriarch will condition the visit to her satisfaction or the visiting male may even pay with his life. I say this as a result from knowledge I acquired as some game-farming friends of mine has elephants in captivity. Crossing electric fencing is hardly an issue for the matriarch because she takes a teenage male place the young male (and it is always a male) between her and the fence and let him walk unsuspectingly alongside her as she deliberately holds him in a position where he walks between her and the fence with the current. Then at a moment she decides on she thrushes the young male allowing him to plough through the fence and take the electric current shocks while he goes on his way braking the fence altogether. After that the fence is open and clear for the rest of the herd to cross. She will never act in this manner towards young virgins but the young males get the stick every so often. With the advances the African elephant show would the African elephant be a good pet. I would hate to find out because they may set the rules and not me.

I know for a fact that a Nile crocodile does not make a very obedient housebroken pet. Should any one have an idea to keep one in his swimming pool be warned the pet would not distinguish very well between his owner and his next meal. And his love for children might be somewhat different to that which a good pet should have as is the case with dogs. With dogs man had thousands of years in breading good pets but it is unlikely that the first relations were as timid between master and dog as that we grew accustomed to.

1. Through many generations of exclusive inclusive breading did we finally manage dogs to have become what we wish them to be. In this lies another fact to analyse. Many different breeds make many different dogs and the one race has characteristics setting that race apart from other races but in the race itself the variety of characteristics find more prominence in the race than in individualism. Characteristics of dogs connect more to the type of dog than to the individual dog and therefore some races have inborn hunting skills where others may have guarding skills. It is the breed that brings the selection and not the individualism in personalised characteristics. Therefore it cannot be said that a dog has a conscience but it is better said that a dog has a better breading line.

2. Some evidence suggests that when Cro-Magnon – man arrived agriculture replaced hunting as the feeding method and we are confident man exclusively kept that dog for it's hunting and sharing abilities. With the arriving of agriculture man then extended his space-time manipulation not only beyond his physical abilities in hunting but also his physical strength in working with tools. This must be the biggest leap of all even much bigger than the leap of the electronic age but such comparisons are extremely difficult to make.

What would be man's drive to not only manipulate his personal surrounding but also manipulate surroundings of other forms of life to their benefit but moreover to his benefit? What would give man such judgement as to select species beyond him and feed them to eventfully find more benefit from their feeding than they did benefit from their feeding. Genes it cannot be. The orang-utan has 98.2 % of the genes man has and the gorilla has 97.8 % of the same genes man has. The Gorilla still lives in woods and is destined to disappear while the orang-utan lives in trees and holds no better future prospects. Genes would at least give the species having such close relation in the gene pool with man an idea to follow the trend set by man and copy some of the abilities. Genes it cannot be and that just about excludes the last cosmic or natural physical explanation from the list of possibilities.

There seems to be a massive gap between what man became and what ape became. Science makes a great singsong about chimpanzees with the ability to use tools for their benefit but man has surpassed that so long ago science have no tracking record about the time and the way that came about. It seems as if man was not, then man was with agriculture and all other providing the manipulation of the other species under the control of man and by increasing all benefiting that the animal enjoy man could bring benefiting all around to benefit man. One day man was ape and the next day man became super-specie-of-the-world, the world champions in space-time manipulation or in other words of controlling life. Not only the life of man but also life of others to some mutual benefit slanting heavily in the favour of man. Still the benefit of the other species holds so much that only species that benefit man started to dominate world population with man. Mankind evolved not only more but mankind evolved much better than the rest and the evolving must be a point that should be investigated in sincerity. I wish to draw your attention to the differences there are in gender behaviour of mankind and the worlds separating gender in mankind.

In modern society it seems as if man and woman are in a dual instead of a marriage. Women are forever complaining that men drag them down, block their progress, and deny them their rites. Is it true or is it what they were told too believe by the persons that holds the influence. I would never go as far as declaring woman as being children of the lesser part of God, but from a mans point of view that is not that far off the mark. Woman wants what men have. It was Sigmund Freud that gravelled with the question about what women want. The problem is so simply solved. Woman wants what woman cannot get. Women want to have the greener grass they see on the side of where men are but they do not see all the shit that makes the grass greener. Once they endure the shit on the side where men are, then that too is the fault of man. They feel their needs and see men with other needs. They turn on the image men portray. Then they turn around and blame men for not caring in the same manner that they wish to be cared for. But this stems from our new civil order. Men are not from Venus and women are not from Mars because that would bring about such simplicity, but the problem is bigger than the mind of Freud because the solution was even beyond his grasp. We must do what we did in physics and backtrack the line. We must retrace the line we took to develop from our time living in trees.

We must once again halve k to the point where we find k^0. We must investigate the point where men and woman separated in the psychology of being women and being man. In this way we might stop confusing our gender that will eventually lead to our civil demise. This warning is serious and if not corrected it is coming to destroy us in the future. One cannot go about messing and changing civil order that came into place during the past hundred thousand years and just replace what we think we do not need because we now are so wise. We know much less than man knew say five thousand years ago but we can harvest energy better and that is the only advantage we have...and we are brainless enough to think that improvement makes us wiser than our past.

Men are as guilty in the same sense but men cannot offer the excuse women have. Women received intuition where men got reason to think with. But I shall come to that a little later. Science always carries on about how man came to civilisation and when did this process start. When did man become more than the animal? That is easy. It was when man could see and appreciate stars. When man could somehow convert information about starts man became man. When did man evolve into the early signs of civil order or civilisation? That also is very easy. It is when man and woman parted their roles and parted their goals. When the tribe started to place certain tasks onto the shoulders of men and other tasks fell upon women. It was when men and woman both had the task of bringing home food but each brought food of a different kind. That was the point civil order separated man from beast. That was when man had the time to start developing communication by leisure and not by order or survival.

When men presumed the responsibilities of hunting while woman gathered the food and the two groups shared the result amongst all the members in the group civil obedience started. Splitting the duties and splitting the responsibilities gave much better opportunity for social development and the establishing of order. Men had their role to for fill and women had their duties to perform. It was when woman walked in the fields and smelled the roses for cooking purposes while men did the chasing of dear or rabbits for hunting purposes. Women had to select in order to provide the best for her community while the men had to hurry and kill to bring home the food for survival.

This was the biggest era of human development and from this humans established the functions their development allows them to enjoy at present. Yet in anthropology no one recognises this as an era. The one group being the women established a calm and selecting atmosphere and worked in conditions of approving and approval. Men on the other hand performed their duties by shouting, running, selecting in stealth at speed and killed in swift to become successful in the role they had. This was the order for a longer time than man can think back. The imprint still totally dominates gender behaviour in the present day and age. It still dictates social behaviour and understanding the way we think. It is the main issue dividing the way men and women come to conclusions and make discussions.

Look at women doing shopping today and look at men doing the shopping. Little can frustrate any man more than to be on a groceries baying expedition with his wife and at the same time hide his frustrations to please his wife. Women smell and taste and look and feel. They judge and decide never to come to a decision. Men rush in; grab what they want without paying any attention to the products they don't have on the shopping list. They want what they need and ignore all else there may be. They grab and run. Even selecting a cashier is done in such a way as to find the shortest line and pay in a rush. It is split second deciding all the way

See how men select any TV channel and compare their methods with how women select their channel of choice. Men grab the remote and blast through the channels with much indifference about any content of the programs. They use the TV remote to kill and not to select. Woman on the other hand look and test using time to find sense of the program before deciding and then move to the next channel. This behaving pattern still persists as part of our genetic developing culture when we as humans decided on how we were going to create our future. This behaving pattern is stronger than our recollecting of time. It is the printing of our genetic patterns. It forms the basis why we have gender. This is so strong that it is this, which the feminists hate and try to destroy because it is this that which lesbians cannot mimic in their pretending to be men. Whether the alliance of united lesbians world-wide and all the feminist orders would like it or not, they are woman living in a society shaped and dominated by a man's brain and this includes all societies, nations and cultures as well as science disciplines. It is due to the brainpower of the man that got man as a specie out of the tree on the ground living in the protection of houses. It is the brainpower input of man that got us away from an existence of gathering and hunting while surviving on a minute-by-minute basis to living in high-rise buildings and skyscrapers while the entire society is running on electricity and not the back breaking labour input of the human body. It is the brainpower of the man that got us wheeling in cars running on fossil fuel and living in excellent comfort. The brain of the man that got us living off shop shelves buying what you need when you need it as you need it and in a variation of choice no species could ever dream of. It is due to man that mankind is the domineering species on Earth. However, correctly or not, this has a downside in our culture. Every aspect of science is going through the brain of the man.

All psychoanalysis procedure goes through the observation of the man because even when woman do it, the method applying in their training was from the man's perception when it became excepted procedure and then the woman would approach any procedural treatment accordingly and the resulting conditioning is with keeping the male approach in mind. Even Sigmund Fraud was baffled by the non-compliance of the woman in the psychoanalysis investigation. This is because the man wants to approach the woman as seeing the woman from the other end of the man's view. This will not work. One has to see a woman as another specie formed in another Universe because that is exactly what happened years ago when man developed as a specie.

The man became the hunter, providing the meat while being out in the field all day long. He only returned after success brought completion to his task. He fought a hard battle and was bruised and hurt and when he came home he wanted softness and a tender appreciating side. He wanted something pretty and soft to take away the scruffy image he had about his hunting companions he shared time with on the hunting expedition. He wanted to get rid of all the competitiveness he had with all the other male companions that joined him on the hunt. In their company he had to be the toughest or he was downtrodden like a dog. The last thing the man could ever do is to complain to his male companions about an unfulfilling or problematic sex life because there will always be men ready willing and able to step in and help him...by satisfying the needs of his wife with or without his knowledge.

Therefore talking about sex problems or any other form of regression in his male rankings has never became part of the male culture and ask any male and you will find the most virile sex animal that has the ability to go on through the night with as many partners as there are willing partners. The male is as tough as any lion tamer can be and the male will wrestle the common garden-variety elephant to death at any spot where it needs to be done. There just never is any limit to what the man can do or what he can achieve. This is the result of the competitiveness there was in the male rankings and any hint of being inferior will cost him is wife and most probably his life. He had to be the toughest, the strongest and the most powerful to scare off all other contenders because a satisfied woman is an unwilling woman and by approaching an unwilling woman the other male companions will only find they are unsuitable as they are matchless in the face of his male lust. On this we all agree and this is the male approach. The man must love his woman or many others will love her more than he can. It was part of all ancient cultures that a man had many wives while a wife had (officially) one husband and in that view he had to be the best or become one of the others tat is part of the wifeless bunch living lonely in the tribe. From the man's stance he saw a wife having other qualities he was after. The wife must be soft, apprecative, pretty, adored by him as well as all other men because man wants to come home to something others would wish they could come home too. No man wants a wife that no other man wishes to look at. If she is not wanted by all other men she is not worth fighting for and to hunt is to fight with his life being on the line so in order to provide for the love of his life, his wife and his children has to be as special as the man thinks he genetically is. She had to cuddle him and cook for him and be ready with sexual favours at his commands.

As long as his family see him as some sort of semi-god coming home with the spoils they yearned for all the time he was away, his is as happy as a lark. He will take on the world and hunt every beast in order to provide for his wife as long as all other men in the world would envy him because of his adoring wife. This is what everyone knows the man is. The woman is there to fill the needs of men. On this information women are judged and every man falls flat on his face because that is not what women want…in fact that is least on the woman's list of priorities.

A woman will not wish to have a pretty and cute looking man that takes all the attention away from her person because her man is better looking than her. She would not wish to share a position within the clan that thinks her man is more soft-spoken and more feminine than she is. She would not look at a man that looks as if he is the flower next to her and she is the bearded cactus or be in a situation where the man is better clothed with a more colourful garment. She will not tolerate a man that will take her place as the pretty picture. She wants a man that compliments her femininity. She wants a man so bad looking that even at her worst, she still is far more feminine and pretty and pleasing than he is at his best. She wants to be the one that draws all the eyes as they walk alongside each other. She wants to attract the attention and have everyone envy her over her good looks while they do not even noticing the ugly bearded slob next to her, and the more ugly the bearded slob next to her is, the prettier she would seem to be. It's all down to relevancy where she is being prettier as he is scruffier. If he is totally bald, that will suit her the better because then he hasn't even got hair to give her competition. If he looks like a something the cat dragged in, she becomes a beauty queen although she herself might be no Mona Lisa. He must be there to fill her needs while he is stupid enough to think she is there filling his needs.

The woman has to be this way and think this way. She has to be ready to substitute him for another as quickly as dropping a hat. She would have mouths to feed and she would have to be ready to replace her husband with the next willing, ready and able man that is seen by all as some strong man to secure her immediate position in the clan and to care for her children or else all men in the tribe will use her as they please. There always will be a fair chance that the man does not make it back from battle or the hunting party and if that happens then she can't weep for weeks over her loss before replacing him. She has to feel about the man in terms of him providing her security and if he is unable to do that, then he has lost the reason she wants him. If there is anyone that think I am harsh and that all this I mention does not apply in modern times, then go and see how long it takes a widow to replace hubby after death when hubby dies when hubby had no life coverage. A wife will give conditional life long love, but her commitment is acceptable, permissible prostitution and if he fails to make her feel secured, needed, wanted and special or that he is the best she will have available, and without having that security, he is gone.

She will provide as long as he provides but if he does not start to provide she will easily get rid of him and take the next in line, even if it is just for an hour or so. If the man next door will appreciate her more, she will have him and will carry his children. That is why woman are the poison killers. They will feed hubby mushrooms as quick as he does not fore fill the requirements she seeks. She sees in hubby the provider for her needs and that is what the hubby is. He is there to compliment her and to make all others appreciate her more. Men, you are daydreaming if you walk around with any idea your wife loves you as much as you love her. All I can recommend in that case is wake up and smell the arsenic because no woman has the ability to love any man. A woman loves a man because he helps her love herself even better. A man is a compliment to the wife and that is all. She will love you only when you make her love herself more. She will love you only when others will love her more because of your influence or your position or your status within the clan. She loves you on the condition you take care of all her requirements and no less will satisfy her. If not then she will stop loving you as soon as she sees someone better.

A woman can never love a man. A woman loves a man on the condition that the man makes her love herself better or more and thereby she feels everyone adores her more. A woman loves a man when he makes her feel as special to the degree she can't feel better and in a way that she will love herself even more than she does without him. A woman can't miss a man's company but only misses him when she misses the way she feels about herself when he is in not her company.

Men and woman are not from different planets. They come from different Universes where they share little but the responsibility in the roles attached to their gender. A while ago I explained to my sons what women want in a man they choose to have as a selected partner. Women do not choose a pretty face as men do except when they are teenagers without brains that they had not developed yet. A woman can never come to love a man. Women are incapable of loving a man. That is not in their nature and that is not in their brain. A woman loves the way a man loves the woman. They love the tenderness a man shows. All women want to feel important and special. They want to be the best choice there is and when the man shows this affection they then respond by falling in love with this affection. They love that what the man brings and not the man. The man that makes them feel most important and overall special will win the prize. Women want to be cared for and made to feel they are appreciated. Woman want to feel they are needed more than any other woman is needed. That is how and why a woman loves a man. It is what the man can be for the woman that the woman loves. Woman will show love in other ways than men do. Women show their affection by caring for, by doing onto, by serving with feeling, by caring when care is needed. Women show their affection in the way they treat men. The food and the drink they serve the man is their affection they show. Woman only respond to tenderness but does not serve with tenderness and soft touching.

This does not go for men. Men think that woman will fall into their laps if they are as beautiful as a woman is. No women would like to walk next to a man that looks more radiant than she does or smell more feminine than she smells. No woman would appreciate a man in her presence that talks more softly and has feelings more tender than she has. The woman will not appreciate a man that has a softer inner feeling than she has. She may enjoy this for a brief period but that man will be dismissed in no time at all and will find him being substituted even if it is for a few hours. A woman feels woman when she is in the company of a man that makes her feel woman. She will not feel comfortable with a man that challenges her effort being a woman. The man must provide a platform from which she can stand out just because he is a rough being and because that will bring about a relevancy placing her as dainty as she can possibly be. The more the man is man the more the woman will feel woman and in that she will be appreciated as coming across being beautiful and tender and loving. The more robust a man seems to be without being vulgar or scruffy the more tender and fragile the woman will seem in that mans presence. That then is what she will come to love. She loves the softness she has in the presence of the man she is with. Women want to feel special while men want to be special. That is the crux that puts women where they belong and men where they should be.

But women think they want men to feel tender and understanding because woman think about such matters in terms of loving and appreciation. That is the manner woman show affection and they incorrectly wish to find that in the behaviour of men. But that is not what women truly want because as soon as she finds that in a man she immediately ditches such a tender being. Women want to be secure because men can give them security. Women want to depend because such depending relieves them of the other side of the Universe. Woman wishes to provide that which she is provided with. Women want to love back while men want to feel love. Men want to be admired while women want to feel admired. Men want to be special in the eyes of their woman. They want to feel appreciated and not to be appreciated. Men want to be appreciated and not feel appreciated. With such differences going unrecognised by both gender men see woman not appreciating them while woman see men not appreciating them and no psychologist ever realised the difference there is. The words used are the same but the approach comes from different Universes. Woman think men will show their appreciation by doing some of the chores once in a while, while men will feel appreciated when woman admire their bodies and their manly strength once in a while. The woman is not developed to appreciate the body of the man while the man recognises that as love. He sees it as love because that is how he appreciates her with the feeling of love for her. It is the way she looks that blows him away and it is the way he looks at her that blows her away. The man admires the body of the female while the female wants to feel protected partly by the man showing such admiration. If he likes her he will fight for her and then he will fight to protect her. The woman thinks she feels appreciated when a man cleans the pantry but when the man cleans the pantry the woman feels threatened because back in her mind she feels he is about to replace her. She feels threatened because she feels no longer needed. Just as the man hates when a woman stands up to fight another man on his behalf. Then in that situation the man feels she is taking away his reason to live. She does not see it that way but to a man having a woman protecting him leaves him undesirable and worthless. He appreciates such behaviour, as a sign that she does not need him any longer and in that woman as well as man don't know what they want. The man is not developed to do house chores because back when the developing started men had to rest to preserve their energy for the next successful hunting trip. Then the woman fussed around and about them and treated them with care in an attempt to lure him back after his hunting. Now woman want to work and earn what men earn but also want to abolish all other traditions that brought civilisation to the present form. Women now see them as fit as men doing what men do and having the capability of even going to war and fight men but that will rebound and what then will happen to the women on that day will be such a sorry affair I hope never have to be an onlooker of that. Taking the woman out of the house is into the work place might seem a bright idea but the price we as humans will pay in the end is horrific. The best is that it is so few that can see what is coming our way for this behaviour we have in breaking with traditions that took us thousands of centuries to establish.

It is a result of the Hoggenheimers and Mammonites wanting slaves but not to be bought as property because then that will rush their capital. It will give the slaves property value while the Hoggenheimers and Mammonites can have all the slaves they want without putting a price on the commodity. By not paying they set living standards every sucker wish to follow and then agree to give a little of the blessing for a lot of labour. If a person work such a person may afford a drink of water and if a person work hard he may afford food. By enslaving his sole he may have a roof over his head. By telling the woman how miserable their lives are going to be should they remain at the house and raise the kids as a housewife they the Hoggenheimers get at least fifty percent more slaves to use and pay them with favours that belong to mankind in any case. They give them human rites and that rite is not the natural rite to eat or drink but is to serve with being able to vote. Democracy is a human rite. Voting is a human rite but eating and drinking is an affordable commodity that one are privileged to have because one can afford the luxury.

How senseless did propaganda become! By getting the woman into the work force they cut the salaries of the men by at least half and as a bonus they get a large group of sex slaves. The Boss at work is the Master in charge, which sets down the rules and the woman refusing him is out on the street. I know about the lawsuits and the laws against sexual harassment but those Hoggenheimers is in charge of the politicians voting for the law as much as they are in charge of the civil servants writing the laws. The laws…all laws are one huge smoke screen to favour the powerful and dominate the powerless. My human rite I have is democracy and the rite to vote. Good lord people are fooled that easily. That means I can go where another sixty million other suckers go and turn the table on politicians because I have a say but that say is one say amongst sixty million others. The Hoggenheimers lifts his pen and the politician ignore every promise he ever made to the democratic voters because the Hoggenheimers and the Mammonites will withdraw his campaign funding and leave the politician out in the cold should he the politician not play to the rules the Mammonites and Hoggenheimers lay down. The politician will have all his morals intact but he will be jobless. He either gets bought or killed. The politicians underwrite the laws that favour those in power and the laws are meant to fool the masses and protect those in charge from the masses. That is democracy at its best but it can get much, much worse with legal slavery going onto all which is much worse than the standard Chinese form of slavery as is the case in America but I do not wish to enter that debate.

We have placed all the blame squarely where it belongs because the poor women love to see them in the role of the blameless victim. They love the fact that the powerful exploit them and placing a relevancy of unblemished innocence in their court. Facts, however, tell a much different story. Leaving for work they are painted in an array of colours to match the ancient warriors going to war because it is a sexual war they are going to. Their faces they paint in the best arrangement they can afford. The close they where takes a wide chunk out of the peanuts they are paid and that should in a sense go to better their children but they spend it to better their chances in winning their sexual war. What is this painted glamour for if it is not to fight the others and beat them for the chance to be the one that is going to be sexually exploited that day? When at home with their husbands they look dreary and pail and as grey as mice. There is a blatant lack of trying to find sexual interest from the man she is married to but those at work is her challenge to combat. She has no and shows no interest of trying to find the enticing in her husband's sexual gratification or try to stimulate his interests.

Woman at home with their husbands alone truly try their utmost to look as unattractive as they can be but going to work where they play the sex games and where they put on war colours to fight all the others for the favour of the boss. I have heard some say so what...who gets hurt. The children get hurt. The children pay the price for every divorce. The children pay the price in every divorce and the children become the price of every divorce. The children pay the price for mother and father being at work or travelling too or from work for the best part of sixteen hours every day. The next generation goes wasted with skew morals and no standards. Where are the parents or guardians of the fifty five thousand children in America alone that goes missing every year? What did occupy the parent when that child goes missing? Where was the parent the minute that child vanished? They were hunting and scavenging for money that those powerful Hoggenheimers and Mammonites so gracefully allow them to have. In exchange for all the democracy they the mindless masses have the Hoggenheimers and the Mammonites allow them to live and eat as long as they show their appreciation, admiration and gratefulness for such generosity as which those in power bestow on them. By giving the woman the vote meant that those in Power cancelled fifty percent of all the votes the men had. By allowing women to vote they did not extend democracy they cut democracy in half. Now twice the number more can eat from the same cake allowing every one to enjoy half of the nothing shared previously of the cake that was already shared by so many before the vote was extended. The voting influence did not get more the voting influence reduced because every citizen then had that much less of a say.

Believe me, I do not think woman should be kept barefoot and pregnant back at the ranch, but woman evolved different from men and if women received their voter's role opposing the voter's role of men and not sharing the voters role with men the system then was less about bullshitting the mindless and more about sharing and distributing power evenly. Women find clarity on other issues than men do and women observe differently from men. But those in power saw to it that men and women would share conflict and not share power. Once again the stupidity of the masses lead to the dismantling of society by granting the masses a substitute they do not understand and cannot enjoy since there is nothing to enjoy...and again the children pay the price by offering their future to go to waste. The Hoggenheimers fight and grant privileges in order to set the table so that the next generation would become better and more obedient slaves. To allow this to happen they then must create less educated slaves that are better suckers to please with a lesser ability to think what should please them. In this process women lost the respect and admiration of the men that they previously had some fifty years ago. Back then men offered women their seats without thinking twice about it. I remember the time when it was scandalous for a young man to sit when there were ladies that was not seated in any public place or public transport. A man would never enter a door in front of a lady and a man will never sit in the presence of a lady standing. Now there is a gender war and the children goes wasted. Woman should care for the upbringing of the next generation while men should care for the woman who cares for the next generation and from that a balance come through which mankind developed on the road that brought him to where he now is. The children would once again find morals and reasons to have morals. The genes they built over so many centuries will then not go wasted by abandonment. It is not the politicians that will lead us back because the politicians have gone rotten. It is not the Hoggenheimers or the Mammonites because they are selling mankind out in their quest for power. The only persons we can trust with such a task are the scientists. He must bring morality back in place because he can judge (should he choose to do sow) with history and science bringing on our future. We must see why man evolved so much further than the ape did and Newtonian scientists' must stop trying to bullshit the masses by telling them there are only one and a half percent difference between the chimp and man. If we wish to entrust Newtonian scientists' with our future they must present the truth they are committed to. I have in two or three pages showed the difference there is between man and woman let alone between man and ape and believe me I did not even come close to the way a novice such as I can scratch the surface in showing different behaviour patterns there are between man and woman.

Of course as usual and as with most of Newtonian scientists' findings, I question the accuracy of the gene pool percentages strongly and I am of an opinion that such percentages are in use for political issues more than scientific proof. I prefer leaving it at that. I am a very small fish in a very large pond sharing the pond with very powerful other fish that can destroy a small fish like me with one gulp. Going into the development journey as man followed the trail one has to look at not what man achieved by own ability but with own measure in manipulating others in life. When man started with chips and flint it was progress but it was also very limited. Only when man acquired the muscles of more powerful animals and took their ability in measure with mans manipulative power did success arrive at a level that brought progress in leaps and bounds. It is not mans hands or legs that brought man's domination and control but man's brain that brought response from other life to benefit man and find benefit in shared life styles where mutuality brought about safety and mutual prosperity. It is surviving more than anything else that means good or bad to animals and most of all surviving of the species and in that the animal only find man and man's company good because man holds its safety. When a lion brings down it's pray the rest of the flock will start grazing immediately without showing even slight remorse to the victim and for the loss the close relatives are faced with. By the death of one the rest find safety and to animals that is good. To the rest it is about surviving and if one pays the price, little concern goes to who paid the price. That is what annoyed me about our all-famous-pop-star-wife. She truly go beyond what nature puts down as rules applying her liking (and selling her book with cheap bluff) while other sheep walking on two legs like only humans ought to cheer her stupidity as if the stupidly were they're own. They have not even got that much brains to acquire that much stupidity but has to borrow to get their tally that High. The biggest annoyance is that with such stupidity those can vote and choose my future because of democracy. They prove almost all ways not to have the thinking power of a mouse but they have the rite to choose my future and I have no say in the matter but to follow what such morons may wish upon us. With the novel idea (novel as it is only man that uses mutuality single-minded and still provide beneficial good from all angles) of widening the use of abilities provided to different developed species, man gained extensively in progress and comfort. After all it is much lighter work riding a horse than walking all the way.

But gains in comfort goes both ways as the horse find protection against predator attacks while finding good nourishment in winter and the best hay to feed. Such a diet provides the horse with the strength to carry the rider and enjoy own comfort with the fact of much reduced fear and anxiety. By having male and female and promoting mating the good in the life of the horse becomes better. Did man rob horse from freedom? Did man take what was not his to take? Should man get some conscience attack leaving him with sleepless nights about his cruelty in robbing the horse of a natural life of freedom? Well the humanists will tell you with teary eyes and running noses that the horse should have its freedom as all are born to be free. But this emotional outcry comes with the comfort they, the humanist enjoy of secured sleeping a good all year round food supply and breading safety. I have not seen one humanist go running into the mountains never to return to civilisation, to enjoy the freedom they should enjoy as much as they wish that upon the animals in captivity. To the horse, after the initial fear of the subduing and the ultimate realising that the subduing is not life threatening as he can live with that, he finds comfort and even enjoyment in that. I see on a daily bases how horses get jealous when the owner takes one to ride when the rest wants to go first. They come and nestle with a desire to connect and in jealousy push each other away to be the one receiving the owner's attention. It is a bigger issue of the conscious to decide the likes of others in what they like and what you think they may like. No one of sane mind will hurt an animal in your care and when slaughtering that we do in the most humane way as described by law. No one cuts of a chunk of steak while the animal is alive. We humans have civil norms and values and by using our brains we can live and let live with more dignity going around sparing the animals huge cruelty than what the animals would have come to face if they're fate still was in freedom and being hunted down by wolves and hyenas. Such is the difference between those having bleeding hearts and brainless skulls and others that can think. Now we arrive at an interesting question as how do we think and reason. I am sure all humanists will have as much to say about my way of being correct only as they will find many arguments as proof of the fact that by dislodging my logic they can prove me being beyond the norm of classifiably insane. There is ever a clear definition about rite or wrong and all principles we find appealing or appalling is within the brain

According to an article I read the brain holds more connecting lines than does the Universe and I may even accept such a statement on the grounds that life has much more complexity than does the universe. After all life takes the dimensional barrier as far as the Universe does and then beyond where the Universe stops. This does not make the Universe simple because I cannot see how any person may ever come to understand the flow of light as the light uses both the straight line of singularity, the half circle presented as the Roche limit in singularity and the Titius Bode triangle making light representative of every aspect which connects space-time away from singularity with singularity as light where Π meets r to become the value of C. But in the brain this is only one function as electricity holds an equivalent of light forming electricity as the messenger to whatever energy is above life and then in the human capacity above even what forms the barrier to life. The arm is not human life because a human can loose the arm send still be alive. Therefore what ever is in control of the arm is in control of life, which puts what we find as life at a higher dimension than that of life. You may argue that in case of animals such thought also control life because a dog may lose his legs but not his life. But even as complex as that may become there are relevancies between life and the physical because where the physical uses pain as a warning system the mind uses fear as a warning system.

By following such a line of argument one can freely deduct that an insect as our corncrake, which we discussed earlier on, is representative of life as the same life we find in the arms or legs of our body and the life we control but is not truly part of the energy "me". Clear to all it must be that life we find in mammals are advanced above and beyond the development the insect arrive at. If that is the case then I may claim that human life has more developed than what other mammals did because with my manipulation of space-time I may manipulate other mammals to harvest some of their manipulative abilities in benefit of our mutual relation inclining more to my benefit. The cow does not seem to mind when I milk her but can any person imagine experiencing a milking session involving a crocodile? Well, fortunately crocodiles have no milk but if they had I would never volunteer for the honours of being the first to train a crocodile how to behave in a steady manner when in a dairy session. You may have or may not have noticed but I am telling you that they have a sharp side and they have a blunt side and the sharp end holds rows of teeth they surly know how to use. Even the blunt side hits like any whip never can and I am sure a fully groan specimen may kill with that tail. Going down the order of evolution we come to bacteria and viruses, some of the lowest forms of life. Do bacteria and viruses count as animal and if not then surely they count for life because life they are. In that we find the equivalent of bacteria in higher developed species as we find that the insect may have the developed life mammals use in their bi-products included sustaining their superior life development. We can see evolution by applying a relevancy of devolution to siphon and separate life from life. I am trying to indicate that life becomes a compliment where the lesser developed formed a mutuality and aided the supreme form of developed who is controlling the master brain in that form of life wherever the master brain may be attached. Life is above and beyond the cosmos and surely even the most ardent supporter of atheism must grant me that much. By that grant must the atheist then add the fact that life cannot only fall onto a category of to be or not to be but there is a range in life forming a line of development and superiority. Life is more than life but has status of being lesser or more and that is the point I wish to address after all the talk. Within one body a range of life values combine in making whatever accomplishments the life form accumulated by extensive development that range in development. It is appreciable in concluding from a range of facts I mentioned but mostly from human common sense we all know that on top of the range being the model best manufactured and with all accessories all other models also having life envies and fear is man. What will make man that special?

In 1905 a case was reported for the first time of a woman that had a hand, which attacked her every night by trying to strangle her. In the manner the hand acted it was clearly out of her control as it was clearly out of control of whatever controls the brain have over body functions. She would wake at night and feel someone strangling her but the person strangling her was something she ought to have under her control. Imagine waking one night feeling someone squeeze all life out of you with a murderess motive. Even the thought of that will make most people get up and bolt their doors and windows just in case. It must be awful having a murderer wake you with such a horrific intention. Go one step further and think that person may be one of your house members you trust with your life. A thought of that becomes rather preposterous! Then for the ultimate in revulsion; think of the chances something acting in such a horrendous manner is something you know with every grain of your body as that thing acting is your body. There is nothing worse to be scared of than being scared of "you". How do you fight such an act? You cannot hide and you cannot go without sleep and as you go to sleep you know that there is some part of you yourself hat is after your life. If this is not enough to drive any person into hysteria I do not want to know of anything worse.

This flabbergasted the doters and no one seemed to make any sense of such phenomenon. I suppose if such an incident had occurred before it would have been denounced as an act of a demon of some kind but fortunately for medicine the art of healing had abandoned forces of nature as a scientific accepted fact unlike the likes of physics still clinging on to such madness. If my memory serves me correctly this case was in Germany. Please remember unlike our distinct academics I have no extended library to find all kinds of information but have to rely on a failing memory being destroyed by my diabetes. Then in France later on another case became known about a woman that had a hand also out of control where in this case that hand tried to forcibly turn the steering of the motor car she was driving to force an accident of a serious nature. This manipulation was seemingly as much out of control as was the previous recorded case. The common factor about the two cases was that in each case a person had an arm that was intent on destroying that person without the person aiming to do so in free will.

Later on in America two neurosurgeons planned an operation procedure where by they aimed to relieve patients having chronic and continuous convulsion attacks caused by epilepsies in the brain. These cases were dire and with the operation as a last resort all the serious after effects became a secondary factor to the superior motivation of saving life and improving the demented quality of life by the patients. The operations involved only the utter most serious cases that left no other option for improvement. It was this or death and not choosing death the patients chose the intended operation procedure instead, but still it was extremely serious and dire options in the choosing. They reasoned that the epilepsy was a result of the brain having vibration and with the vibration stimulating other vibrations through the brain in some cases the one caused the next vibration and it was more a reflex of the first causing the next as the symptoms was going on a prolonged non stop convulsion. To stop such reflex by the brain tissue they held the argument that when cutting the cortex the two lobes attached will not have the reflex and thus the continues convulsions will loose the continuous effect.

By separating the lobes the nerve attack coming about in the one side will not transfer to the other side thus it would not cure all elliptic attacks but the prolonging effect will be reduced. One vibrating lobe wills then being separated from the other part not cause the response in the next lobe and it was diagnosed that it was more a response to the reflex allowing a reflex to the respond and this brought about a never-ending cycle. The idea was that it would result in reducing the severity of such grand mal epilepsy

According to American law the doctors first had to show a high degree of success by operating on rats in order to prove that the consequences of such an operation is in acceptable levels before starting such a procedure on humans. To obtain the rite by law for the granting of the operation many rats underwent the procedure and the procedure then were extended to many other species. Every aspect of recovery and side affects must be documented to an exact accuracy with no exception to the rule in the slightest. The behaviour of the animals before and after and the general physical data then goes to excessive detailed scrutiny by the finest the medical profession has to offer in America. Accuracy in the process of accumulating data and other relevant information is beyond question especially in the country with the highest standard in medical care. The after affects the procedure had on animals were indicating no serious side affects of any reason for concern. Many species went through the procedure eliminating defects if whatever possibilities there may be.

When monkeys went through the operation procedure our primate cousins had no side affects in any way. There was perfect hand eye co-ordination and the nerve system had no complications with the motorized operating functions in any way. This confirmed the surmise the medical profession then at the time had that this third lob was just fibre with no function of distinction. The fibre was in position to stabilise the two lobes and had no connection with the lobe in a functional manner at all. All the indicators brought about such positive results that the American government granted the licence for the first experimental operation conducted in a human in absolute confidence the procedure went about and with a very good outcome. But shock was looming to all medical experts. In every case the patients had one common disability. It showed a horrible disadvantage no one expected in the least.

All patients showed the science of a phenomenon later named after a movie Peter Sellers made famous. It was named the doctor Lovejoy syndrome because on of his characters in the movie was an eccentric half mad all crazy German general that had one arm always trying to strangle him. This was meant to be funny in the film but the patients suffering from the reaction of the aftermath are not so inclined to the humour. They all had one arm that went out of control and the arm showed serious signs of having a mind of its own by doing the most annoying things the patients had obviously no control over. The one arm had life apart from the person free will doing things that would embarrass or even threaten the so-to-be owner of the limb. It might have been portrayed as a comedy, but underneath this is the foundation of man. This is what makes man distinctly different from all other creators carrying life.

Well I am no brain- surgeon although I am inclined mostly to form an opinion of my own that may not always stroke with informed opinions by professionals. The test operations were conducted on a variety of animals including monkeys, the so-they-say close relative of man. Well primates might be as close as they can get but I am of the opinion there are other species being still closer to man, but that is somewhat off the point. From all the facts I mentioned the past pages I drew a conclusion of my own because if any person is experienced on the matter, then in that case I am more schooled than most ever were. I did this on the grounds that although I might not be highly schooled by books on the matter, I have more experience than most experts. I have been fighting what is thought of as depression for over forty years. I have been forced to study the subject because I have been forced to fight the subject. I have been in a constant battle daily / nightly / hourly and no doctor will teach me what there is to know on this matter about this matter. I am even dreaming about the subject at night and I know what it is to fight suicidal tendencies, because I am in struggle even in my sleep. No animal has morbid suicidal tendencies. Animals will beach inexplicably on the seashore and some deer would go into a semi conscious state when there is drought, but no other animal has the fight in itself to kill itself as a normal behaviour principle.

I showed that man has a higher evolved form of life than other animals.

The bad hand syndrome is the rage and the retrogression infested in society brought about by every person.

Taking what I see from people's behaviour including my personal tendencies I concluded that what we think is normal living patterns is in fact very unmoral and what is accepted as being thorny behaviour issues is quite everyday commonplace behaviour. We all talk about a death wish and associate that behaviour with risk takers such as stunt men, hells drivers and other total professions committed to indulge in reckless actions with death defying consequences. However those I just mentioned are the mild-mannered amongst us because they only give some minor outlet to their normality whereas the rest hide it under a thin coat of paint called acceptable social and economical vices. Every person excluding none is a self destroying suicidal freak bent on total self-demise and is not untypical or outré, even amongst the most gentle amongst us.

I will present one case as a scenario keeping in mind that the person I am referring too is the most peace loving, docile, scared-of-her-own-shadow type of person I have ever encountered. This story I am about to tell is as true as my personal past and I see both the persons I refer too on a daily basis and therefore I must hide identities or live to be sued. In the case of Daisy, I will admit there has to be more advanced do-gooders, the save-all-persons than daisy is or more perpetual rescue-all than she is, but I can't think of any one outclassing her on that front from the top of my head. Yes, let's call her Daisy. Remember that Daisy is the biggest do-gooder South of the Sahara and Daisy has a brother that is a banker but moreover he is much more a loan shark than he whatever else he could be doing, going by any other description occupationally. He runs many loan firms lending money to the downtrodden, those amongst us that has no credit access or those that couldn't find any ability to be helped by the normal institutions available on the open market. Let's call him Shark. But Shark invented a means to sidestep legal procedure and charge extra absorbent revenue on money lending to the under classed in South Africa because Shark detected a loophole in the law. Before the law could catch up and be changed Shark became a multi-millionaire and I know of some cases where local banks borrowed cash from him when they ran short of supply. In the event of that happening even the banks paid through their necks for using his stack of cash for a few days. If ever I knew a

Hoggenheimer living as a Mammonite on a lesser scale than those charlatans in Switzerland that call their profession bankers, then he is a fine example. Shark would have carried on eternally breaking the monetary backs of the financial-spineless but the large institutions that are even bigger crooks running banks than Shark could ever be started to copy his system and eventually the government had to jump in and stopped the entire scam when it became too wide spread and commonplace. This I am telling to paint the picture of how big Shark's love for money is. His love for money was eclipsed by none other love he had.

Daisy and Shark is half brother and sister, sharing a father but is divided in ranking by a mother. Daisy is about ten years the senior of Shark and Daisy lost her mother in the early years and subsequently was raised by Shark's mother as the one filling the spot holding the lesser part of life while Shark was brought up as the God's gift to humans where he could stand strongly alongside Jesus Christ in the overall importance viewed in respect considering all men that ever lived. What Shark got as a child, Daisy was never allowed to even dream of. Daisy grew up not only to respect Shark with every atom of her body, but to fear him with the fear equal to what she should have of death and even now at present as a person in her fifties she was and still is never not reminded about this state of affairs every living second of everyday.

Daisy and her husband were normal farmers and as all farmers in South Africa, which included me, were living on the verge of poverty, having a daily battle to just have bread to feed our children. Then the final crunch came as Daisy and her husband had to sell out because poverty forced them just to come and scrape some living in town. They were destitute, poor, lost everything and every dream they ever had. With Daisy having children still at a school going age Shark came to their rescue. He had many houses standing empty because the boom of the nineties that came and went left the town empty and with it many workers that were suppose to live in the empty houses while not living in them any more, many houses got vacant. As the people that lived and occupied the houses left because they had no more employment in town, Shark bought the properties for almost nothing and that he did because doing that could hide taxes by supposedly differing cash on renovating houses. With the houses empty and his sister in destitute he had it in his heart to be merciful and so he had Daisy, his poor sister, with her family, living in one empty house that was very close to his mansion of many million Rand. Daisy and company had no money, no income of any sort and therefore he gave them a house at three times the going rate as what applied when the town was still bustling with workers and she grabbed the opportunity because no other person would allow them to occupy any house in town, also because they had no money, no income of any sort. The payment for the house was exuberant but Daisy was desperate so Shark had Daisy sign over the amount she was suppose to pay for the rent every month and added 25% per annum in the event of falling behind in payment as this was coming from her part of the inheritance should their parents one day die.

However this merciful gesture did not put food on the table so he had his brother in law become his personal gardener and paid him accordingly…and in South Africa a gardener's wages don't buy you your daily bread even for one person, let alone a family. Still this was part of the housing agreement whereby they could occupy one of his many houses. …And still Daisy and her family had needs and had to survive so shark had Daisy work in his money parlour, but he paid her less than he paid the tea girl. These conditions Daisy had to accept or she had to starve with her children.

There was Daisy working with literally thousands of Rands of Sharks money, lending it to all that was as desperate as her position left her and all this happened while she herself had such a generous nature with an all consuming bleeding heart. Daisy at first was checked many times a day about the conduct of her work but as the months went on she was trusted to the full…and why not, Daisy had the fear of God in her for Shark because without Shark she and her family had no money, no income of any sort and no livelihood. Daisy depended on Shark in every way that fills the nightmare in which she finds herself.

With all this gossip notwithstanding, there is a motto behind all this and I am coming to it.

Then Daisy started lending money to some destitute friends of her that was not cleared by the department that does the screening and clearing and the reconciling of pay back procedures. The more favours Daisy gave to friends, the more friends did Daisy suddenly have. Daisy never took one cent for herself or for personal need but the more desperate her situation became, the more she could understand the desperation of the needy that became her friends. The snowball effect resulting from her bleeding heart was self-propelled and as perpetual as it was ongoing. The outstanding balance she supposedly carried as on hand cash ran into the hundreds of thousands and when she approached other cashiers they lent her money from their drawers because Daisy was the sister of Shark and Shark was bent on helping his sister because Shark was forever telling anyone that had time to listen how soft-hearted and kind he was for helping his sister so generously. His generosity shown towards his desperate sister almost had no bounds, or that was what he had everyone believing and therefore his sister surely had permission on distributing funds without his direct consensus. Therefore, with all his generosity he is showing towards his sister, why would they question when Daisy dipped into their drawers…she just had to have the permission of Shark or she would never approach them, with all the brotherly love floating all over and after all she could put in a good word on their behalf next time she was socialising with Shark. Little did they know that Shark would rather socialise with his well-bred dogs than he would socially talk to his sister and her husband.

One day all the floating cash ran out. Not one of the clerks had any more money to give to paying customers or then even to any customer because Daisy in all her understanding had hundreds-of-thousands-in-cash worth of friends all promising to get paying back as soon as their financial dilemma turned favourably again.

Hell broke loose and I mean hell broke loose. I never saw Shark that angry because I had a ringside seat when he caught Daisy in front of my house when she was visiting my wife. That is how I know that much detail because Shark was never soft spoken but that day he was even louder than I get when my temper gets out of control. I couldn't miss a word even if I was as deaf as my grandpa was when my grandpa was still alive. Things he said to her ranged from when they were children to when she would eventually die and the conditions he hoped under which she would eventually die and he hoped for her to experience the utter misery she will experience every instant in-between and not once did he have a kind memory about any memory he had about her and their shared childhood. I couldn't miss anything because it was happening out loud only meters away form where I lived.

Shark kicked her out the house that very evening, which is most illegal in South Africa, took her car illegally. He took position of their furniture, again illegally and had the kids and dogs and all, on the street by nightfall.

Then for the first time Daisy's father came to her rescue and took her and her family in and gave them logging. The only person that Shark apparently respects is his father and when the desperados were in their fathers house Shark let go of his per suit. The father drove to Sharks second hand car dealership the next day, bought his daughter a car from the lot at cost and they still drive the very same car after many years. The father then got them both jobs through his influence and saw Daisy and her husband get to a living standard once more. Shark was not satisfied but fortunately had that much respect for his father that he came to a conclusion that satisfied him as well as his father.

Eventually Daisy singed over millions and millions of Rands of money left to her in her father's will. The amount she singed over became so much money that her Father did not even have a portion of that amount in his entire will! Shark could not run to the Police because an investigation of any sorts was the last thing he could endure. He could not claim theft from insurance because insurance insist on a Police file and Police will adore the opportunity that will give them the chance to uncover and see what Shark's books would reveal.

So, what's behind the cheap gossip? In the partnership Shark had the destroying hand but the destruction was aimed at his sister from childhood and not at him in person. He started destroying her instead of himself from a very tender age. Daisy had the protecting hand and she protected herself by protecting her friends from the life she had to endure under the destructive hand of Shark. They both destroy and they self protect at the same time. Some would say Shark got what he deserved and other would say Daisy should be in prison. Some would say Daisy should be granted a stature next to Robin Hood and others might say she is a sole mate of Al Capone. Others would reckon Shark is a do-gooder with a most kind personality depicting a tender under belly and that he just took what he was due while he protected his sister and she unlawfully stole from him in a reckless manner showing her un appreciating nature. It is never right or wrong, good or bad but it is wrapped in relevancies applying.

Taking into account my contemplation of human behaviour about drug abuse, child molestation, alcoholism, reckless endangerment and the list will never be completed, I realised the behaviour of normality was self-destruction and a serious battle by forcefully fighting the self destruction. In every person there is inherent urges to become self-destroying of the person that should be protected for deer life. We all carry this urge in our brains and we all are in a battle to either contain or to allow the urge to control our lives everyday. This is the ability to fight self-destructive behaviour that gives each one the personality we are caged in.

I am not referring to the symbolic little horned devil sitting on the one shoulder and the smiling angel sitting on the other shoulder where each is suggesting either a bad thought or a good thought which will have us capitulate to temptations or guide us away from temptations. What I am referring to is real destructive issues there to kill or to destroy physically. I am referring to a person within every person that holds the urge to kill or maim or destroy the person and another person also within the person trying to defend and fend off the destructive person. I call this retro-aggression and psychiatrists call this having depressions. I am talking about a real killer.

In every person there is the hand that has the urge to kill the person and there is the hand that fights off the hand that wants to kill the person and all the while the person sits in the middle playing a game of control between the two hands.

Some persons have the hand holding the non-aggression alliance well and truly in control of the self-destructing hand and in other persons the hand of self-destruction is totally in command. Sometimes and mostly the hand will destroy others well knowing that in the social arena that is the best way to self-destruct. They put you in a barred cage and lock you away from everyone else but yourself and when that happens the self-destruction can aim fully on the person that the self-destruction tendency really aims at. But in most case the more docile versions are more commonly practised. Such persons will become wife-beaters, alcoholics, drug abusers and the more common social misfits while others more progress to forms of serious criminal activities or have uncontrolled criminal tendencies. They could even become the serial killers where they have no ability to stop the overbearing self-destruction, even when realising the police are on to them and they will be caught with the possibility to die, they still keep on killing until they are caught.

They are not fighting off a demon but they have the demon we are all carrying. The demon is as much you and me as we are the demon within. We all have the one hand trying to destroy and the other hand fighting the hand that wants to kill. The three personalities are one and the persons are all in one mind. The three personalities link by all sharing equal responsibility in front of man and in front of God. The seriousness in the situation is washed and bleached by the press and Hollywood as the yin and yang eastern culture. It is so much washed away it is downgraded to a state of becoming a minor threatening concept or leave people with the idea of a person choosing between bad or naughty thoughts and good or sweet thoughts. That is putting the problem equal to a fairy tale. This raging war we battle everyday, which I sketch, is all out self-demolishing, self-destruction of either the person or committing explosive and uncontrolled aggression towards another person. We all have it and no one is exempted. It is only the shift in control that applies either more or less control with some having more aggression than others having more control over the aggression within. In most cases this is what drives us to become exceptional figures and great Kings and Important Generals among which are Alexander the Great, Julius Caesar, Napoleon Bonaparte and all other history fondle and is lovingly revered in memory of human history. They destroyed nations and cultures and yet they form the beckon of human memory. Their murdering routs they caused being the angel of death to hundreds of thousands became the light points we regard as culture as development. That is why we so tenderly make an effort to remember the monsters that killed and murdered nations and civilisations, sparing no grief to woman and children, toddlers and the aged. The greater the King was, the more blood the King had flowing in rivers representing human despair and yet, those are the heroes we enjoy to remember. Kings are not remembered for not having war and not killing, but because of war.

This war rages in all of us with different levels of control leaving humans with having different personalities. It is an inborn nature we all have but some psychologists give the personalities pretty little names, naming each as if they become personalities that are detached from us. Then the psychologists and his inferior feeble minded victim acting as a patient can have much fun while feeling intellectually superior in unravelling the concept by complicating and confusing the issue totally and have the poor sucker pay for the treatment to un-name what was nameless to begin with. The psychologists do not vaguely know how to respond and alleviate the suffering so they further the establishing of the personalities and by doing that they exacerbate the problem to such an extent it then become beyond managing which makes the problem Titanic and above what any mortal man could cure. If they don't know how to kill it, they cultivate it to become something that kills the holder of the condition in any case and then the psychologists can acclimate the seriousness of the multiple personality syndrome...in South Africa this process is called job creation. The politicians make a mess of everything and then ten times the number of people has to be employed to avoid a catastrophe that comes in any case. This is then one form of self-mutilation brought on by confusing even more that which was what no one ever understood before. The bad hand is the cause of any form of self-mutilation, be it bad habits or serious crime.

In prison we find many convicts practising the art of self-mutilation and those that to my thinking who should know better is confused by diagnosing it as a form of depression. By considering it as a depression the doctors prescribe anti depression drugs that should suppress the anxiety that they regard is being associated with depression. They think when a person cuts him or her with a blade the person does it as a result of depression, which is altogether crap to start with. By suppressing the emotions the doctors help to further the actions of the self-destructing personality because the personality they suppress is the controlling part or personality that contains the self-destructing personality, and with medication they exacerbate the situation as they suppress the controlling factor. There is no medication to suppress the bad hand self-destructing personality because if there was, one could give a murderer a few Bob-Martin pills and he or she then afterwards was as rite as rain. With suppressing medication one does contain the containing personality - side by depressing and then suppressing the feeling of vulnerability, which is the guard against the self-destructing personality, but when suppressing the feeling of vulnerability, that increases the violence of the self-destructive side. In childhood parents promote the development that would favour either one or the other side by exiting the fear in the child and allowing uncertainty to reduce the behaviour of the violent character. Giving the child a good old fashion spanking for instance helps the child suppress the self –destruction the child allowed to surface when acting socially unacceptable and this destructive feeling then will be contained by the child by giving the child motive in supporting the self protective feeling. One look at the way youth develop currently should tell the liberals filling the offices of professional psychologists that they are heading for disaster very fast and are railroading society on the wrong track that is heading for social collapse within fifty years. Criminals are not born, but are bread and while we all have the self-destructing murderous tendency we all have the ability to control that, which will give us self-protection. If a child is not taught to control the tendencies, then in later life the tendencies will control the child. This is called irresponsibility leading towards criminal behaviour. We have to learn control by applying self-control. This is how our Creator created us, so that we can create ourselves. We all have to learn to behave socially acceptable or become social misfits.

In the Bible it reads that man was made the last but far from the least with more superior qualities than all life combined because man is all life combined and then added more than the fare share. Any one thinking of our Creator, as a magician is mad and that I state without excluding even the Pope or preacher of whatever denomination. The Creator is a building architect applying mathematics and physics we can never come to appreciate. If there were any one that has an opinion that God spoke a word and magic was the word that person would have another opinion when thinking with some clarity about Creation as a whole. The Creator is Creating and by creating there is a building process involved. Every person starts an individual process of building a human just after birth. Looking at tribes living in regions far away from civilisation as one can still find in the Amazon River we find those individuals being adults by body still play games we find our children play with much amusement and childish enjoyment.

It is far from incomprehensible to make some sort of comparison between our children at play and those adults at play and the similarities are astonishing. Racist remark it may be but the grown ups are more child than the western child is child. They defiantly are backward in mind and mentality. The most logic deduction is that the child in us represents our development phases through a long journey. Humans at birth are animals. Babies can make noises to convey their needs and nothing more. As the little life grows it develop not only by growing but also by culture and what the parents put into the culture. It is more than likely that the developing pattern children follow is the same developing line humans evolved through as the generations brought more insight and better understanding to the following generation. It is only of late that there is some pattern of devolution taking place especially in the western world but that trend is set by a culture of greed. The parents are chasing a good life and instead of placing morals they push money in the hands of their children to get the children out of the way so that the parent will lead the life they choose and rid them of the burden of children while conveniently blame teachers for the children having unacceptable behaviour as much as they pay psychologists good money to correct they're mistakes. It is moreover a fact that parents do not actually pay the professionals to fix but pay the professionals to rid parents of responsibility, guilt and of course the children. The price society will pay for such luxury is far more expensive than affordable.

I so many times wished I could have my life over again. There is one study I shall conduct and that is follow the pattern the child indicates how the human developing process came to pass and draw the parallels from that to man's evolutionary path. It truly must be a study worth one life. To know man that well must be the ultimate there is to know and afterwards death can bring no regrets for wiser no one can ever be.

I think we should now return to what the Bible refer to as the tree of Knowledge and reflect once again on that verse. The quote was and I quote in firstly Afrikaans: Boom van kennis, kennis van goed en kennis van kwaad". Directly translated it reads as follows "tree of knowledge, knowledge about good and knowledge about evil" and that is where my argument starts in my attempt to indicate the possibility of the sixth dimension

From the stage of toddler I found eve dice of three characters in all persons. There could be more but not less. Let us call the three persons three entities of the good the bad and the ugly, but not in such a specific as being a saint a pleasant and a demon. The entities are rather more under cover that that straight foreword by definition. In every person's brain somewhere there are the trinity of entities ruling our lives.

The trinity are one person and not individual persons but the same although very integrated they are also very separated. It is not a question of schizophrenia with multiple personalities because I do not believe in that. I think that was made up to convince who ever needed convincing about something to do with nothing and has no base or then has the same base as physics hold gravity responsible for a variety of facts they otherwise have to admit they no nothing about. The three belongs to one person and in fact is the same person. In the Afrikaans book I named them Ek, my and myself translated as I, me and myself. To make matters more interesting and subdue confusion somewhat I wish to keep the Afrikaans as that would make explaining slightly less complicated. Ek is I. My is me and myself is will you believe it myself. The "My" one pronounce just as you pronounce the month of May in English and that would make the pronunciation of myself as you would say May-self in English with the pronouncing of self in English and Afrikaans exactly alike. Ek my and myself are the same and there are no distinguishing between the characters but at the same time they are as far apart as three that never met before. Every of the entities belonging to the same body, the same brain one has a character as unique and as far apart as another being on different sides of the universe.

Every personality has different likes and different dislikes and feels a different purpose in life as much as to life Once again I wish to press home the fact that this is (to my view) as normal as breathing and has nothing to do with the mental instability known as schizophrenia. Although only one personality claims occupation of the body at any given time, all three take responsibility for action all the time because all three are the compliment of one.

Any one such personality may claim occupation of the body at any single moment and normally do not relent occupation easily but of the three one is always in charge and the other two take position when the main personality loses concentration or relaxes guard of the situation. Every personally has own motives being apart as far as the north may be from the south. One may even think them in classes of being the person's personal god and personal devil but such a thought may place boundaries that are unfair. I would rather describe them as one is the charger being scared not even of the devil himself and the other will be the cautious the guard, the one always on the lookout for trouble coming. They form the one that is in charge and the one in charge cannot be excused because it was a totally foreign entity pushing in charge in the direction it never wished to go but was blackmailed in doing that wrong! The one in charge takes responsibility to the full for each dead the body did and every wrong committed. In the end they are only identifiable but not that clearly deniable and they always appose each other in complimenting one another. In the American cases where the neurosurgeons performed the operation in cutting the cortex to split the brain lobes of the operated patients a condition became a situation where the persons found the uncontrolled motorised motions of limbs under their control supposedly, but not under their control at all. Then for the first time the phenomenon became a syndrome with a name. It became "the alien hand syndrome" but also find referring by the use of the Doter Lovejoy syndrome named after the movie. There are cases known on record to result from a variety of brain damages and severe apoplexy. I can confirm as a witness that in the cases of my biker friends their faces changed with mood swing. Not bone structure or complex feature but the facial expression changed the way the muscle form the face. In cases they looked somewhat alien, but my referring to as normal is not about such extremes.

The alien hand syndrome observes cases where the one and the patient refer to as the naughty hand and the clever hand. The naughty hand seems as if it had a life and mind of its own sometimes acting to embarrass but some times even to endanger. The main connecting issue between these cases is the hands control not being within the owners authority and that the hands will obey as nothing ever change but then on occasion from the blue it will act on own impulse and with a motive clearly never matching the owners intension. I wish to underline with no exclusion to whatever intension that the alien hand syndrome does not prove the trinity that I refer to and the alien hand syndrome underlines two facets where as my view state clearly three identities. The reason for this mismatch holds a most intense connection to my personal religion, which I never share, with any person out side members of my immediate family. The alien hand syndrome only confirms (to me personally) a connection of some sorts in some way.

Many years ago after I had to overcome some personal problems at the time through which I was admitted on occasions to an institution. I spent time in nerve clinics where I had to recover from some brain disorder called endogenic depression. It is a condition that the patient will get severe attacks of depression and the cause of this is totally inherited by nature. The disease comes from a bad gene carried from parent to child and as much as my father suffered from it so will my children and is not very scares or very serious. I would say it is as serious as you make it to be. My regrets about suffering from this condition is minimal because of the extent of learning I had the opportunity to come by that otherwise would never have come my way. But then from what I saw in the times I was admitted to the clinics I mentioned, my (unprofessional) conclusion is that in just about all cases the prognoses depend almost entirely on the patients willingness to recover and how serious the patients are about recovering as the recovering is al about the relevancy struck between obsessions and true problems. In that there rests a balance between fighting for the sympathy they wish to evoke in others and fighting a battle to achieve oppugn from the problem. In hindsight and after all the bad is forgotten it was worth wile because the learning of the human mind and the way others think and feel was enriching beyond my suffering. At the time the suffering was almost overwhelming but to escape ones own problems one can listen to others and by learning from them one get perception about their state of mind and your own state of mind. It was in this period I came to the conclusion about the trinity within us all. By increasing my personal learning curve I decreased my personal discomfort.

At birth those coming into life receives a spot in space holding singularity. Life will have the opportunity to defend and maintain that spot to secure the surviving of the life in that spot for as long as possible. Life is an energy alien to the cosmic Universe and foreign to cosmic principles. In contrast to all mainstream physics members' eager promoting of life being a normal factor of the cosmos in the cosmos, there is was and never will be such proof. Mainstream science has one goal with such promoting and that is to raise tax paying interest and arouse the cheap thrill seeking attention of the weak in mind and weak minded tax payers, whom does through the paying of their taxes also fund the other wise incomprehensible expenditure of space research (that is referring to the mind set of the mindless masses, those brainwashed to believe in democracy). I know it is the normal mentality going around because in the past I have been asked the question that if there is no life there in outer space and other galactica, then why is the cosmos there. For what purpose was it invented! Well to me that question presents the mind pattern of the mindless masses. But the mindless masses also has a function and a spot holding singularity claiming sunshine since they are the ones burdened with the task of paying money into coffers where such money is directed to some sensible causes other than sponsoring the ventures of the murdering military. People are most willing to fund devises that will maim and kill persons they have never seen or met but show resistance to funding plausible causes such as for astronomy research.

Life is not part of the cosmos. Don't bullshit the weak of mind with spreading the idea that life is to be found across the Universe going at a dime a dozen, crawling all over the cosmos at every point there can be and if you are part of the weak of mind believing all the bullshit about life being available throughout the cosmos, then don't let them bullshit you. Life is only a secured part and a factor on the atmosphere of this tiny planet and also found in the oxygen richness of the water of this planet, the third dot circling around and being away from a very ordinary star in a very ordinary galactica. There are so many criteria the cosmos have to meet to support life on a specific planet that the solar system failed the meeting of such conditions on all other solar structures. There is such a slim possibility to find life even in the realms the cosmos present that I am sure we are the only life in the Milky Way. If the next galactica has life somewhere any communication between us and that life will depend on intervals of two million years to give any responding reply on what ever is said in an ongoing and flowing conversation between us and our next door galactica owners. Even if there is life then for all useful purposes there is no life as far as we are concerned because if they sent a massage on waves travelling at the speed of light over the vastness of the space between us some two million years ago the massage will reach us now but the answer we send back will take at least four million years to confirm to us what they replied to our reply. Two million years ago there were no humans or any intellectual means to find a massage some foreign entity sent and the way we go about there will definitely not be any humans or be an Earth holding and supporting human life four million years from now. If we go on wasting out Earth recourses as if there were no future generations coming as we are doing at present there will be no future generation four million days from now. We will then succeeded destroying through money chasing all that can be precious. With our smallness that we find so big we do not even envisage that less than one lifetime ago e were incapable of knowing about the possibility of a manner by which we may receive or send massages to other galactica. The papers were cracking jokes and laughing at Goddard for his ideas about travelling to the moon in a spacecraft. Now all the members of the establishment has totally gone overboard in the other direction and is dishing out madness too everyone wishing to participate in such madness confirming space-travel and aliens visiting us just to have sex with the most unlikely candidates one can imagine. The whole aspect of time through space being in space makes such thinking of communicating with other possible life as insane as speaking to ghosts or finding demons running free to torment Earthly creatures or devil worshiping or any undesirable fantasies such as presented by witch hunting cases. I explain all this in detail in the Seven Days of Creation.

The life we know is only on Earth but on Earth life is so abundant we cannot anticipate a cosmos barren with life. Life in our cosmos and with our cosmos being the Earth we cannot spare a thought that does not include life. Even a volcano spewing fire in our vision connects to life as that action endangers life. Life is the mainstay we come to realise. Life is the ability to manipulate space-time by occupying space-time and displacing occupied as well as unoccupied space-time. Surrounding such a spot occupying singularity is also the engaging of many other spots in singularity filled with life. All the other spots receiving life is part of the main singularity holding command but that is the same way as a star is filled with atoms being separate but joining to secure the unit. The life under the control of the singularity depends on the development of the life in command of the singularity dedicated to life in control.

My guess is as good as the next person about where specifically the life in control is situated in the body but with a mind in reason we can exclude obvious places. There cannot be space between "me" in singularity and where I hold singularity since time cannot apply to distance "me" from myself. Using logic presented the fact of singularity and then produced singularity. Now the same logic may present life and singularity and life in singularity.

If I was my body there could be no time applying to a massage sent to space holding life in my body. If the life I wish to convey a massage too was part of the "me" in singularity no massages was necessary to send because then the I that was receiving the massage sent by the me in awareness of the need for such a massage would then be aware of the massage long before I was sending such a massage to me. Only if the part that was receiving the massage was some life under my command but not part of me was such a sending of such a massage necessary. That part that receives the massage I send would not need any massage because being "me" was then also being part of the immediate information chain. The fact that I am in a command centre and I am sending commands by which I am ordering occupied space-time to comply to certain of my wishes and the sending of that massages went across space and took time to reach the position I whish to command, the area under my command as well as the structure through which I send the command was enslaved by me but was not part of me. With this ability to control and manipulate space-time to order space-time under my control makes the body having the position I am referring too as part of the manipulated space-time and not the manipulating singularity. The body falls in the manipulated sector because that massages is part of the manipulating process. The manipulating process is having wishes to control and manipulate space-time to further the wants and needs of life within the commanding singularity. This sending of massages through space and time is proof that with the massage I am distancing "me" from life I control through decision and managing such deciding.

If I were part of me occupying singularity then I being "me" would have to adhere to space-time rules because of my body being part of space-time occupation. My body would have some form of heat display, which I then would present as say electromagnetism of a form of space-time displacing. I am not the life in my body but as all other space-time I take charge of the life in my body with some affection. When born I receive a motorise body well developed and a mind that has me in four personalities of which I am the centre non-thinking and emotionless no body serving one purpose and that is to secure and control the other of my three personalities. There is the singularity holding part that has to fight for control of the mind and that has to identify, judge, inspect and approve or disapprove the thoughts and the wishes of the other three. In total I am four but since the centre is the dominator I will concentrate on the three I have to control. My function on Earth is not to allow the other three to dominate and control me but I must secure control and dominance over the three.

You may wonder what gives me the authority to claim singularity. If you think back I have already showed that the cosmos acknowledges my point in singularity as the cosmos is sending all light crossing space and using time to meet me in the centre of the Universe. With me finding that all light is rushing towards me I can make the claim of being in the centre and being the centre of my personal Universe where from I have my point in singularity. Where ever I move too or whenever I shift the centre that is registered to me follows me and places me in the centre of my Universe. But it is not my body that is in the centre because I can use time and space to see light fall and the light then shows bright spots on my body and my body prevented the light from acknowledging my centre position. My body does not form singularity and does not share my singularity but is only close space-time supporting my singularity. My eyes are not the centre part because if I wish to close my eyes their centre function falls away but my centre function never falls away. Even my thoughts are not part of my singularity because thoughts are attached to time and time secures space. I am the part that is without thought but under the influence of my thoughts and my thoughts bring about my actions.

All humans are very aware of this, which I am talking about and we are aware of this for as long as we are aware of ourselves. When one tell another person to concentrate and not to let his or her mind wonder in thought you are telling such a person to be himself or herself and not to allow "outside influences" such as thought provided by your other characters which you acknowledge to be present and is providing emotion to detach her or him from him or her. The person commanding the concentrating is sending an order (which is the manipulating of space-time) and such an order will prevent your entities to distract you from yourself. By any outsider sending the order to concentrate is to have you agree that you are aware of the other characters and are making a significant effort to fight them in the short foreseeable future that the enduring will bring benefits. For you to be yourself without yourself denying you to be you is your effort to manage and control the characters you have been aware of all along. It is a way we always had in the past going as far back as speech will take us and recognised we are fighting our self by means of controlling our minds and preventing thought to take us away from being ourselves. By concentrating we confirm the knowledge of a more intimate person within ourselves and of acknowledging that you can be another part of yourself but that part of yourself is not you in the pure essence. I have to show this because then I can prove that what I refer to every one knows about but never thought about.

Athletes and sports people are most aware of the control over the commanding of the body they have to practise. Practising is learning to control the body. It has little to do with the body because when frightened witless any person can command his body to outperform the best efforts of the brightest and most accomplished athletes. Practising merely means that we are repeatedly going over a routine time and time again to find the ability to exclude any of the other characters from participating and blurring our control over the accomplishment. I have the suspicion we learn to control one of the characters as we involve such a character to help us with the positive control over our action in that very specific trial we were working for.

From birth we are in a teaching process and a learning curve where we learnt the teachings in order to establish behaviour acceptability's of the other entities. With the discipline our parents control us we learn to appreciate what is accepted and what is unacceptable behaviour as we learn to control the entities hat control us. But while this is going on and influences are shaping the teachings we undergo, we still fall under the laws applying to the universe. We are in space-time occupying space-time and commanding space-time. That means Kepler's formula has to apply to life while life is in singularity that is in the cosmos. The three positions we find as three characters are always shifting by motion through emotion. The characters are changing their approach to match and fit the approach to that which life in command post singularity has. The attack by them on singularity is constant and even present in our dreams. The attack is meant to destroy when not combating the good and evil they enlist to detour our life route. I strongly do not believe there is criminals or else we all are criminals but I do believe the criminals opt to become strongly influenced by desires of the other characters they are part of. By this I do not say criminal behaviour are acceptable and not punishable. I say there are control or lack of control present and as such not good or evil. With that I admit that I think punishment must be much more severe that it is at present. But there is no clear cut good and bad to every one running on a thin line. Even by preaching the Gospel such preaching becomes shady because the question larking in the shadows is what is the motivation driving the do-gooder. The pretence may cast a shadow on the sincerity and might overshadowing the issue as to what is the intent behind the goodwill teachings are. Many times I have asked persons that was in the process of trying to convert me to some or other form of faith they think is bringing them eternity a trick question just to measure their honesty. I ask them to instead of giving me their faith, which they are dishing out so freely to instead give me money. I ask for the money they have or the car they drive to find what is truly important to them. When finding reluctance in them to share such earthly commodity but an eagerness to share their faith I point this out to them about what they appreciate as not sharing and what is worthless enough to share. I show them that the faith they hand out for free is just what it is; free rubbish they dispose of in an effort to reap the glory of becoming my messiah and the saviour of my soul.

They bring salvation to me on the condition that it will make them feel like god and that feeling they attached to this dishing out of the teachings of their gospel. In this handing out the gospel they are elevated to becoming my messiah bringing me eternal life and feeling like god in the process. It has little to do with my salvation and everything to do with securing their position of feeling godly. They reap the glory and give me their disposable shit. Even such seemingly precious outpour of goodness and salvation has a very shady background hidden from the unsuspecting customer with good faith intended. This is still the part of the good-hand bad –hand syndrome because they know they bring me their worthless shit and enjoy me basking in their glory because I am stupid enough to enjoy their toxic waste. If you have something precious, you keep it and spend in extremely carefully. When you have rubbish you throw it at any person stupid enough to want it. They will not share or give me what they consider to be valuable as is the case with their earthly positions but what they discard as trash they share with me under the pretence that we all will play the game pretending the morals they teach me is of worth the effort of receiving. There is no good and there is no bad because the good and the bad come from the three characters prompting me to form an opinion. That opinion is my relation to what I observe in their behaviour to my behaving. That opinion is based on past experiences that brought gladness or sorrow. That is the motion through which my characters evolve. Where my characters are a^3 the evolving and the setting of moral standards are forming by $T^2 k$ or the motion of emotion in the space of previous experiences good or bad is just factors forming the six sides or six borders that the three personalities will use to launch the onslaught. Another case that I can mention is the British nurse that murdered patients to free beds for less serious patients standing a better chance on life thus being the more deserving patients. By killing those critical ones the less critical ones receives a chance on life because the less critical ones may use the chance to cheat death far better than would the most critical ones or that at least is the way she saw it. Again in the most evil there is compassion and the most compassion there is evil. It depends on a^3 of the characters that will bring the other three sides $T^2 k$ forming the borders I observe. Life in the Universe is physics applying.

This mans that from Kepler's formula we even can find much more than the basic physics in cosmology. We can read the establishing of man. There are the normal recognised and appreciated good and bad but there are two other characters never before noticed. There are the person locked in singularity somewhere in the mind that control body and sole which the singularity quite correctly use as disposable space-time. That one character secured in the ruling singularity one may see as k^0. Placed in relevance there are the three characters that form the "I" that makes up the "me" within me apart from "me" within the private singularity. Within the mind we can also find the parameters set by the identities to the entities. It is the borders of movement and the parameters of allowing good and evil to come to terms with appreciated and accepted rules. Some may murder but find cannibalising much unacceptable while other will rape only after murdering and removing the victim's eyes. Each sets his or her personal standards of rules about what can be correct and incorrect, accepted and what should be denounced or tolerated and rejected. In that sense there are the a^3 but in appreciation to the three characters we find the movement we allow the characters to work in giving the acceptable tolerances of $T^2 k$. We have within life and our minds the six dimensions controlled by the seventh. We have $k^0 = a^3 / T^2 k$ and to each there are individual acceptable norms applying to each person according to private boundaries. Analysing and categorise conditions is essentially a meaningless way of job creation.

Lets not kid our selves about rite and wrong as moral issues. If you kill a person while not being employed to do so, for instance serving as a soldier, the law stipulates you should be hunted down and hanged for that callousness. If you have an army uniform and you are dispatched to another country serving under the flag of your nationality and you start killing, then you are a notional hero.

If you rape a woman you should be in prison. When you rape a woman that is a member of another nation in an act of war, then you helped to pacify the enemy and no criminal charges will ever be brought against you for committing war atrocities. If you steel property while being in civil liberty, you are a thief. If you steel while warring in an enemy country, then you only claimed personal trophies and memorabilia. Lets not start the bullshit about expectable and acceptable behaviour and what forms human moral criminal behaviour. In Afghanistan soldiers walk rite past poppy fields and cocaine factories without giving the illegal substance a second glance because they want to hunt down the Islamic extremists. If someone is torturing any person while not ordered to do so as a soldier, you are despicable and need to be locked up for life. If you are a soldier torturing to get information from the enemy, you are admirable. Good or bad, rite or wrong, acceptable or barbaric depends on who is benefiting from the aggression unleashed. If you are Iraq's Sadam Hussein and you killed five thousand Kurds, you are a war criminal. If you are by the same measure George Bush and you killed a million Iraqis by bombing the living shit out of entire civilian suburbs from American war plains, you are fighting the axis of evil and those actions are justified, even killing Sadam Hussein in an unlawful and unjust court proceeding, you then are acting according to all western standards which is the method whereby you then are guaranteeing civilisation and you then are a rock being the pillar around which western democracy spins.

There is no criminal behaviour spawned by brain damage, where the person was abused as a child, or have to have some form of mental disordered, so that they then could to be classified and categorised as some mental disaster acting out some demonic anti social criminally insane act when murdering or stealing or raping…murder, torture, rape, plundering is within everyone and it is the role in which you and I play that makes you or I hero or foe. We all carry such tendencies but it is society that will judge us in relation to how the national pride (in other words those very rich and powerful) reward or punish us for such actions. The unacceptable acts were the culture of every person in ancient times and became part of our developing genes. Those that were not like that did not multiply and did not become part of the gene pool that later formed the basis of all generations that followed. Finding psychopaths and schizophrenics and all the other mental disorders are just labelling what is in every person walking the Earth. It is a question of some controlling the feelings and urges and other like to pamper the desires by acting on it. If there is no control over such murderers impulses, hang the bastard from a lamp post and rid the world of such human waste. To supply reasons why some are as such is to give motive to those to be as such. Some of us are tall and some are short and the tall one can reach the top shelve while the short ones need a ladder. It is a bigger effort to fetch a ladder every time but if you must, then you have to. Some has to put in a bigger effort than others to maintain normality and those that can't, should be terminated by ending their inability to care before they are a danger to innocent people.

Those Neurologists and Psychiatrists coming up with fanciful labelling of the categories of the criminally insane and trying to identify behaviour backgrounds not acceptable in democracies and classifying those misfits accordingly where they try to find reasons why such persons still prevail all the while studying the persons with criminal behaviour patterns, they should rather start conducting their studies with using politicians, bankers and the most powerful individuals that holds the authority to order and allow members of the general public to become soldiers in order to protect their riches. Politicians, bankers and the most powerful individuals are the worst criminals with no moral convictions except to have others die to protect their belongings and their status. Again I repeat, lets cut the crap, there is no good or bad as formed norms in society making behaviour acceptable or not but only reasons why anyone acted in the manner in which the act was performed. All wars are totally un-defendable and all countries have armies willing to murder.

Some person will cannibalise but never rape and judge such behaviour in disgust. Others kill to keep company and others kill to part with the company that they murdered but keep the body as surety that the murdered cannot come back and haunt them. There are people that with best intentions commit unspeakable horrors as acts of kindness. To them torturing others may present a show of goodness and is quite impropriate to consider the wrong in the deed. They commit criminal deeds as well intended behaviour not for one second minding the consequences of their deed. To them their deeds are well intended and those thinking of it as wrong are the ones being wrong. They cannot see how their behaviour can be disgusting criminality because with such crime they try to secure kindness. We try to standardise offences and criminal behaviour but many judges send criminals to jail for pornography while they keep themselves entertained with pornography and prostitution. Other judges hand out stiff sentences for drunken driving but they drink and drive most every night. Standards are so personal and individual we may never come to terms with the reality behind the variations there are. Every one holds the position retaining singularity in life and by life we apply motion but what we do with such motion makes each individual independent from the rest of the cosmos. It is in the motion from limit of goodness to limit of badness that we derive in our characters as we place the borders of behaviour of our three characters as acceptable or deplorable. It is because of the variation in the tolerances of the motion within the character limits that standards will always fail. There will always be those opposing others and never once will absolute agreement come to be. Rules vary in diversity between individuals that are in agreement with each other.

Inside all of us lurks to personalities above and beyond Ek and they go by the names of My and Myself. I fight them for control and sometimes I win ad sometimes they win. Most of the times they behave well but sometimes they can be opprobrious without my consent in the matter. One of them and I leave the choice up to you to decide which one I am referring too are truly obnoxious and spiteful more to yourself than to others. The other one is normally timid and can be classified generally as your conscience. Between them you have to take the control because you are the boss.

Being the boss and demanding control you are the balance as much as you take blame and shame when not being the boss and not in control where matters go out of control. You carry the consequences always, as you should because whenever what ever goes wrong others will pass the blame on to you. And so they should because you are responsible even in such times that you are not responsible.

Ek finds himself in the middle of My and Myself and as Ek is in the middle Ek get advise from My and Myself. In accepting or rejecting comes regrets and jubilation but ultimately the final choice is with Ek. The characters are strong at times and are weak at times limbering and dominating whenever opportunity presents. In the Afrikaans I identified the character by names but found it somewhat complicating the issue in the English and so I did not name to identify by character.

Ek takes full responsibility for all actions for Ek is My and is Myself by only being Ek. We all have this in us and some find ways to fight it better and some of us are in a desperate fight for sanity and survival. The issue is not the guilt because every one is guilt ridden, as we all have to cope with this behaviour in some degree. The stronger any one denies this the harder the one is in a fight for self-protection. Cases are sometimes most serious and other times just under the skin but it is there in every one. If you are human you are fighting. Being the one finally carrying the burden then becomes self.

It could be where the father feels he as person never accomplished anything and with that self-detesting he puts all his blame, guilt and rejection for what he is into the passion he feels for his child. The passion he feels is carrying the burden of hate he has for his image and that drives him to expect from his child what he never could achieve. The child has to be many times better performing with many times more positive results, be a champion, be a scholastic genius, become the school president and make a mark as a pillar of the community although the child is only a child. The child has to outperform all others on all terrain in a fashion fitting the image of a champion, which the father never was. It could be as serious as I indicate or it could only be in little suggestions made to better his child.

When I am speeding on my motorbike or driving my car at great speeds, I am not trying to commit suicide but quite the opposite applies because I am trying to fight for my life. This is not an act of depression but an act of surviving. It is not the bad hand trying to destroy me but I am fighting with the protective hand trying to stay alive. That is why I feel alive after such an excursion with speed. By fighting to stay alive I am restricting the influence of the bad hand by exuberantly enabling the good hand for the task of fending off the bad hand's murderers plot. That is why stuntmen and other risk takers are happy and feeling alive. Evil Knievel was no depressed slob trying to fight off a sinking feeling of failure, while fighting against his tears as he felt sorry for himself. He was addicted to the rush he got from jumping busses on his motorcycles and approached every event with the most skilful planning possible. He was fighting the bad hand all the way to the scaffold from where he jumped, while nursing the good hand every second of the way. His planning involves his good hand in combating his bad hand. He was fighting with his survival hand to control his attacking hand. He was staying alive and fighting off the fear of being killed. Sometimes the bad hand won and he did stupid self-destructing things going way over the limit like when he went and attacked the person with some iron rod. The entire act was totally self destructive and the price he paid was enormous. But the limit where he fought everything was on the edge and that is where he played his survival games.

If one takes this back to how man began this makes sense. Man had inside him the something that tried to kill him all-day long and was in a fight with him within his self every living second of the day. This kept him alert and aware of dangers not only from being attacked to become food for predators, but also from something within trying to destroy him. That made a being as vulnerable as man super alert in defence and that gave man with his most limited defence an approached to danger that surpasses all other beings. Man had to employ his own wit and intellect to survive and the best way to improve your potential is to fight yourself. It is a fight you can never win but you have to fight constantly to survive. By winning and staying alive I am fighting off the urge to kill myself but the urge to kill me is not a result of depression, it is an anger aimed at me that I could turn to other things with immediate anger to call on as an aid for protecting oneself but when in idle ling mode it is aiming also to hurt me by switching a thought. It is not a feeling of getting into sack and ash and feeling sorry for myself all day long. No, it is readiness to fight at the drop of a hat and most times this aggression is aimed at myself, where I am able to control it in order to apply it at whatever tries to kill me for food. I am defending myself against myself attacking me so that I can switch my attacking side to attack the danger attacking me as food and in that I have an attacker defending me as well as a defender defending me by also attacking whatever attacks me for food. This triple awareness gave me the edge to start applying a mental awareness and with that I could depend less on strength and more on mental wit.

It is again seeing the facts not as they seem but taking the facts in relevancy, as the facts should apply as need call for. It is retro-aggression that has roots not in despair and depression but in survival and fighting to stay alive. To come and diagnose this, as sitting in ash is the most counterproductive discipline man could think of. It is madness on the same scale as war.

Sometimes the fighting goes to far and sometimes a child never learned to correctly control the fighting. This to my thinking is what stimulates and drives gangs and control gang members holding such skew purpose in life. Sometimes the child feel his parents are on the side of the attacker and then that pushes the balance and sometimes the child feel the parents are over protective and that will get the child to go along with the self destruction part. In these cases the child never learns to control the balance and always feels a failure because he can't get his balance rite. In many drug addiction cases stems from where these feelings of self-preservation and self destruction gets confused and the child also brings in feelings of sexuality dabbling somewhere between the two feelings.

Those that think suicide is a result of depression has no idea what depression is! Depression is a feeling of feeling down trodden, in the depth of despair, it is a feeling of sitting in sack and ash filled with desperation and killing yourself is nothing of the above. The feeling being in the depth of despair may lead to suicide but it will not be the desperation that ends in suicide. Suicide is a feeling of changing from what is a wall of misery to be the feeling of victory in having an answer, driving your problem to a conclusion, leaving the hand that wants to protect and giving in to the hand that wants to kill. It is changing sides within yourself, coming from the side that fights to keep you alive to crossing over to the side that drives to end it all. The ending of it all is not an act of desperation, but it is a total feeling of complete aggression, it is using the aggression you are always fighting against with aggression where that aggression you then no longer fight, but you use to it fight with it against the situation you find your life has turned into.

As it is in all cases with all things in the Universe, there are no one conclusion ending in one result and with that result it ends all further conclusions, but it is many conclusions working hand in glove where one conclusion might be the strongest fitting one and some then it might be pushing the other reasons also applying into a lesser obvious pertinence that then might be inclining slightly more towards a factor forming the background. However that might be, killing yourself is not a feeling of despair but it is a swapping of alliances. It is the most sincere from of brutal aggression that there is. If you can't kill someone you think is to blame, then instead you kill yourself to put the blame on the one or something you wish to kill.

People play Russian roulette not because they feel depressed or because they taunt fait, but it is because they challenge the hand of protection to the ultimate. By playing with death they are alive. By pushing their luck when odds are against them they are completely in control of life. This has nothing to do with suicide of feeling a desire to die. People looking on see this in an opposing relevancy because the onlookers see it in view of possibly dying while the person spinning the barrel see it in defiantly staying alive.

The same goes when suicide is contemplated but this then is in reverse order. A person feeling attacked and depressed about feeling attacked does not commit suicide because they feel attacked and wrongfully accused. They commit suicide more likely to get on the winning side. They feel they are losing the battle with life and against life and then they are depressed. Then the feeling they have changes from depression to aggression, but they can't attack everyone attacking that suicidal person. Just as all persons are, the suicidal wishes to get on the winning side. The person wants to become a part of a winning team, which are normal and a tendency within all of us. On this basis sport and therefore money in flourishes and the Hoggenheimers are making billions in raw cash. That person wants to also win and being alone on the one side will make the person wish to swap sides and go on the attack, joining the winning brigade. Then the suicidal will attack the one the person feels everyone is attacking. The person will relieve the protecting hand of all further duties and will charge the attacking hand to get into maximum force. This change is a form of the most aggression any person has. It from then on never is about depression because it progressed into violent aggression. When this change enters the scenario it has nothing to do with feeling depressed or feeling sorry for yourself or being downtrodden, but it then becomes a feeling of God damn me because everyone is damning me in any case so I too am damning myself. It is getting aggressive to show everyone I, the suicidal too, can be a winner by fighting this miserable shit I am and the one everyone is fighting.

If the medical profession goes on fighting depression the way they are doing at the moment then they will keep on losing the battle as they now do and getting the precise results they now are getting. Depression has risen to become one of the major illnesses in society that cause deaths of thousands and the medical profession can't start to realise it is the medical profession making a mess beyond words of the situation they can't begin to understand. The reason is that the whish to depress what can't be depressed and by not depressing what can't be depressed they are depressing what should never be depressed. While fighting depression as suicidal tendencies will be as if trying to curb a fire with diesel fuel. It is not only suicide in the manner of taking a gun and blowing your brains against the wall. Drug abuse is suicide. Spouse beating is suicide. Criminal actions are suicide. Yet, like science does, they pack everything into compartments and label the lot differently so to ensure it is a mess as to make sure it is then a form of job creation

I wish to tell about a situation that is very typical and again it borders on gossip, but with this gossip there is much to learn because this gossip can be about millions of persons every day. This story is as factual as anything in my life could be and I am making nothing up, but the names.

In a town a person has a garage where he repairs vehicles. Lets call this person Jack. Jack had a son working for him and with him so lets call the son Jackson. One Saturday Jack did some overtime and while working with no one customer going to come and supervise or visit, Jack had a number of beers while working. I guess Jackson also had a few but I wouldn't know about that fore sure. They were toiling and the beers got them getting rousing. Then Jackson had a fender bashing with the company breakdown in the shop on the lift. This got Jack furious and the booze got him excited. Enraged with semi drunken anger Jack came up to his son with no kindness in his heart and Jackson new all about his dad's temper so Jackson physically ran away. Jack could never catch his younger son and was beaten by youth, so he enlisted cunning. Then Jack started soft talking Jackson about it being an accident and it wasn't that serious and Jackson should come and inspect the damage with Jack.

The very second Jackson stood next to has dad Jack had Jackson have it with the fists and he did it in a big way. Jackson was only a boy of sixteen and had no credible chance in an exchange involving fists much more than words. After the confrontation Jackson got on his motorbike and drove off feeling very bruised and very rejected and very powerless.

Jackson went home and hanged himself in the garage, in such a way that when his dad opened the door, he saw Jackson swinging. When Jackson did this action, he was not depressed. When Jackson did this action, he was fighting mad and furious. Jackson wanted to win the fight he had with his dad and was so intensely worked up that he thought of nothing else than to get back at his dad. Jackson wanted to give his dad a beating that would cripple has father and give his father a life long scar. Jackson saw his chance to get back at his dad in a way no man being a father would be able to endure

Jackson got in a position where he totally abandoned the protecting hand and embraced the destructive hand. He did not think of suicide from a point of being depressed but he wanted to give his dad the ultimate hiding he could ever give. He wanted to win the fight at all cost and show his dad what his dad had coming. It was the most aggression he could command and muster against his father. He knew the suicide was the biggest punishment his father could get…so he did it. Jackson killed himself in rage. Jackson could not beat his father so Jackson committed suicide and knew in that act he would destroy his father…and Jackson was successful in every way. Depression was as far away from his mind as an iceberg is from floating in the Sahara.

The true problem is with treating the disease called depression. Jackson would not have done it, if he never thought about is before. Jackson would have contemplated suicide before and Jackson would have mentioned this to someone, most likely his mother. Jackson's mother would have taken him to a doctor about the suicidal tendencies and the doctor then saw this as depression. Being the depression the doctor diagnosed, the doctor would most likely have prescribed some emotional depressing drug to suppress the anxiety Jackson had. This was the big error because the anxiety Jackson experienced was his fight against his self-destruction by trying with anxiety to use his protective hand in the battle against the destructive hand.

The emotions of controlling the destruction then were suppressed and when the bubble burst, all the control went flying away at the speed of sound as the destructive side came out while simultaneously the protective side left the building because all that was suppressed was his ability to think clear and have the protective side help him combat that which he fought all along. By his father going against the protective side and working in alliance with the destruction, he also used the destruction to help him fight not only himself but also fight his father by destroying his father by destroying himself.

If Jackson was diagnosed correctly before and, if Jack was given the correct treatment by controlling his temper and not having booze to suppress it, all this would not have happened. If Jack took control of his senses and not take alcohol to "relax" him, he would have been a father to a boy that was alive. If the medical practitioner saw and then diagnosed Jackson's condition as aggression and not as depression, Jack had Jackson take over the family business later in life. Jack loved his son more than he could love life but his self – destruction was aimed at his boy to aim it at himself. But everyone saw depression and no one was able to recognise there was no depression anywhere within a country mile and everything that took place that day had everything to do with retro-aggression. You don't blow your brains across the room because you feel depressed, sorry for yourself and tearful of the miserable situation you find yourself in. You don't kill because you feel sorry; you kill because you are enraged. Doctors have to recognise their failings in diagnosing the problem. There is this lifetime battle within one and all and the aggression does not hide behind depression but the depression gives way to become reckless destructive aggression. Don't treat the depression with a pill as a quick solution, but teach the person to control and use the aggression productively.

These same feeling can bring about the very opposite where the father tries at every chance he gets to run his child into the mud because the father fears that the child will push him out of his role he has to fill but lack all abilities to fill the role. He thinks by destroying the child he is protecting the child because he then is securing his position as the father figure. Knowing well the chid carries his genes and believing the genes the child is carrying is not worth much because the father has a low sense of self worth, he protects the child by showing the child what he (the child) is and allowing the child the realisation before the child will one day find out for himself in the cruel world. This cruelty is about protecting the child from the cruel world. There is no good as much as there is no bad, and in the same breath everything is wrong.

The same behaviour may come from the father in fortune, the one with success, the pillar of the community. He is the top judge, the successful attorney, the town's top businessman, the city mayor or the one carrying the admiration of others. He hates his child for that child will one day inherit all the goodness he now has and that makes him sick. He might feel the child will never become the doctor he now is but through his reputation that he worked so hard to achieve will become even greater than he now is. That makes him to push the child to live up to his personal greatness or destroys the child to show the child how much the child should be great fill for the admirable fortune the child has to have a father such as he.

It does not have to be about someone you love. I shall be very frank about my case and millions will recognise their fight. My personal struggle involves money for one. I have not the slightest idea how to administrate my money affairs and sometimes I know I have this inward hate towards money. Whenever I have money (and that is not very often I do admit) I allow people to sucker me with a sob and crying story about their hard life and the bleeding heart dashes through, the knight within me with the glittering armour takes full control and helps me to give away sometimes even thousands and tens of thousands, knowing very well notwithstanding all the promises of repayment at the time, I shall never see the money again. It is not I being the bleeding heart and then yea it is I the bleeding heart but that is one character. I the bleeding heart am on the background where I the hating bastard am rite on the dot standing on all fours and then some shouting and urging me to help the others, or to go on and buy what ever I feel I need but I know I can't afford. Forever the urge to buy is ensuring me that there is thousands more coming my way in any way so what the hell, let some goodness flow. This will always come where for some reason some money fountain runs dry just afterwards and I drop my family into financial surviving periods allowing them, the ones I love most to suffer the hardships. Not once did this occur and was not the prelude to personal hard times! I am a middle-aged man. I realise I waste the money because I hate feeling comfortable so being without money makes the fight I then have to fight mostly unbearable, but that is the good part. Then I have to fight of depression, brought on by being without money and not fight the suicidal aggression. I know these characters. I recognise the precise feeling accompanying each one. I have scrutinised and analysed they're being part of me decades ago as being the I in the me and the bastards being the I in the me still catches me again and again, sucker punching me over and over.

The first time I went gambling I recognised the one personality of self-destruction being very aggressive and taking centre stage as it came to the forefront just lying under the skin, filling me with anxious excitement and then and there I realised gambling was not meant for me. In that sense I have beaten him hands down by not starting the habit. Old positive me is about cars speed and going crazy in my mind and I know it is the strongest one because that one am the biggest I. A middle-aged man tearing down some street on a massive motorbike showing some youngster what the bike can do when there is someone on it that knows his onions are grossly irresponsible. How childish can you get, how in mature can one be. That is the I and that is the very me and that is Ek and I have as much control over that as a drunkard has control over his drinking. Saying that is also admitting I do not wish to fight him as I do not wish to beat him. I find his company very pleasant stimulating and destructive and when he comes to the foreground all other characters harmonise within me, giving me all the different feeling each one of them should supply. I am then self-destroying giving the negative character his day, I am childish giving the neutral character his day while the positive one and I am the same person to let all power loose.

My characters bring me in conflict whenever I bay or sell or demand a price for my services rendered. I would bay the biggest shit at the highest price in the belief I am helping the poor slob. When I sell I get all guilt ridden when trying to make a profit because I get this idea I am cheating the poor fellow by insisting on the price I am aiming for. In all cases I feel so guilty to ask any person money that actually belongs to me I get a nerve attack or a running tummy. It is not out of fear because if any person gets violent or aggressive I know how to defend myself. You have to know that if you are a devoted biker because they always get drunk and strong at the same time and fortunately I was in motor racing during the age my mates learned to drink. But with racing cars always braking down and crashing there is never money or time to drink so I never got around to start the habit and by the time I stopped racing every one accepted me as a teetotaller so I never got around to develop the habit. But asking for money or insisting on a fair price is more than I can achieve.

I shared with you this Ek, My and Myself part of the personal me-story so that you, every you will know what I mean because the objects may change but the objectivity and feeling and the motives never change. One example how the process works will be when a mother that hates her child for birth pains, unwanted pregnancy, feeling unfulfilled through a bad self image or carrying on where her parents left off with them treating her as a child in the very same manner she treats her child. The negative character prompts the mother to reject the child as she detest the child and loathe the child. The Positive character saddles her with a tremendous guilt punishing her as anxiety in blame riddles her.

In self-punishment aiming to destroy her because after all that is the purpose of the negative character the positive character punishes her as she fills her with self-loathe while neutral character tells her about the wickedness in her. She gets reminded at every opportunity about her love she must have for her child and her duties as a good mother to love her child. It is her responsibility to love and protect and because she strives to be the good in her the punishment is severe.

If she gave in to the negative character and start enjoying the hate and blame she feels towards the child the positive character let loose the pre-historic maternal instincts with a flow of torturous guilt hidden behind such strong emotions she deflects the hate onto her self. Then comes the neutral character reminding her that no one should ever know about her hate towards the child because as she detest herself so would the world detest her if ever someone became wise to her feelings and she will be driven from society with hammer and tongs. With the conflicting hate polarised and swinging between her and the child she knows that all of human kind will see her for what she is and with the hate she feels towards herself the negative character takes that hate and turn it into fear for others finding out about the truth. In realising that no one may ever know about her true feelings toward the child and know about the loathing she feels about herself she takes all precaution to hide it from the world.

In this mind game of rocking emotions the positive character supply her with advice in how to take charge of the situation never to the benefit of the child but to her benefit in protecting herself from the outside world. The advice will never have any concern about the child because she carries that burden by herself. The positive character continuously reprimands her of her evilness while the negative character fills her with hate running between the child and her feelings about herself. The neutral character reminds her constantly not to allow any one to find out about her feelings for her child and demands protection of the outside world finding out about her and her child. The neutral character pushes the fear to match the severity of the onslaught by the other characters in order to maintain equilibrium and equilibrium means almost insanity

As the insanity at times almost become intolerable she reflects the blame for her situation onto the child being there and making her life hell. Again this diverts the relevancy from her taking blame to the child having to carry the blame and therefore correctly must suffer for the blame. The relevancy of carrying blame shifts from her to the child and this makes her suffering under blame less heart full while the child carrying the blame more correct. This the negative character grabs by prompting her to punish the child as severe as she can and through this stop the child torturing her. The cycle leads to the next cycle and in this sanity guilt love and hate becomes one flowing emotion of disturbance. With this conflict within the mother the child's developing personality receives knocks the child cannot stand and less understand. The child starts behaving rebellious and unacceptable to the mother's neutral side and in the eyes of the community. This chance the positive character grabs and being positive only as far as the mother's well being goes the positive character advises the mother to leave the child and let be. This will show the world how much she loves her child by refusing to even punish the child when the need arrives.

In this advise the negative character joins by advising her to let the child become out of control establishing the fact that should her hate towards the child ever leak out, the world will not blame her for every one in the world that has contact with the child will hate the child in any case. When every one despises the child the negative character swings into action by filling the mother with more hatred towards the child and when opportunity comes and they find themselves alone the hate comes to the foreground and then she punishes the child with most cruelty. This can be as part of actual criminal prosecutable child brutality or it could be most cunning and devilish in conspiring but the brutality is all the same. With the presence of such severe child brutality all others in the community turn a blind eye not to get involved where each outside person will find some excuse not to become involved. As every one has a struggle of their own they too are in self-protecting not feeling the urge to come into the open and defend the child. Being in the open will unveil their personal fight for survival and they then will become the target of the community. All this is in the very distant back of our minds never in front where we can kill it but present in the way to be us and not to be us.

All three personalities agree on one thing and that is that the world must with its entire people have a hate in the child as much as the mother. But as every one find the child unacceptable and revolting she can feel better about her feelings because now she is part of the crowd. Being part of the crowd will bring sympathy from others with their understanding about the hell and the torment this child inflicts on her every day. Such a feeling soothes the aguish of the wrong she feels she is committing as much as the wrong the child is committing to her. By allowing the child adverse behaviour and defending the punishment of such behaviour the child will become more unacceptable and the situation heads directly for the disaster she hopes to accomplish. She directs the child's personality in that direction while she feels all the torment others see her go through. She allows the child to go to nightclubs doing drugs and commit self-destruction while the mother is merely a spectator because after all that is the child she hates.

When the child is out at four in the morning she can feel good about not having the pest around. She can feel good about her hating the child. She can feel good about all the sympathy she receives about having such an unruly child. She can live the life she claimed, hating the child, feeling sorry for herself and good about others understanding what she is going through. Should the father, a teacher, a policeman or any other figure try to stop the madness and bring order to the child she will attack that person with all the hatred she feels towards herself and towards he child. She will destroy the prevention of her self-destruction because any body trying to discipline the child is fighting her aim to destroy the chid and after all then the world will see how she can fight for her child's protection by almost putting her life on the line.

This becomes the Sudan affair where the bleeding heart buys guilt relief and be god while the philanthropist pushes guilt as hard he can and be god to collect money on behalf of the Hoggenheimers that then can be god with such wealth distributing it to the Mammonites who can be god by baying from themselves as much as selling to themselves with unscrupulous profits making him god and allowing the Mammonists to be a slave driver and being god to the slaves. It is this sickness of society no one cares to see because every one gets what they want, even the luckless get what they want with the minor condition that when the luckless suffer most that becomes the region where most profits are for every one in the chain of gods. So the luckless must be in crises starving as they are dying to gain most profit for every one. The profit has little to do with money but with being god. Any attempt to stop the situation will never be tolerated by any party and therefore my remark that every one will press for my castration because of my suggestion to rectify and bring a solution.

The mother will fight any and all positive solutions with tooth and claw and no one should dare to lay a finger on the child because she know how successful she is in destroying the child and it is so easy to shout child abuse when someone wishes to correct the ways of the child to the benefit of the child in the interest of the child. Her devoting love will protect the child from such brutality as what a good hiding on the backside will bring if the hiding is done in love and the child knows it was on behalf of care. But that will stand in contrast to the mother's brutality and punishment and the child may recognise the difference and wise up to the difference.

The social worker will never accept such brutality after all it may cure the little brat and put our social worker out of a job. The lawyers and judges are on such a big job creation drive by minimising penalties and getting criminals back on the street for the next cycle in crime they will fight any interference that may reduce the crime and decrease their chances of money making. With so many to loose so much on the one side and only the child to gain on the other side all brutality in the name of positive punishment will become child molesting and will never be tolerated by those with influence in society.

The mothers behaviour becomes a reflection on the disease within society and it is in everybody's interest but the child's not to admit to any knowledge about the foundation behind the scenario, after all it is only a child going down the tube and to top it all it is a child no one cares for. Is there anyone out there that can see the parallels running here between the animals not caring for the unprotected infant crying in desperation for a mother while every one ells in the species cannot be bothered? May I now comment on the fact that we are going the way of the animal and devolution of our species is in progress?

Weather you care to admit it or not but greed and money is destroying man to the fullest while man is enjoying the destruction with all its lust. The drive in society after W.W.2 became progressively to feed the children to the hyenas of society, which are the crime bosses, the prostitution rings, the drug pushers because after all, they bank the money at the Hoggenheimers to the convenience of the

Mammonites and Mammonists. The ones that are caught in police action are the ones not part of the official system and they become the offers the politicians demand in protection to show the public the system is doing what it can but unfortunately it can only do that much and if it is not good enough it is because we all are human. Hundreds of thousands of children disappear through out the world and no special task force has ever been set into action to get behind the problem This problem has no boundaries in as much as it is going on in every country there is world wide. I have an awful, awful feeling and please consider the next remark as a thought with no substance but that internationally oil is bought with children as payment because no politician through out the world shows much concern. But let three banks get robbed in one day, then a special task force comes into action and gets the culprits with extreme prejudice. This is the symptom while the mother's behaviour is the condition.

The positive character advises the mother to defend her child against disciplinary measure, the neutral character reminds her of the image problem and demands protection while the negative character sees the child slip into the ditch and everyone is happy. Every aspect is in line with almost one aim and that is the destroying of the child. What ever may bring positive results everyone shout down by making the connection where the punishment links directly to what may be extremely negative because from the onset it seems cruel and negative. Giving the child the spanking of his or her life driving the fear of god into the young person will have extreme negativity but that must stand in complete contrast to the love the child then must receive before and after the spanking. The child must know with one hundred percent certainty that the parent is and will always conduct the child's care with one aim and that is to ensure her or his well-being. But if the child knows that every aspect of the child care swings around the drive as far as getting the young person destroyed the yes, the child will find all connection to punishment intended on the destruction aspect but it will remind the child of the similarity and lack of contrast in the usual treatment because of the absence of the love and caring aspect is in harmony with destroying.

The whole aspect changes around when there is another person also punishing the child but with prejudice intended. When the punishment comes from despise and not from care the mother sits back and allow this to happen. If the father is sexually abusing the child the mother will suffer greatly all in silence all quiet not allowing outside intervention spoil the situation because then it is the father who is to blame. It is the father that is destroying the child and it is the father that is the devil. The other parent can then take responsibility and blame and the mother becomes the second blameless victim in the case where she does not participate in the abuse only because the father then plays the part. That is the only aspect that changes where as otherwise the scenario remains the same. All characters play their part as if she is doing the destroying because she is doing the destroying by helping to provide the perfect environment for the destroying and not allow any clue get outside the close knit intimate family circle. She takes a part in the abuse and takes as much enjoyment as if she was acting although the blame and the soothing shifts somewhat but all intensions still encircle the destruction of the child.

The conflict within the woman may drive her to protect the child she hates as much as the father does. Her neutral character tells her that now she is no longer to blame therefore what ever happens she can stand in the shadow of the male and he has to take the blame. But the negative character will not tolerate such idleness. The negative character wants the child's destruction but moreover the character wants her destruction. He advises her to action. When the father is at his most dangerous being overwhelm with cruelty she will jump in and save the child by physically protecting the child while knowing full well that she can do as little protection as the neighbours budgie can. This action will satisfy the positive character by her showing her unbound unlimited care and devotion as the epitome of true motherly love. With her actions she know she will unleash much more anger and the father will loose all control. The negative character finds stimulation in this action and supports more involvement to unleash more violence all the way. In such a rage she knows the father will then beat the daylights out of her before he turns onto the child with more rage than he had before. With such reaction all three characters are satisfied The positive finds a way where she can become good, the neutral character knows that society will condemn the father and the negative character will justify the cruelty as the correct way to go because the child the father and the mother is bent on destruction. Now the positive character can tell her she did what any good mother would have done, the neutral character knows the beating is the same as what she endured as child in any case and that did not kill her so this beating cannot be that bad while the negative character will enjoy the situation to its full as she and the child is being destroyed.

The balanced behaviour of a person with the trio not having violence to promote would run outside and call outside help from any source available at that moment. A woman beater and child abuser is always, always a coward and when real trouble arrives he will stop immediately. But if the trio were involved in violent provocation exemplifying hatred to the child, she would take charge in a different manner. Even if she does act in this manner she will not call the police or get the husband behind bars because she argues that the family will suffer with the father not providing at the time. No one receives any money while being locked up. The excuse she uses is that there will be will no one to provide for the needs of the family living expenses. The father's inability to provide is all but the truth as she wants the situation to continue because she is enjoying it as much as the father. Should she truly admit to the seriousness of the crime she would have the father locked up as if he died and never allow him close to any member of his family again. To her and her child the best will be if the father is dead because the father will destroy where ever he involves himself. The only sane thing to do would be to declare the father dead as far as the family concerns go because when he is released from prison she would have a new life to live that no longer will depend on him or his providing. She would recognise him for the monster he is and not see him as the senior partner in crime, which he truly is. Keeping this partner ship in place by using a lame excuse like ninety nine percent of woman finding themselves in such a situation uses would satisfy her three characters and once more the money matters more. Now the stage is set for the beginning of the next round because with the father as with her the creation of the next climax begins and develop until the next time hell brakes loos where the father then has even more hate and a lot more to prove and correct. His fury and outrage will cover his hate but also the negative character in him will demand revenge as compensation for lost pride. He will repeat his role and she will repeat her role and the child will have no role but suffer destruction and again and again the process goes on and on.

By throwing her body in between she knows no woman can stand against any man in a physical fight. Her excuse for aggravating the situation is that it is what any good mother will do and that also becomes the advise of the positive character. She knows she will outrage him blowing his week self esteem out of reality because now she takes him on as a man insulting him at the area he feels the weakest because he acts in the manner that he does because he knows from his weakness he can never manage to be a man in the company of men. The outrage she unleashes within him will satisfy the negative character. Reminding him of his coward ness and weakness will bring the monster in him to its full potential and that is exactly the plan. By her action he will be reminded of his weakness and that will be in response from advice given by the negative character as for the violence part and the shift in the blaming will be in responding to advice given by the neutral character. The child will now see how much she cares and that she does not hate but love the child to a point where she will sacrifice her life to protect the child. That will please the positive character. Then afterwards when the beating or molesting or whatever cruelty is completed she runs off with the male as the lioness did when the lion killed the cubs. The mere fact that she still remains with him runs parallel with the lioness accepting the animal behaviour because the human litter may not be dead but that is not her fault and in any event the fun can continue in full rage the next time around. Why call a halt to all the fun because everything will be all right after the husband pushes a few hundred dollars in the hand of the child or bay him a brand new whatever that will sooth all pain going around. He does not care for the chid as he can bay the child's silence. She does not care for the child because she allows the situation to continue regardless and the child does not care for the child because no one cares for the child in any case. After all, the father did show remorse when he bought the child a new whatever. The theme is about money. If the father is caught he would most likely get a fine because the penal system does not encourage incarceration for such minor crime. He can bay his freedom even by penalty of payment and money wins the day.

In the event where the child stepped out of line and the father (or mother) comes down on the child as hard as he can, to shock the child with such force as to scare the child so much the child will fear any thought of repeating the incorrect dead ever again, the other parent holding the hatred will then come in and phone the police, contact the magistrate, get the executioner in preparedness and go on with all dignity going to madness. She will never allow such abuse. She will rather see him dead than punish her child. She will make such a fuss and such a scene that should anybody not get involved they will become assessors to crime and child brutality. This again is a charade where her characters enjoy every second of her instable behaviour urging her on to over react. Now the moment has arrived where she can show the child who is the devoted parent after all She can show the world just how much she cares for her poor little baby that only went about to destroy the child she was destroying in any event. She will never allow the destruction to stop because of an action a balanced parent sees fit. The law comes down so hard on this man as they wish to scare off any other parent that will ever try to correct the behaviour of a self destroying child. After all where will the next generation of criminals come from if there is a well balanced society and how many lawyers and judges may become jobless.

Every one in an influential position throws their weight behind the mother's instability destroying the caring parent for caring. The judge himself has three personalities to fight and no remorse but to uphold the law to the letter. The politician that helped creating the law realises there are many more unstable persons out there to vote than stable persons and with the majority being mad it is clear with whom he will side when writing the next law into the law books. And besides he also has this little fight going on inside himself. His negative character has an enormous advantage because to him was not only given one life to destroy but so many it can keep him busy as long as he can remain in office. His neutral side tells him to remain in office because that is such a lovely place to be and his positive character now with him in office can be the god he always new he was.

The madness runs deep as it runs wide leaving no pillar in any community in strength. The description I give may be mostly in exaggeration of the truth but the truth it is. Even if the slightest way of is applying or of finding such evidence that this is taking place it will show that with the least provocation the balances is shifting towards destruction and the way society is, is an indication pointing out more than just strongly that the human race is on the decline. The truth stands out as a sore thumb showing that there are massive problems waiting for man on his spiralling way down the devolution ladder.

This is part in all layers in society from the very rich to the very poor and every one puts the blame squarely on the others without accepting any blame. When this child becomes an adult the process not only continues but worsens. For the sake of argument let us make the child a grown up male. The child as an adult misses the mother that he had but also never had. From this he acquired the loss he felt with the loss he does not recognise. Now he is ready to find a mate. But being human we humans have the culture that mating is a lifetime commitment.

The wife he is looking for must be someone like his mother but with a slight twitch. He hates his mother by now as much as she hated him all through his natural life. He confuses love and hate, compassion and punishment, caring and rejecting, in a way that allows him the freedom of becoming a most confused person. With all confusion running wild and seeing love and caring in the same light as not caring and running wild he goes on the prowl in search of a wife. He longs for stability, which he totally rejects. He wishes for companionship he found to smother him. He hopes for security he does not care for. In everything there is a threat. What should bring devotion brings hatred because that is what he recognises but does not care for. Will this young man's three entities have a wonderful time. He came into the world unwanted because of free love, he was raised in anger because of a free and fair society, he was neglected because of democracy and now he wants to employ the culture that brought him destruction as a child. In loving his wife he hates his mother. In wishing for her companionship he fears for his life. In pleasing her he understands only rejection. He is even more bent on destruction than his parents were. Where there might have been someone that tried to show him rite from wrong he got that show of caring as massage that that is a person is out to get him. Any person alive including his wife that will make any effort to show him the wrongs in his ways he will reject to a point of committing violence. In that he will find the rejection he hates so much and he will kill to destroy that. He may or may not administrate violence when at home. That comes with the role of the dice.

The wife he loves he has to hate because to him that equates mother-love. To him loving someone is destroying that person. The one he cares for he has to destroy. That is the love he was taught to accept. This leads him on a journey of more self-destruction than any attempt his mother ever made. His yearning for the mother he never had pushes him on to every woman he can find. His wife being at home does not please him because she is not the one he was looking for. His morals are mingled like a mixture of concrete. The good and bad are so intertwined he cannot see light from darkness. He starts a life of adulterous affairs partly to destroy his wife whom he confuses with his mother and partly on the hunt for the mother he has a desperate need to find.

With devilish cunning this sets his three characters in motion where they can join forces and destroy at will. His positive character allows him to bestow the love he feels but cannot share onto the woman he is with just because he does not love her. His neutral character finds her acceptable because it will last but a night and the negative character helps with the charm because he will once more destroy himself and the woman at home, which as a matter off fact he truly and dearly loves. Unfortunately he does not recognise the love as love because he does not know the feeling of love. The love he recognises as love is the feeling he feels for the woman he is having the adulterous relation with because he hates her as she reminds him at that moment as the female figure representing his mother. If he really hated his wife he would be at home destroying her but as he loves her he does not whish to destroy her so he destroys her by not being at home. The positive character puts all his attention into charm that he throws on the female he is with. But because it is the positive character it also helps him realise that he can have fun in destroying the adulterate as the adulterate is the one destroying the one at home which he loves. The neutral character tells him to carry on because he has to punish the one at home for not being the mother he is searching for and the negative character is in heaven as every one around goes to hurt.

He believes that his wife at home truly loves him and for that he does not wish to hurt her but in the connotations he has about love he also knows that if she truly loves him she will try to destroy him. After all that is what love ones do when they show love. But because for the simple reason that she does not try to destroy him the neutral character holds that against her and make him believe that she is acting in such a way simply because she is not caring about him. Such behaviour stands totally in contrast to what he thinks love is, while his positive character keeps reassuring him of his wife's devotion therefore he should set his mind at rest as he will not lose her. The negative character finds this unacceptable and reminds him to do onto her before she can do onto him. The outcome is a vote of three to none in favour of the affair and another round of cheating starts for another night.

While the cheating takes place the positive character will remind him to love the one he is with as the one he loves is not there and the neutral character will tell him that as long as no one at home finds out then no one gets heart while the negative character tells him should his wife ever find out she is in any case getting what she deserves for not being the mother he wishes to destroy back. He will not enjoy her company and may not even enjoy her sex. He is in search of something and that something she will no be able to provide. During the relation he will get bored and then dismiss her as a dirty rag.

The worst that could happen to her is to let him find out that she has true feeling for him. That would place her in his power a place no girl will wish to find herself. The mother hatred will come to the forefront and he will start destroying her as he then can inflict all the injury he does not wish to inflict on his wife. By chastising her he will find some accomplishment and relieve and seeing her anguish will fill some of the need he finds in repaying his mother. But that will only bring some satisfaction and it will only last for short periods. But he will still yearn for his mother and in that there will still be a need to run more woman down. Because he does not have a clear image of what his mother was and what love is the characters can play mind games he will not understand. We all get some notions when one with a clear image of a mother and love between mother and child come across a female. We all have thoughts about what may be but we discipline the thoughts because we know we love the one at home and do not wish to sacrifice what we have for what may be. Hell, there is some woman I have met that is as attractive as any creature can wish to be but the very last thing I ever wish is to spend even one night with her. It is not because she is unattractive but to the contrary. She knows what she is and she knows how she excites men and she uses that charm to get men to dance around her with pleasing delight. The worst fate that can ever come to any man is to get involved with such a woman or even worse than death will be to marry her. She is the female of the male I described and she is bent on having men flirting with them and then just throws the verbal cold water on them to enjoy the reaction they get. In a marriage the first signs of trouble will drive her into the first bar where she will pick up the first victim and destroy her partner and the one she is with one blow just because she did not get her way in the argument she had with her husband. The poor slob that someday lands her as his wife will have enough information to write a book about hell.

With the conflict another situation with another disturbed child may provoke the complete opposing figure of which he is in search of.

In the next scenario of the Don Juan now in discussion holds the neutral character in place that will not have a clear picture of the woman of his dreams. With the image of his perfect woman being very vague any of the other two character will come in and pour their versions of the perfect woman in his mind. The reference picture that he has about the woman he wants will be completely out of focus diluting his perception completely. He will want the woman every one desires. The disco queen or the brothel bitch or the bar tender goes out every night with another guy. He will wish to find the woman that treats him like dirt. Being treated this way will so kindly remind him of his mother and with his wish to please his mother as a child he will transfer that to his wife to be.

He will forever find some tart he wishes to please that has no wish to be pleased as she is in search of a man like her father. Her hopes are to find a man, and usually with success is one that beats the daylights out of her. Her first her second her third husbands will all have one thing in common. They will be woman beaters without exception. Or they will be drunkards, or womanisers, irresponsible persons but that is precisely what the woman has in mind although she will die before she admits it. If someone that truly loves her for what she can be to him turns up she will hate such a person because his devotion confirms her rejection. To her that man is the representation of her father and she in her twisted mind thinks her father was such a nice and devoted man because her characters will never allow her the opportunity to see her father for what he truly was. As her attraction and his attraction does not meet the requirement of their characters he will follow her like a lost puppy because her rejection of him is what tells him of her true love and devotion and that is precisely what she find so revolting about him. She wants a man that loathes her and here is a man that adores her. That throws her characters into disarray just as much as it throws his characters in disarray.

She hated her father as much as he loved his mother and because her father was the personification of brutality as was his mother they had to accept what they received as parents. But since neither had a real figure to relate to their characters turned the image they built around that parent around as to make them very acceptable. The characters they have will for the rest of their natural life bring to them the opposite of what they had as parent. That leads them to the very opposite of what they are in search of and what they find is what they wish for although it is exactly what they do not want. They both are condemned to one life of misery and if there ever is a hell, that place will be a merciful relief when they die compared to the life they have.

I do realise from the examples one must deduct that these cases are only the mental cases and anything more serious would find the person a patient in an asylum kept under lock and key by the President's special request ordered through the highest court in the land but it is not like that. The extreme I underlined because the extreme is the easiest to understand. But the destruction the parent has in mind for the child could be baying the child a very expensive pair of shoes only to let the child feel slightly important or giving money for a movie you would not wish your child to see but from the expectations your child has you cannot deny the child. It could be that you allow your child to go out with friends knowing the next day the child will write an important test. I do not wish to go into detail why that is part of the destruction or why it may be destroying the child because that is not the issue. The issue is the personalities lurking and being you. There is no slip of the tong with some wrong words slipping out. It is a deliberate intentional conveying of a massage to the other person about the true nature of your personal feeling and thoughts concerning the other person. What is the slip is your allowing one of the characters taking control for that split second while your guard was down. The turning of the cars steering wheel landing your mother-in-law in front of an oncoming vehicle while you were looking the other way and can swear under oath you never saw the oncoming car or had no inclination to go that way in any case. Why that happened is a total mystery because you will never do such a thing intentionally or other wise while the truth was that you were quite enjoying the nagging old witch's' company and her on going tormenting in her criticising you in the way you handling her precious daughter. That calling your employer by his first name when you were actually out to impress him. Calling the young girl in the office my lovely in the presence of the biggest gossiping bitch in town and realising this will lead to direct link involving a phone call by informing of your wife within seconds and knowing there will be hell to pay that evening while the truth is that you truly never even noticed how beautiful and smart and lovely and sexy and gorgeous this young girl was. Any thought of her appearance never crossed your mind for one tiny second. In another case your unintentional looking at a girls legs as she gets up and finds herself in a very embarrassing position for that split second while she is glaring at you for being such a dirty old man. Your looking down the blouse of a very breasted beautiful girl as she was bending over while your wife has caught you with your hand in the cookie jar. This is every day incidents with no intension of ill on your part but happened when you did not have full control of that situation for one instant. It all happened by accident but you believe me the deed was as deliberate and intentional as any of the cases I mentioned. It could even be as serious as your kicking a business competitor on the shin while you slipped and almost fell. It is the one character shouting your innocence as the other character is calling on your record always showing good manners and polite conduct while the third will never let an incident slip by.

This can and does even go as far as a nation. I am an Afrikaner Boer and being that I am not blind for my people's mistakes. When four Boere gets marooned on a deserted beach of a desolated island far from any other culture you can rest assured that within the hour of landing between the four there are, they would have started five different Christian denominations and six different political parties. It is a well-established historical fact that during the Anglo Boer war at the battle of Ladysmith the Boere had twelve thousand able generals with not one soldier amongst them. That is quite typical because we listen to God through His Word and no one ells. This last remark does not exclude our personal characters acting in the capacity as generals when promoting twelve thousand times three different opinions.

....And then there are the Bible punchers...the ones that will convert you weather you need converting or not from ....well seeing that they never met you before it does not matter what faith you may hold, because they hold the opinion that only they can bring you absolution because they not only personify what ever they believe their god is but they see themselves as the direct extension of God, a finger or hand of God controlling life on Earth.

They are hoity-toity, overbearing and haughtiness rolled in one god given container forming the BIG They. They can recite hundreds of Bible verses for minutes on end and that they believe is the key to their absolute presumptuous claim to God.

They walk with God and they talk with God and they discuss with God matters of mutual concern giving God advise where needed and as they see fit and where God is in their opinion straying from decisions taken at their previous meeting.

I am not referring to the normal God-fearing pious person that goes to Church and feels his thirst needs quenching on Sundays. I have no rite even to discuss any person's religious thoughts and belief. What I am referring to is religiosity to the extreme, a mental unstable drive of laying on the hands to heal, involve every one in religious debates every second of the day, starting to pray out loud as to draw attention of all persons around that should observe how their closeness to God has become as they are in constant prayer and will grant you some time between prayer because you should remember, they are keeping God on hold and the line is busy. Only they have the rite to prayer because after all when ever they get hold of you they wish to pray for you as if you have no connection to God, God has only given him a direct line and all others have to go via the switch board and wait their turn if they get a turn. Normally and in most cases, actually always I let them be but times arrive where they interfere so much you have to put some perspective in them. When I take them on issues their argument has the same logic as that of a pregnant pig and I have learned not to let them off easily. You press home the point and make them as big an idiot in front of as many as that wishes to listen, and destroy their mental thinking ability for the rest of the day. I have had situations where they tried to run away from me and I would run after them taking with me as big a crowd as I can possibly gather at that moment and destroy their image to shit. After such a session they are normally so annoyed with me they ignore me flat where I then leave them alone. I would never do that to other people for no reason can be important enough to humiliate your fellow man. But with them they leave you no other choice.

Not once and I repeat not once in my life did one person ever come to me with the introduction of: I come to you in the Name Of the Lord, and that bastard did not cheat, swindle, rob me, or steal from me. I have reached a point where I decided that should any one in the future introduce himself and refer to "I come to you in the name of the Lord" I'll chase him from my property like a bad dog. They are the biggest crooks and con persons walking on earth.

They too have a massive trio rage where their characters use religion to go out of control knowing very well all people will respect God and their referring to God always bring the other person in obedience. That is when they hit home with the most devious cheating and swindling you can imagine. If someone of their likeliness offers to pray for you don't close your eyes, grab your purse!

They're knowledge of the Bible is astounding but they know nothing about the Bible. They learnt a thousand or two thousand texts and recite them with speed, throwing the one after the other without making sense or allow the meaning having any connection. That is only an eye blinder; a way to astonish you so that you will lower your guard and that is when they hit home. There is no such a thing as a free ride and they always want from you ten times more than what you are prepared to give and when finished with you in the very last paragraph of the small print area it is only all about money but of course "in the Name of the Lord".

You wish to spread a gossip or any untruth, well be sure to use their channel. Normally those services they provide for free but then it must be juicy, unfounded and completely void from truth. Their trio works on the basis that the positive character takes them to personify God, therefore they can do no wrong in the eyes of the Lord. The neutral character advances the notion that they may convert you and that may be useful in some future schemes where you then can fit into more devious plans while the negative character is of the opinion that by your not believing the way they do, you are doomed in any way so robbing the convicted bears no shame. God put soles like you on Earth to be useful to their likes where after you will go to hell anyway so what the hell, they might turn some profit before you meet your final demise. After all if you cannot see the light they give you, you may as well be blind and being blind you don't need more than what you can see. By them taking from you and giving onto them they are receiving with a self help scheme what the hand of God on Earth deserves and where you are going to lose everything by your departing to hell you might as well start at a point where you are still useful to their blood sucking. Should any reader not believe me take some time and start discussing non religious issues with intent on your part to learn some angles their characters maintain as informed opinions. They are not hard to find. As with all criminal hoodlums hanging out at places of criminal conspiracy they normally hang around at the tents of the evangelistic preachers commonly referred to as the "Happy Clappies".

When I shared time in clinics with persons having some psychological problems this was the beginning of my theorising in this direction. I wish to state once more it is merely an observation of a layman fighting to analyse his own condition and took time to see where similarities were between different humans in the same boat that was sharing a mutual difficulty. Some were alcoholics where I am a teetotaller but still behind the condition I came upon similar causes giving one person one crutch and another person another crutch but it is the crutch one has to loose and behind the crutch you have to find the pain causing the person to grab for the crutch. What ever I share must be taken as not even an informed opinion but as merely another opinion where we all have opinions and it is worth the while to share opinions of an assortment and a variety.

In the last part I indicated persons holding the righteous views about their religion to advance and use as an excuse for their almost and sometimes definite criminal behaviour and malice intent. On the other hand I have seen suffering where these characters dish out what no one can bear. The agony and torment some people go through is of a much higher pain than I ever suffered when I came off my bike. The pain is more real than physical, the fear is stronger than death, and the confusion is louder than not understanding. It is horrible because some of them feel the anguish moreover than they would if a genuine murderer was chasing them. With a genuine murderer you can try and escape or hide but in their suffering there are no such luxuries.

The way they suffered and reality of their hallucinations had put the fear of God into me and made me more than willing to get over whatever small difficulty I had because my luck is worse than the Irish. (To my mind no one can have worse luck than the Irish because they got themselves in a spot on Earth from all the places they could chose next to Brits, where the forever meddling and interfering bossy Brits is occupying the very next island) With such bad luck going my way I might just find my problems increasing. I had electro convulsion treatment on several occasions and to my opinion the treatment has a healing affect as the neutral character loses some dominance with the loss of memory through the electric flow. The patient then loses confusion by gaining perspective where the dominance of the entities reduces. As the generating of electric flow brought about by the life factor decreases electric tension in the brain the location of the entities become affected as well

Gravity, electricity, time is all the same thing and in a more or lesser manner influence life and moreover life in the brain. By reducing the electric tension the mind stimulates and as the convulsions allow electricity to escape from the sells stabilising the brain activity and helps sorting the influence the characters have on the person. I must admit that directly after such treatment I don't feel such an excessive urge to speed. Fortunately the condition normalises quickly and I can get to enjoy the exhilaration of my crutch once more. That proved that influences of such characters do vary and can be in dimensions of interpreting and it seems that influences from outside sources can be a major consideration

You may believe it or not considering my poor academic background but I have an inquisitive nature and a need for mental stimulation by acquiring facts and information and that was my academic downfall. My positive character always urges me to test another person's knowledge base and interpretations of facts. This was present even as a scholar as I did forever test my teachers. My neutral character will then classify him in filing order typifying and classifying ranging from brilliant to shit where my negative character will fore ever test the teacher in relation to my personal abilities and from that stance supply the necessary admiration or animosity. I never allowed my teachers at school to escape and when I became suspicious of their depth of knowledge hell would be upon us, moreover on me because they had the cane and always knew how to use the thing. However in cases where the teacher became a source to quench my thirst for knowledge I would eat from his hand. My positive character would take charge and push the others to a silence where no one knew they existed but in the other events of me growing suspicious about the teachers abilities the negative character would destroy any form of harmony that may develop between us and that was the normal in all but a few cases.

In primary school the teachers thought my behaviour was cute but in high school it became intolerable for all parties concerned. I make this remark to indicate that outside influences does play a part in younger minds and through positive stimulation the influences on the characters can be directed to a positive outcome for the child. No one lives in a tight cased cement container never to have an ability for change. It is the duty of the teacher to recognise and direct the children's interests to the benefit of the child and that could lead to the benefit of the class.

I established my theory with all relevant information based on my personal case. While this was going on I also realise there are nothing about me being exceptional or unique and if this applied to me so would it then apply in other cases. With a clear objective I started discussing other people's situation with them to draw similarities that would match my case. It was similarities I was after and not parallels so every time I got behind the whole issue by befriending the person and in that way I could establish a confidence that no professional could. As there was no malice intended on my part and I did not brief the person on my theory or tried to offer remunerable advice I could see no harm coming to anybody. The information was never brought to paper establishing personal files and since gossiping is not one of crutches no information of any private nature slipped past me. If ever any advice came from me it was certainly not on the grounds of my theory so I could not harm any person in any way.

But the more I came involved with other persons the lesser the importance of my personal issue became and the more I detected some golden thread running along lines undetected. In some people I could even detect which character reigned supreme that day by remarks they made or the moods they had and in limited cases there was differences in facial muscles, as the mood swings occurred. But in the Afrikaans book which up to now a very limited number of readers had access to, this is the first time I went as far as mentioning my observations to any person. I can even remember the precise moment the light of understanding went up when a psychiatrist Dr. Steenkamp, which is still treating me explained about the mind and the free will of persons' personal thought, the way a person react on they're dissensions and the total absence there are of demons or other spirits that may influence the mind. I state this categorically I have never commented about this theory I have and least of all to dr. Steenkamp so I do not wish for any person to conclude he or any certified medical practitioner had any personal opinion about my conclusions. I merely said this because only a few incidents stand out in my life as very memorable and this was one such moment.

What I found was that it was as good as a human trademark apparent in every one, slightly more apparent in some than in others, they are in every one all the same. The entities are mostly absent but come in when a person has his or her guard down or when a person has emotions with an influence stronger than the person's ability to control. All actions man make is with intent. Some might be under the guidance of one or more of the characters but every one has full control over all their deeds, without having an excuse for conduct. They rule your life as much as they are you and will promote your true intensions whenever you do not wish to. Fighting the characters is fighting yourself but you can and you must find a way to recognise them because their intentions are to harm and never to uphold. The characters come with certain emotions bringing along certain feelings. The feeling may be an excitement that does not match the situation or an anger that does not fit the occasion but if one is vigilant you may catch the feeling before the feeling catches you. It may be very slight in irritating handling like for ever pressing the wrong "t" being the "y" on the keyboard or turning a cup of tee over on some important guest or having a dislike in someone you never met before and should not have any special opinion about.

One incident to try and prove my point I wish to raise is from my personal recollection and I wish to share it in order to avoid other peoples' affairs. I suffer severely from acrophobia. Putting me on a double-decker bus is about as high as I can go. In all sanity taken seriously one cannot have acrophobia. One cannot be afraid of heights when steel bars inches thick will prevent your falling. One cannot be scared to look down a glass window when it is closed and you cannot fall through. It does not make sense and yet I can assure you it is a fear greater than the mind itself.

The fear is irrational but should any one try to loosen my grip once I grab onto something he is not only endangering my life but he is seriously messing with his own life. I am aware of the problem and believe I do have a rational mind until it comes to heights. There is no thought, there is no reason and there is no arguing about matters. It is instinctive irrational animal-like behaviour where I go into a survival mode. I cannot fly and yet I know flying is the safest form of transport so much in fact it may be a thousand time safer than my cars or bike, but that is the rational and it disappear when I look down and see something small down there realising it should be big. Even just the realising that I am about to leave the Earth is more than I can control.

The fear is almost if not an obsession and becomes uncontrollable. Then one day i stood on an exceptional tall building (well exceptional tall for me a person coming from Ellisras the true one horse town in the middle of a semi desert) of about eight to ten storeys high. I did it purposely to see what the emotions was accompanying the acrophobia because the attack must have some prelude. It can't just hit you like a brick that is nonsense. Something has to form a fore play, a sign of what is coming. Even if it takes one second it still is there. Nothing can just overpower the mind instantaneously but every thing must be about a collective of factors and facts coming together.

Coming out of the lift as I was walking towards the corridor where after entering it I could look down for the first time I intentionally was waiting for what ever to come first and announce the shock. Then as I came to the open I felt the feeling of fear but it was first another fear, one I was use to and knew. It was one of the characters coming to the foreground as if called. That made me realise that it was not the heights I feared but the negative character. I feared the character might take control and make me do what I did not wish to do. The fear of the heights is there, and that is no maybe but that is an extension of the problem and not the problem as such I may not fear the character and I may feel uneasy about the heights and it could even be that I become insecure but the problem was outright the negative character coming to the front.

As my dominant personality losses confidence the negative character comes in. It is not a case of him pushing me or my jumping but it is something going on in the realms of my mind where I do not understand all things all the time. It was a fear of what I may do to myself and not of the heights. Under all of this was the presence of the negative character lingering almost like a shadow feeling not present and not absent but just there. Then came the shock of the actual height and all logic flew away like a little bird. I was clinging and grasping for dear life.

If I can take my mind back as far as I can go back my very first recollection I can recall is a scene where I was on this huge tractor and it was far down. I was definitely under two years of age because I know which farm it was and my Grand Father sold the farm in the Free State that had the tractors before my second birthday. When we moved to Tzaneen he farmed without tractors so it was definitely before my second birthday. I was on this enormous tractor (enormous because of my youth) shouting desperately as I was crying hysterically for help because I remember the thought that I had no chance of getting off that tractor all by myself. I do not know how I got off or who helped me off and being where I am now I must have gotten of because I am not on the tractor any more but that day my negative character and my acrophobia met and got mates. Of that I am sure and if I am correct, then parents should take care not to scare their children in an innocent prank of unintentional fun with their child's fear. It could have lasting consequences to the child. It is not the heights I fear but it is the character taking control even if I know there is no chance of that happening still there is no logic as far as the phobia holds ground. If that is the case with me it should be the norm. People are not scared of objects because objects hold no threat and every one knows that. My humble opinion is that one of the characters is dominant and the dominance is so much that the person feels threatened by that character. The character is in control so often through a depression or an anxiety or a mania of sorts that when situations arrive where the scene should be normal fear becomes the norm.

The negative character brings the threat the neutral character bring the warning about dangers and the positive character joins in by bringing the fear. The positive character brings the fear in to dislodge any attach the negative character may launch. The positive character on the advise of the neutral character disables the body and disables the ferocity and fierceness of the negative's dominance. By pumping adrenalin the body goes numb, the legs go weak, the arms shiver and the body has no strength to function while the person who is in the middle of the junior civil war see object as the reason for the attach but the object is only the trigger and not the show.

I too, am of the opinion that these characters and the way they perform their balance on the day and in the situation makes the hero or makes the coward, depending on the balance at that precise moment and occasion. Phobias connect to the negative character and mania to the positive character and by allowing un- protection through some situation triggering the mini civil war inside the mind; the person becomes a bystander where the person should be the controller.

A kleptomaniac may put something small in a bag and swear by the fact the kleptomaniac did not know about the action. That may be as much the truth as it is a lying because the actions was not deliberate, but the actions were intentional. It is easy to ignore the compulsive behaviour and claim non-participation when participation may have been semi unintentional but still enjoyable. The one character may distract the attention of the person but it is done with full participation because in the end responsibility is with the person. One would often hear the remark: "I knew it I knew it was going to happen" when something was going wrong. All humans can read situations and your mind told you something in the situation were desperately wrong. While the positive character does warn you of events coming it is a deliberate action to allow the neutral character to distract you while the negative character can play for time for whatever occurrence to take place. The whole scene was a deliberate action by the person to gain a negative outcome to produce some suffering or hard ship to some degree because that is why we are on earth. But I shall get to this last remark later on.

This takes us back to the bleeding heart baying off guilt by paying the philanthropist to collect on behalf of the Hoggenheimers dishing out to the Mammonites paying the Mammonists for some slave driving. The actions are deliberate but the true intentions are deliberately unintentional. We are bullshitting our conscience for gaining our mistrust. It is the lye of culture and all participate but some participate to a degree that does not please others. The degree might be to some extend not serious enough to bring commitment and the persons would stand on the side line and criticize without direct involvement because of fear of own guilt uncovering or even of a want to participate while others would come in and rescue but not to save but out of spite because of personal yearning for participation that the person knows would not be permissible.

Another scenario is where a person is drowning. The rescuer comes to save the drowning victim. The victim is exhausted beyond normal mind control. The lifeguard reaches the victim whereupon the victim tries to drown the lifeguard. The victim has lost rational thinking and is then in a mode of action versus reaction. The positive character grabs and clutches at the lifeguard in anticipation while the neutral character tries to survive the conscious and the negative character wants to save the situation by taking the lifeguard with the drowning effort because after all it is the responsibility of the negative character to destroy and destruct as much as possible.

Some gave these characters names. The negative entity goes by the name of a death wish. The neutral character goes by the name of don't care. The positive character goes by the name of optimism. By naming them we found once more a way of avoiding our duty to recognise. It is much easier to dismiss than to admit because admitting has to lead to prevention and prevention is no favourable option.

We say the drowning person got panicky but that is another word we use to escape from reality as much as an excuse in avoiding responsibility. By being panicky we deliberately excuse behaviour and responsibility about actions we may commit to explain irrational behaviour. The negative character wants to punish the lifeguard and even make him pay for his life in interfering with a situation the negative character is enjoying thorough rely while we others only recognise the efforts of the positive character because that will be the nice thing to do. The next time we behave in the manner as to kill our saviour we too can be acquitted on grounds of incompetence. The action of trying to kill the lifeguard is as intentional as the trying to grab onto him to be saved and as intentional as loosing control through the neutral character.

When I first read about the Lovejoy syndrome it brought to my attention that not all mortised control of the body is in the domain of the person all the time and sometimes there are some part of your life that can take control of your actions when you are not in absolute control. It is all a mind game you play with yourself in diverting responsibility with the compensation of enjoyment but the avoiding of dismay about yourself. There are no excuses because the final responsibility is in your power. It is in your power and much more even it is your birth duty to fight the characters but the moments you lose the fight you take the responsibility for the actions because you momentarily lost the fight. It is a win lose situation where only you walk away with the prize as much as the punishment. The muscles are under your control and you are in charge but the entities are little pests being you and are thorough testing you by grabbing control whenever they can. The entities are not only and absolutely negative but are positive as well. We have all been through situations where we admit afterwards we came through by the grace of God. We always hear some one remark that he or she does not know how they did "it" but "it" came through far better than "it" should under "normal" circumstances.

All our phobia, all our desires and all our hopes in achievements we pin on luck or the role of the dice but luck and the role of the dice has nothing to do with it because or achievement good or bad as our accomplishments wrong or correct, and our thoughts being acceptable or not is within us, in our control as much as it is us.

A paedophile should be hanged from the nearest tree because he gave in to the want of the characters and not because "he is not in control of his actions". He wishes for the characters to take charge and even deliberately set up situations where they may take charge, because he enjoys the dead as much as they do and more because they are he. They are in all of us but for some certain behaviour are unacceptable and for others it is not.

The judge bringing judgement knows in his grain that the molester will commit again yet he sentences the criminal to a few years of incarceration and is fully aware that their is no chance of rehabilitation because the culprit does not wish to be rehabilitated. He will find his rehabilitation as a death sentence because molesting children is keeping him alive. When the judge do not bring the death penalty he, as much as the culprit participates in the next cycle and therefore must take responsibility and participation in the next round of child abuse. But the judge sits there with the idea " there am I but for the grace of God", which is true in a way but also is not true in a far bigger way. With my fast driving I do not wish for a cure, but I know something is going to go wrong somewhere some day. That is my chance I have to take. He upholds the same argument and goes to the molesting because the sentence is the chance he has to take, and is the chance he takes. You can bet your bottom dollar that should I know before hand the next road race would kill me I would not participate, and the same goes for the molester.

With all certainty that the hangman's loose is waiting he will have second thoughts about his next molesting session. Murders always fight for their life by fighting the death penalty. The underlining is that there is no black white and grey. There are no clear-cut defining borders and sides. The neutral character can be as destructive as the negative character can bring a positive out come. When the mob comes out to lynch the actions of the mob are negative in lynching but as they do not wish the continuing of the criminal's behaviour it becomes positive when doing the demonstration. That it will bring conflict to the child's guilty feeling and in that sense their actions are neutral to the child, the participation is positive in preventing the repeat and there by positive in the negativity of lynching where the police prevention is negative by protecting the paedophile and that is positive by upholding justice as it is neutral by delaying another criminal's relapse in crime because relapse he will.

The atheist does not go on a disgraceful child molesting campaign because he thinks there is no God and if he does not get caught there is no punishment. He avoids indecency because he is human. The paedophile may be the biggest Christian around but argues that since he is only human and humans are sinners and sins are alike he can maintain his behaviour until judgement day. Such a line of argument is very typical of the trio being in charge. The responsibility is always with someone or something else and the person never pin it to specifics but shifts the blame to wherever is convenient. And so does the Judge! Hang the bastard and judgement day to him the sinner will come a little sooner. If he does not wish to control his characters help him by eliminating him with his characters. By molesting the child he starts another cycle producing one more child having little control over the characters the child will fight when he is an adult. Stop the violence by stopping the cycle by stopping the one not controlling his characters. If death waits as a surety he will mend his ways or seek help to accomplish change. It is easy not to change and difficult to change but we all can change.

One may have be opinion what I refer too is about the good and the bad, about the saint on the one shoulder and the devil on the other and think well…yea I've heard that before… But it is much more than that. People stop at serious motor crash sites not with the intension to help. If any one admits to that that person is untruthful. They stop to feed the urge. People watch blood sport, not for entertainment but to feed urges hidden deep within the mind. When an armed robbery takes place with possible killing spectators run to the scene. When there is a fight on the schoolyard the word spreads like fire and little else can generate more enthusiasm. The most brutal serial killer always receives the most male. Woman would throw themselves at the criminal misfits with marriage proposals; coming up with the excuse they have enough love to concur the beast's evils, but that is a hideous lie and they more than any one else know that.

They wish to share in the darkness of evil, find someone that could lift the veil covering the beast within. Stopping at the scene of a bloody accident evokes prime animal senses covered by social upgrading but is still very much lingering within. It serves as a reminder about the days when a feast came from such human blood. At some time all of man was a cannibal, man-eating monsters that feasted on the flesh of the enemy after a conquering battle brought victory to some and victimisation to the others. When faced with starvation it is a normal sense that kicks in where people will start eating human flesh. Shocking, as it is… the shock is about realising the urges more than revolting.

We all have the darkest desires of committing unspeakable atrocities and most barbaric acts that will shock the normal mind into panic and frenzy. Some people dislike blood sport, not for what they're purist of heart tells them to reject but the rejection comes from the craving they fear. They fear the need for such beastly acts may linger and run out of control. A story about a mass murderer is bestseller weather it is fiction or truth it is popular. It sells and there is a valid reason for that although our minds reject the reason.

When thinking about these crimes we put it in the basket of the negative because it is where it belongs. No one can ever be positive about such behaviour and with that we create fabrication of truths we wish. It is as negative as it is positive…it is neutral. The good / bad character is not the danger; it is the neutral that is dangerous. It is the neutral taking our minds on fantasy journeys. It is the neutral character we most easily identify with and mingle with. It is the neutral character setting our morals. When Ted Bundy killed the many woman that he did with the brutality that he applied he was not negative, he was super positive. To the woman "forgetting " to lock her door, walk alone down the dark ally at night, stopping to give a strange man a lift or hitching a ride the deed is not negative. They are super positive in their expectations of what may come to them. Sure, they do not ask to be raped, or beaten, they do not beg to fall victim to crime, but lower down in their minds they know of such a possibility and very, very deep, deep down they stand neutral to such an event because they do not mind the excitement or the sympathy afterwards. The bank robber, the pickpocket, the shoplifter, they are not negative they are supper positive. It is being neutral that is the negativity.

The animal cannot see the Universe from any other position than the one the animal holds. Singularity placed divinity in the centre of his Universe and the animal has no means or brainpower to translate his mind to another position other than his centre where his needs and desires are. From there comes the neutral holding positive or negative as well as the he. Negative means he should fear and positive means the other should fear. In this religion can be one form of atheism and atheism another form of religion. Atheism is the inability to see anything except from the centre of that person's Universe and the inability to transform to another position seeing it from another viewpoint other than that persons Universe centre. When Albert Fish sent the parents of the child he cannibalised a letter informing them how he devoured their child he was positive and the deliberate pain he caused was positive. Peter Kurten the vampire of Dusseldorf only became neutral with his last victim, but he always remained positive to his wife so much that in the end he forced her to betray him so that she could claim the reward. By her collecting the bounty he realised his final act of being positive as he always felt about her…but only to her and so his final offer he gave her was that the bounty that went her way. To him that deed was as positive as his killing and eating of three children in one night was positive because he was unable to translate his centre point to their position. That is being animal in every sense and animal on two legs. The Boston strangler was always neutral working his way to becoming positive for days although that meant the utmost negative to his victims.

Being human and religious is the understanding of other concepts forming a centre way beyond your centre of the Universe and moving away from the animal neutral to a human stance Not eating the flesh of the sheep does not in any way make you more positive but it makes you too stupid to see the universal picture of the totality of Creation and its wider meaning. That is trying to prove you are not the animal you know you are but cannot seem to separate from. In the neutral character is the one that no one propagates because that character is much to close to us to be comfortable with, that is the animal, the beast or civil the morally accepted or rejected but still nursed by all because of pre man mentality. From that the other characters stands positive or negative but the neutral character holds the centre and the key to light and darkness in the mind of man becoming man. Claiming the position all other characters including the self occupies a part of the mind. The neutral character sets the tone and the rest will follow. In the neutral character lurks the animal and depending on the person, the darkness of the animal comes to the foreground or stays in the back ground but is forever present in the mind. Identifying with or identifying the neutral character places us in the realms of man or animal. Setting us apart from the neutral character is what produces man and not animal. In the same manner as not eating sheep is the saving of one criminal life where every one involved knows that criminal is destroying dozens of future lives and expanding the problem by creating many future criminals where they then would create some more criminals at a ratio of a dozen to one. In three generations the one criminal is then the cause of hundreds paedophiles walking the earth. That is degenerating civilisation and it is all because the law enforcement from politicians and judges through the lawyers and civil servants down to the bleeding heart and the cop on the beat that wish to prove them not being the animal that the criminal is.

By creating the environment and breeding ground for the animal and forcing their positive ness in neutrality they destroy the future of the following generations of man. Law not taking blame for their actions in the neutral stance is as much being a criminal as the criminal's neutrality creating his positive to negative relation. The law enforcement' officials excuse for not wanting to be as bad as the criminal is more destructive and a much bigger misdeed to society than what they do to the criminal because after all it was the criminals free choice to commit or not commit and as much criminal not to wish to pay for his deeds. Either we teach the criminal there are billions of centres to the Universe or remove him from our centre but allowing him to become forcefully our centre proves as little as it is destructive to every body. But by throwing him in prison where he shares time with others the same as he accomplishes only that law enforcement may promote crime to establish more cycles for their and all the other criminal's benefit but to the disadvantage of man in general.

The paedophile or any other criminal or anti social behaviour is quite rectifiable where psychologists must teach the criminal in recognising the incorrectness of behaviour, the recognising of the characters and the control of the characters by thought control. Criminality starts with a thought and that is the point I started in my discovery of the characters. Being suicidal as I am and that being my problem starts with a thought starting with a feeling leading to a depression. I am never suicidal on my bike or my car doing high speeds. Then there is only the positive character with the neutral character keeping guard for traffic control. Never do I at any stage exceed my personal limits or endanger lives through recklessness by outsmarting my personal ability where my mind works at the speed matching my vehicle and that is the secret to success. It starts with a thought. It starts with a feeling. That is the gate to progress the characters follow. The remedy is striking a link where one will recognise the thought sparking the feeling and recognising the felling sparking the thought.

There is countering the feeling by thought as much as suppressing the thought by feeling but the BIG issue is that you HAVE to know yourself. The alcoholic feels the urge for booze as much as the thought for liquor but when recognising it he counter acts and that is where alcoholic anonymous has such a great success. If he did not wish to recognise the urge as much as he did not wish to suppress the thought alcoholic anonymous can do little. It is the difference between cure and tramping. But it is fighting day and night and fighting yourself knowing you are in the fight for your life for the rest of your life. It is fight you can never win and must never lose. It is continuous round after round with no victory and no defeat, no prizes and no glory. Only shame to follow defeat and the winning part has no recognition for effort or accomplishment but to you. That alcoholic fighting himself and finding his determination is the human victor. To me staying neutral is staying alive. That is what prison should teach the criminal in recognising thought and control thereof. The big courtroom confessions these murderers make are about showing the world how big their Universe are, its about how they are the centre of other's Universe and the control they had in destroying the others with their Universe but it is never about apologizing to show they can understand what others have in the Universe they destroyed or that in fact are others with an own Universe to have. Should he think he can outsmart be unable to rehabilitate he is animal and should become destroyed like all other raging animals being out of control. He should therefore fear death or find death. In the modern penal system rehabilitation is a word used in the courtroom with no other place to have. Life should be about improving and not sustaining.

Alcoholics and all drug addicts lose the neutral character when intoxicated. The person becomes super positive loving the world for the world loves him right back or becomes super negative by getting aggressive or drunk with remorse hating his parents for what they did or never did. In this the addiction of the parents always plays a part when the child of the alcoholic also becomes an alcoholic. When the addicted becomes sober the neutral character takes control with vengeance, as the addict needs the next round of substance. To the neutral character surviving means finding the next round of intoxication. There is never a balance and when the positive becomes positive, drunk or sober, it shares a spot with the negative and of course the other way around also applies. The drunkard is prone to mood swings that is apparently out of his control, but that is a fairy tale. When sober the same applies, as his moodiness seems to follow his every move. The drunkard uses the substance to escape from himself and his own adequacies he feel about him in his self-portrait and therefore allows the characters a free hand when intoxicated. This we in South Africa call Dutch courage and will have different names in different regions but all names apply to the same behaviour.

Things are normally not as serious as in the case of the child molester or the mass murderer turned cannibal. On average it does not have such deep and intense underlying emotions and fights to the bitter end. It could be a case of the young man is meeting his in laws to be for the first time. The dinner is in great preparation as it is in great anticipation for all involved. Every one but most of all the lover-boy and soon to be in-law is under pressure to impress. He is the new face in the family having all eyes on him because every one knows all the other faces and he only knows his face, which he cannot see in any case. He is out to impress and is tuned for this all out effort of do or die. In this effort he relies without relying on some help from the characters because he can use all the help he finds. He lowers his guard completely to allow the positive character unhindered passage in the situation arriving. In the background lurks another character that is far less dominant in the whole affair.

It will take but a flash in a moment for the negative character to prove a point but as he opened all the doors so wide for the positive character he now is even positive in being neutral. With all the tail wagging he has to go through his negative character takes little appreciation in any possible discomfort on his part. Lover-boy is yearning for acceptance to such a degree this making of his all out effort is rather new and strange for the characters as they show an all out help line helping by the full range of their individual abilities and to use the chance in such an unhindered open channel participation. But one is waiting his chance in quiet anticipation. At the moment of climax where father in law to be wishes to make a toast will be the moment all three characters join the fun by creating the incident all three have been waiting for all night. It will be to the determent of boy-impressing of course. The negative character is very aware that the neutral character may prevent his planned action there fore the strike is lightning quickly. The neutral character sits lapping up all the attention as the positive character is all out helping with the impressing of all around. With every one well occupied the chance comes for the negative character to make his point for the night since everybody was having fun and he had to sit idle and unwanted.

As the tray carrying the wine glasses, which are filled to the top, passes by lover boy, the negative character motorizes the arm closest to the tray and hits it with force. The action holds the speed of a boxers punch and with that lightning speed lover-boy never thought he had such quick muscle movement. The action reaction reflex action reflex reactions is beyond the abilities our boy to be married ever dreamed he had. He is totally surprised at his quick ability in movement and that to his knowledge is miles ahead of his ability to move. In all his life his arm never moved that fast as he turns the tray over on the lily-white table linen.

With him being totally out of place and out of sorts he cannot recognise his ability and that is what his positive character ensures him. His neutral character will be very embarrassed and that embarrassment he places in lover-boy's private embarrassment about what happened. Shouting and proclaiming accident by him as well as the other two characters will bring all members of the in-laws-to-be under the impression that this was indeed an unfortunate incident completely convinced by his very genuine embarrassment. The embarrassment he proclaims are partly his but mostly the embarrassment comes from the other characters proclaiming innocence to him about their involvement and misuse of trust. This he uses to further his embarrassment to exclude all blame of deliberate action on his part in order to impress his in-laws-to-be.

Every one present will feel deeply sorry for him except his negative character that made the point that all the licking may be for tonight but he is still his own man and will have is own way in the future to come. The question is was his innocence really that innocent and was his embarrassment truly that hearty? On the first count no, he was just as deliberate in the overturning of the tray as all his other actions was though out the night and innocent is the last thing he can be guilty of. On the second count, well yes, in a way but not only for all the reasons he proclaims. No one can ever proclaim innocence and non-participation in deeds the person commits. All your actions are all your responsibility. No excuses can be maid without lying through your teeth. Being born means the fight is on and the fight will continue till your last breath is wind. Having trinity around is your birthright and fighting is the option you made before birth and not after birth.

Life of man is about fighting yourself with all the vigilance you can ever muster. That is why you are here having the time of your life for all your natural life. It was your inheritance the day you were born a human. It is not all about negativity but it is all about achievement. It could be quit within your self and it could be in the presence of thousand of spectators. Every sportsman has "on" and "off" days and there is such strong emphasis on the physical that by training the physical no one notices it is the spiritual in training. The spiritual always is the physiological but all preparations are psychological. Practising is about training life to manipulate space-time to the best affect. In all it is only about the physiological. Training is about telling the muscles to obey command and telling command not to obey the muscles. When the muscles shout in agony to stop, command must turn a deaf ear and when command shout to the muscles to go on the muscles must be like a dog and react without questioning or arguing. To describe this we use the name fitness. It is the conditioning of the flesh but it is much more about the mind having control over the body and all fitness is about life commanding the structure in occupation to do what life wishes to be done. Fitness is moreover about controlling the characters than the body. You have to control the negative character's destructiveness to be about the opponent and not about you. You have to control the neutral character to dismiss outside interfering with your efforts. You have to control the positive character to bring subduing confidence. You have to use the fear to your advantage in believing you are fighting for your life and not merely a trophy.

Every sportsman has a story of "absolute brilliance" and always the remark is about the sportsman not truly believing he had the ability in accomplishing what he did. When saying that the sportsman only considers the physical aspect and never the mental drive that pushed him beyond the limits he accepted. On another occasions it is the very opposite when the sportsman declares the day as a disaster because notwithstanding an all-out physical effort nothing went according to plan. His muscles did not respond, his legs were stiff; his arms were not in synchronisation or what ever the excuse for the disaster is. When shove comes to push it are the three characters we find again behind the success or disasters.

The apparentness comes through by him not admitting his efforts links directly to his state of mind and the blame goes to external factors. Everything went just rite or just wrong. It is everything that holds the responsibility for his success or his failure. It is external forces at work and in charge of his luck or bad luck. With that remark he indicates his absence in his actions. His lack of admitting participation and his unwillingness to claim success or admit failure proves that he is relying on something he feels he has little control over.

We all admit that when we are tired we make mistakes. When we lose concentration things go wrong. When being absent-minded we make accidents. That is admitting to the role the characters play. Being tired means relinquishing control, letting go and then the character take control. But also when in fear of one's life you find yourself in super control where you are miles better than ever. That too is the characters at work. It depends on which character takes charge and to what degree does the character take charge. I always teach my sons never to fight a man in front of his girl because you face a man much better than normal. If you are winning a fight leave the other person a way out to escape and never allow the impression the person has no way out. As soon as the opponent gets the idea he is fighting for survival, or he has to fight for position in the tribe in example for the favour of a female you have a monster on your hands and in that case be prepared to fight for your own life. Never allow the idea to enter the opponent's mind he has to win or ells... that will be to your determent. In the instance where you do not allow an escape route or a man thinks he is fighting for a female you have a raging bull and three characters to fight and you do not wish a fight him with his characters on his side fighting against you. That will be your death you wish upon yourself.

It is not schizophrenia I am referring too. I am no psychiatrist but as a complete novice I do not believe in schizophrenia. I do not believe the person is hearing voices from beyond because there can be no voice of beyond. It is his imagination and his characters playing mind games and if a psychiatrist or a psychologist come in and join the fun admitting to such a scenario. With such outside help and sympathy to help the dreaming along the characters then can and will come out to play. The "split personality" in the fight is within every body and can go into rage whenever allowed to do so by deliberate actions, provocation by other, tiredness, fear but also delusions, that is true.

From the article so far one may tend to get the impression it is about big issues like meeting you're in laws the first time but it is not. It is every day all the time things. Sitting in traffic waiting for the light to turn. The positive character slowly shifts to neutral and the neutral slowly joins force with the negative that was in the background all day. Without the person noticing the whole situation within him shifted in frame but he is unaware of it. Suddenly an explosion fitting a war burst to life. Another motorist sits daydreaming with his neutral taking him on long trips because his positive has nothing to do and his negative is in the background.

The light turns green and the daydreamer is just that tad slow in responding because he was in thought. He was in thought as much as he was on Jupiter. His negative character helped the neutral create a situation the negative character can participate and not be bored. So his slow acting is as deliberate as his breathing is and the other person with the shifting emotion sees this, recognises the stunt that the daydreaming motorist is pulling and with the positive character standing on neutral ground and neutral fully in the negative territory, the negative character comes in with a punch like none. Suddenly two very timed persons go into a rage because of the smallest incident. They do not recognise their own behaving and if they do not get their characters on a leach quickly the characters of both men will take charge and blood might flow. This we call traffic rage. It is a very convenient name.

The man sees some one he may regard as rather attractive and smiles at her. The positive character becomes embroiled in the proceeding while the neutral character joins by taking control and urging the man to step just a tad closer, just to see… while the negative character is anticipating rejection in any event and deliberately steps on the girls' toes. Embarrassment is all around even with surrounding crowd because somewhere in the back they in the crowd all no what happened and that brings the embarrassment to their door. Every action on all accounts are very anticipated and pre-arranged. We use the name of an unfortunate incident to file this under. It's about calling your wife but using the neighbours' wife's name. It is knowing you have to cut the lawn but the drowsiness just will not let go so you sit down and close your eyes for a second, just to relax for a second. The one character shifts one position while you were not attending procedure and you miss the opportunity to mow the lawn for one more week. This goes by the name of slipping the tong and nodding off.

Very typical of this is the driving absentmindedness. How many times did we sit back after arriving at our destination and thought how on Earth did I pass this or that town? While driving your car it becomes as routinely as breathing so there is little to be positive with and less to be negative about. That is the chance the neutral character has been waiting for all week and he takes charge by letting your mind wonder to many destinations but the one you are heading to. Little harm can come from this under the normal, but when danger suddenly strikes you are gone, your positive is gone, you negative with all the adrenalin is completely absent and your neutral being neutral does not bother in any case. By the time all the absentees arrive at the scene just that second later, a horrible accident is in progress. This we call driver fatigue. The process lingers on while we are absent minded, not totally in control of the moment and shit happens. But shit happens because we want it to happen for that bit of excitement we do not need, but the characters do because they make misery, To them it is a case to change the situation to something more exciting, more emotional, to press our social standing or just to ease boredom. This goes by the name of the mind is wondering.

Someone may say something out of the order and a rage follows. Why would a rage follow when you know very well that what the other person said was unfounded and if that person do believe what he said, that person is then so misinformed that you should then in that case not even listen to such nonsense. But confrontation is about to blow like a rocket on a launch pad. You will show him what he said was untrue and he will take it back or pay with his life. Once again we allow the characters to shift and completely obstruct our normality.

The best is that with you allowing the shifting you will press old issues long ago forgotten but the characters suddenly helped you remembering this or that and this is the last straw! Under the normal the previous incident hold such minor importance you would never remember about it but at that moment it does not even strike you as odd remembering such trivially while not surprising at all it is clear in your memory that moment. You feel you can murder while never in your life did you ever have a thought about how it may feel taking another person's life. Everything you experience is out of the ordinary and out of order yet it seems to you at that moment as being as normal as discussing the papers. Does your characters enjoy the excitement and taking you along for the ride. They take control and you take the mess afterwards not knowing how you will ever show your face again after such an incident. This we call losing your temper.

You walk down the street with your best suit on feeling as chirpy as a robin in mating season. Then suddenly your foot misses the curb as you were walking and you land face down in a crowd of people you never saw previously or know any one around you. There is not the slightest chance of you meeting any one of the spectators ever again. Yet your world plummets down the deepest mine shaft. You cannot see the light of day ever rising again. You hang your head in shame while trying to conquer the incredible urge to run from the scene as fast as you can. Not for a moment do you stop to think that it is not that shameful and happens to everybody many time throughout they're lives. That the people you hold so important can never be important because you will never see any of them ever again. Again you characters turned a situation around as they turned on you. This we call embarrassment.

All of us meet the challenge in battle without ever realising. A young man newly wed has a friend of his wife staying with the couple for a month or two, just until she can find some other accommodation. This young man is going ballistic with the fight. He is in rage about this beauty sharing a roof with him and with him working shifts he finds himself alone in the company of this woman because she is still in the market for employment. She page thorough many papers per day to find a suitable position but that does not take all day and with his wife being at her work there are many hours pleasuring about unchecked. Now he goes in spin. His positive character tells him of what he has in his wife and that he should not endanger his fragile and young marriage, his neutral tells him his wife is at work and need not to know while his negative went positive by telling him how beautiful this young female is. This we use the name temptation to identify. It is the clearest way we know of identifying the threesome.

We were all at one or other stage young in our lives and young at heart and know the feeling when you are running the hundred meters or you are on the Rugby field and there is that special person looking from the spectators end. You feel her eyes burn in your back and the excitement blows your senses. It is like someone ells takes control and you become that much better to the degree you cannot believe yourself, or you have gone pinching fruit (a favourite pass time amongst the Afrikaner up and until I was a teenager. That was before money and greed came in and made kids criminals for helping themselves to fruit at night).

The other side of the coin was that you knew who ever gets hold of you, will tear of the skin from your behind, but that was part of the fun being the challenge to the danger. Never was police involved one way or the other and even coming home safely did not mean security because if your parents catch you, you have to go back to whom fruit you pinched belongs and fetch your licking. I was a hundred meter athlete in my time and was quite quick, but when caught in the act I saw these real slow guise and can-not-run mates in crime pass me and if Carl Lewis was there they will pass him as well. They beat me by a hundred meters on the three hundred meters and cross one and a half meter fences as if it was hurdles on the track. The neutral character was on guard all-night and got the other two pumped but on stand by. The moment surprise takes over the positive gets negative but charging past the negative to get the body in motion. The negative sees that being negative is a splendid way of getting away from the danger and re-passes the positive while the neutral is pushing both to get out of the way so that the negative can take command of the muscles. With in less than a heartbeat there are four characters (including yourself) that take control of muscles and there can be no fitter and more potent athlete occupying the body than the fear factor presenting the next few minutes. This we call motivation.

Then there is the young person that is the one-week in the dump because he cannot see what the world is all about. His girlfriend dropped him, he is in the middle if an exam he cannot see how he will ever pass and to top it all is the fact that it has been raining for three days while he has this camping trip planned. The positive character finds a few days rest while laying low in anticipation of the coming holiday forcing the negative character to take charge of his outlook on life while his neutral character takes charge of the schoolwork making him take his vacation early while exams are pressing. The next week he finds himself being over positive bout life, about his happiness and about all the opportunities in life that are waiting on him. The exams are history, he is vacating and met the girl of his dreams and have a song in the heart. To tell the world, he is the man of the moment and to prove that he has long uncombed hair, walks with a swing and doing his thing just to annoy his square parents that are living they're life in history. That is the neutral character in charge and being all-positive for one week/ month and then negative for one week/month does not surprise any one. The name we attach to this is growing up. It is all about finding the characters, meeting the characters and blending with the characters in order to fine tune for preparedness for the fight ahead.

We name the events. We know the events. We suffer through such events but never stop to think why it is taking place. Why would your mind start to wonder? If you are in control and you are in that position you should not find yourself out of control or somewhere ells while you were being there. If you were your body why would the mind go absent? If you were your body and mind why would you stray or over react in anger, pity, shame or stupidity. If only being in the body you are in as the atheists believe, then you should be in the body, because where ells will you be but in the body you are. The fight is on and you have to recognise your opponent because your opponent is you in person. Your opponent knows your weaknesses as you know them because you are your weaknesses. Your life is your fight and it is on for the rest of you natural life.

It is as common in every one as miss-placing keys, spilling milk, forgetting some one's name and such minor incidents bringing great embarrassment at the time but is as normal as breathing. It is no big psychological problem forming in the dark side of the sole where the brave does not dare. Yet it can be there. When ever a person comes up with a brilliant remark astonishing the person that made the remark much more that the person to whom the remark was directed is an example of the trio. When trying to repeat the brilliance the very next time we become the stuttering idiot that wishes the floor would dissolve the human body. We find an inability to repeat such brilliance. We all have astonished ourselves from time to time as much as we embarrass our selves from time to time, so there is no exclusion and only inclusion.

It is the person being unwilling to do a task and finding himself repeating an error that he knows he should not repeat but something is driving the person to repast his actions. The more he repeats the error the more he gets annoyed with himself but getting annoyed with yourself must be the most unaccomplished task you can accomplish. His normal reaction would be then to become annoyed with any object or person he may find displeasing and although the object or person has little to nothing to do with his actions it proves the best way to blow off steam. This I refer to as the boss syndrome because that is one of the perks a boss seems to have. Being negative about the task makes your mind go wondering to more pleasant places to be. By your negativity you are suppressing the positive character and that puts the neutral character in charge. But the neutral character has as little interest in you're being there and takes you away on more pleasant day dreaming trips to where you would rather be and that leaves the negative character to be in charge. Who is doing the task with every one gone…it is the negative character, and an unpleasant one at that being all alone and hell is on its way!

The person starts making deliberate errors that can be avoided but with his lack of enthusiasm he becomes absent-minded and that is when the accidents stars cropping up. It could be small annoying things but also it can be very serious injury coming from such little absent mindedness. The negative character wishes to draw all three characters attention to the fact that the negative character is as displeased with the situation and wants to be relived of the duty. The LAPSE in CONCENTRATION is no lapse in concentration and the disaster following (big or small) is as deliberate as the person slipping away on his daydreaming trip. The whole affair is one big disaster waiting to happen and always does happen. All parties involved claims innocence and protest to any involvement accept the negative character that now is relieved of his duties. The net result is that the negative character is the only winner in the end.

Where does it come from? It comes from being human. What encourages the characters to become more dominant in some than in others are the better choice of question? There are at least a thousand possibilities I presume but it is more than just likely that it may present itself as a manifestation of a chemical imbalance in the brain. That does not explain the fact that it is present but may support the fact of more dominance in some individuals than in others. Being alive is about chemicals but the chemicals allow conducting electricity and are not life itself. The chemicals are a conductor but we all know the conductor is not electricity. But when the conductor goes hay-why the electricity goes hay-why. In that sense the chemicals will play a part but not play the part. When there are cross over of wire connections some life will flow in a direction where it actually is intended to be at other outlets. From what I have seen with some of my brain injured friends the control becomes absent but that does not mean the injury cause the characters because I have witnessed the characters in every person I have met. It is as much part of our personality s it is our personality.

The chemicals could be one aspect but underlying fears are the predominant issues that render the characters the possibility of going out of control. An unhappy childhood brings an unhappy life. When the child is growing into a skew adult the adult will bring about another skew child. The main drive behind man is fear. Where fear is absent the man is a danger to him and to society. Fear brings about a conscience. The fear I refer to is not anxiety or being scared but having respect. Having respect for one's parents or teachers or the community and respect for the law. Above all the fear of God brings respect for God and that forms the basis for being man. Persons with personality problems and character flaws may have too much fear or none at all. It is a balance and it is the balance that keeps us upright. Underlying in the balance is respect for yourself and that you have with the respect you have for your parents. When being a child your parents are a substitute for God because only they can bring values and norms that will one day bring about a balanced member of society. Money cannot buy that.

I know that in itself having a conscious and adhering to the conscious does not say much because the alcoholic hides his booze from himself because his characters are in conflict. He knows he is on the road of self-destruction but normally that does not bother an alcoholic that much as they love drink more than life. It normally is the fact that the alcoholic knows about the destruction in his children and his love for his dear ones. He wishes to protect them from him but he has this massive problem that is stronger than his urge for life. To find a way to solve the problem he starts to hide the liquor from himself and in that way he stars to lie to himself. The characters are accusing him as they are destroying him and the escape is the destruction. When he is drunk he has no control and when he is sober he has no control. The characters are telling him he has no problem as much as they are accusing him about his problem and denial is also admitting where admitting then becomes blame and the blame he carries are more than he can carry therefore he drinks to escape the blame of his guilt.

It is not only liquor but also it is sex, drugs, pornography and gambling. Those are the weaknesses that put people in the same situation as that of madness. It is living the life of lust knowing that that life of choice is what you choose and choosing such a life is the equivalent of destruction but modern society makes fighting thereof much harder than ever before and capitulating as easy as breathing. The guilt, fear, anger and despair come in waves and in conflict where the alcoholic is in the middle with his problem out of control and out of his hands. His characters took charge and they destroy him as well as those surrounding him... all that loves and care for him but the biggest destruction is his destroying of the ones he love. Seeing the suffering in the lives of those poor, poor soles make me great fill for the small load I received and the ease with which I have to carry my small burden. In that way we're in a fight but the fight is all in private and we may acknowledge failure where there are great success as we may judge success where there is great failure. We on the outside judge what we see on the outside while the fight is on the inside where we can never see. A rehabilitated alcoholic must be a far bigger success than a successful achiever born with the golden spoon in the mouth but skins the cat in the dark by indulging in cravings of the night, all in the quiet his money can bay. It is about what you fight and not the way in which you fight or how you fight.

I am no philosopher of any sorts but thought about success and failure and my position in life where I as a person will never reach great achievements therefore the question I asked myself is will I die a failure because of that? My success is my children I leave behind and the success they may be. Not in great achievement because in the end even King Solomon declared it was all about chasing the wind. Leaving behind riches in money can bring as much despair as leaving behind poverty.

The only way I may achieve success is leaving behind children that are of a better fabric than I was. The next generation must be better equipped, better evolved and better humans than was the previous generation to ensure evolution. What is taking place to my mind at present is devolution through out the western world. I do not have to bring proof about that because reading the paper or looking at nightlife will bring proof to any one wishing to see proof. How do we western man raise our future and how do we equip our future? Western society removed discipline from schools with admirable success. We removed the authority from the teachers as best we could. The teachers are through out the western world the lowest paid professionals in society. Being the lowest paid brings about the under achievers of society and any successful teacher leaves the profession for better pay in the private sector. This is the cancer of the new age we are facing. We all know my last remark is the truth and by that the parents wishes to compensate, but to modern man compensation is about money and paying to get rid of the guilt. Every nation culture community and individual admits that money is no measure for success and yet that is the norm we live by. Every person living a life is but a building block to establish the next generation. Man is an endangered species and animals are overpopulating the world and walking on two legs is not the criteria for human classification. The position you hold distancing yourself in thought and behaviour from the norms that animals uphold makes man or beast.

Animals are not in a struggle with their improving of the mental but in a struggle with the survival of the fittest, the one that can kill the best, run the fastest outsmart all others. That is not man. Man is about his fight to better his life and not his body or position. Man is about fighting the battle of the best in man. That is not modern man's ambitions. Modern man has ambitions making him the best animal on the planet because atheism propagates the fact that we are all members of the animal Kingdom. If man is that blind it may be best if man returns to the animal Kingdom where he thinks he belong. At least the loss will be small but smaller will be the gain of the animal world.

I wish to take you back to the indicial verse in that there were three things man would obtain when rating from the tree, and not only two…it is the three that every preacher misses …There was the tree…the tree of knowledge…knowledge of good, and knowledge of bad. You determine your knowledge…of good…of bad but above all it is your knowledge. The Bible does not name what is good or bad…that is your choice you make and that is your price you pay in the end…because the knowledge you accumulated will stand you either good or bad, as animal or man.

It is not what you may accomplish or accumulate that has importance but what you learn through being man and standing apart from beast, that is what you take with.

Accept the following or reject the following but consider the following.
Because there's mathematically no nothing there must be a God.

In the educating of Primates by teaching them to understanding human concepts no one can truly present to the mind of the ape the realising of divinity or a Higher Force than we find on Earth. One can teach the ape the position of a divine leader but that will not have any more significance to the animal as presenting a tribe leader or a strong leader but the total realising of a spiritual being that no one can see but still can appreciate is beyond the concept of the ape. The ape cannot pronounce or denounce a Creator because the Ape is unable to relate to such a presence. That makes the ape eternally less than man. The atheists' ability to denounce the Creator proves that the Creator gave the atheists His pronouncing of the atheists as an eternal living being capable of recognising the Creator merely by refusing to recognise the Creator as a cosmic fact. By the animals not being able to denounce a Creator the animals never received pronouncing of the Creator without such ability to pronounce the Creator they can never acknowledge or announce the Creator. Without that ability to announce the Creator, they forfeit any aspirations of becoming recognised as eternal entities by the Creator.

The animals holding life where that life being space-time to be used by eternal entities as the entities see fit and without explaining to the animals why the animals does not require any explaining. That underlines the animals not being able to understand even the mere fact that they are used as a benefit to the eternal beings that has the ability to form concepts about an afterlife. Without such ability of recognising an afterlife makes them denounced and they then can never be eternal beings to the Creator since they cannot admit to a Creators presence. By the atheists denouncing of the Creator the atheists recognise the pronouncing of the Creator, which the atheists choose to denounce. But only if the Creator denounces the atheists' denouncing of the Creator can the atheist become denounced. This must be rather easy for the Creator to establish because the Creator only have to comply with the wishes of the atheists in granting the atheists the life long dream of not having the ability to recognise the Creator such as the privilege all animals do not have and all non-animals have, which will therefore doom the atheists to become usable space-time used by eternal beings that has the ability to recognise a Creator of their choice. We have to acknowledge singularity as a cosmic fact and with the acknowledging of singularity and the Big Bang coming about from singularity we then must exclude nothing or zero as a calculatedly number.

By excluding nothing or zero from the cosmos as a valid measurable factor you have to include God because if there was nothing there was some scope for God being nothing and therefore non existing but since nothing is the only excluded number in mathematics you have to include God as a factor. If you do not agree then go about mathematically prove zero as a factor. Show how you can establish zero as a measurable quantity, What ever you dare to prove or disprove you can only prove or disprove through the human thought and understanding of concepts only human concept can understand. And the fact of nothing is a human concept of not understanding the concept at large.

I do realise with my statement about nothing and God it is stretching matters beyond the argument. When taking the argument that far the argument can include the existing of fairies and other fantasies, that is quite true, but by my placing a fairy and other fables in the realms of insignificance brings only harm to what remained of the child in me and I can assure you there is not much left in that sense. What ever is there is also in the infinity of my childhood memory. Not believing much in the fabric of the fantasy does not affect my position relating to the animal as the animal has little regard in such matters. Fantasy may have significant when I connect ethics to it as the Roman Catholics do with saints, demons, angels and such. When one can disregard such fantasies without affecting norms and values of the civilised, not harming the moral fibre of society there is little harm done by such removal. But when removing such norms for the enjoyment of being superior while the correctness of the argument in its core is invalid and, the structure of society goes to hell, a lot of question marks appear about the morals behind the motivation as to the motive behind the act. Man is moreover morals than life.

All life will destroy other life to its own benefit. A virus will kill his host for the benefit of one life cycle and at that a virus life cycle. What waste we may think. Killing a host only to spurn seems a lot of waste. Not so, because ticks will devour a cow alive without feeling any remorse. The same apply to a lion killing its pray. If there were no chance of the antelope finding means to escape the lion would start eating without killing saving itself the effort. It does not kill quickly through pity like humans do. It kills because of self-interest alone. When one of the herd falls victim to a lion attack the rest will start grazing thanking in that manner the victim for securing their position for one more day. There is no compassion, just self-centred egoistic drive to self-protection. When a person jumps into flames in an effort to save another person he is a hero. He receives praise from all concerned and may even land a medal for his effort. Why would we humans consider that as brave, being exceptional and above average?

From our cosmic position we are in the centre of the universe. Where you may sit or stand is the very centre of the Universe (your universe) because you can only relate to the Universe having your individual singularity and where all other aspects in the cosmos are pointing away from your singularity to all other positions. We shall always be 56 in a Universe of 112. The Earth will always be Π^2 in relation to the sun with the speed of light being $3\Pi^2$ from our position. The Universe will grow in all directions pointing away from our direction, just because we are the centre of our universe. It is not surprising we see ourselves on the edge of the Milky Way. From where we are we will be on the outer edge of our galactica. To the inside everything will be brighter and to the outside everything will be darker. That is a fact, but not a reality. The fact we may appreciate but the reality we can never understand. We are egocentric maniacs coming from the position of our singularity. That is the animal, the trio breathing our air. That is what the fight is about and what one may take with after death changes dimensions occupied. That is wisdom standing apart from knowledge and the animal has knowledge in intelligence but man has wisdom in extelegence. That parts man from beast. Man has to part from singularity's approach.

Us humans must fight to find our place outside singularity, outside self-interest and strive to better of our position not including finance. We have to relate from a position in divinity and not singularity to understand the cosmos and to understand life. If not we may well die as animals, and such possibility is there. If one person spent a lifetime killing his human in him, he may end as the animal he always strived to be.

Being artist's engineers or preachers have nothing to do with it. Knowing the Bible or not has nothing to do with it. Being a believer or not has nothing to do with it. It is your approach to the cosmos that makes you part of the cosmos or that dimension above the cosmos. Accepting that one is not part of the cosmos but part of life in a cosmos puts one above the cosmos, but still in the cosmos. Never lose reality but accept responsibility. Persons believing in fairies and fantasies have the problem of over acclimating the positive while the neutral protects sanity with much of it going in the direction of a lost cause leaving the negative happy for destruction can only follow such obscurity and the arts and artistic are typical in this. This is the line the drug addict follows to the last letter. By pumping the acid the positive hallucinates, the neutral exclude the incorrectness and the negative character are in seventh heaven as destruction is deliberate and decisive. The very same apply to a soldier in war where the positive character enjoys the killing as much as the negative character enjoys the being killed and the neutral character wishes the cruelty on them before they are upon us.

When the soldier arrives back in society, society understands him as little as he understands himself because the fabric of human principles has gone skew. His characters, including him is as mixed as a milkshake fruit salad. You may well ask why is the controlling of the characters of any importance, as they are obviously part of us? There is the middle, a precise middle where I as a person hold my personal relevancy as I see myself. What ever I attach or detach puts me in relevancy to others in life. That is the purpose of my fight with my characters. My precise middle must be very straight and when the middle leans toward any side in particular my middle is out of alignment. To secure a centre there must always be room for others and their opinion. When I say there is no God I go eccentric and when I say there is only God I go eccentric.

As it is in mathematics in the matter of the line and the dot, it is a question of you deciding the relevancy. It is your choice and only your choice to what relevancy you wish to place God as a presence or a factor. By declaring the absence of God your relevancy might reduce God to infinity but in your denial you place relevancy therefore relevancy remain be it in the infinity. It is to the peril of the denier that such a person excludes God as a factor because by placing God at a point of infinity next to zero the denier places himself next to the animal that excludes God because God excluded the animal from extelegence. Such a thought proves the fool. God allowed the animal excluding the admitting of a God because God exclude the animal from the dimension of being human therefore liability to questionability. In the perception lies the norm.

On the other side of the spectrum is the religiosity maniac and such is his idiocy he gives God the role of the animal being on call by prayer to serve the master's call. In all such cases the maniac places himself in eternity by creating a spot in the centre of eternity for himself providing him endless power having God as his slave or animal on a leach in acclimation to his eternal status he then places God just outside eternal big for that spot he reserved for himself. Through his effort by prayer he can heal, bless and doom…and God will obey as instructed. Such a fool proves the thought.

When a person buys a car worth millions (in South African Mickey Mouse Money), because his business is going grate and he can afford to, but has his sister and his brother in law is working for him at minimum wage, not making ends meet in providing for their school going children, battling every day to put food on the table while he is making money like water flowing every one admires him for he is rich. He is treated very softly and no one dares to stand up to such a person because he has money. Such a person as a human is wasting breath because he may walk on two, but he may as well walk on four for the humanity he portrays. Still society regards that animal as a pillar of the community because he has money making a "sharp and intellectual business man". He should be shot at dawn for impersonating man whilst being beast. His positive has only one aim and that is to better his financial position securing a better admiration in society and establishing a front with him being god to the rest he sees as lesser mortals. His neutral sees his money drive as security and that satisfy while his negative is bent on destroying others because his positive is telling him how great full his sister and her "useless" husband should be for his generosity of providing food on their plates. Not once for the shortest and briefest instant will he ever give a thought that it is precisely the other way around. By their self-denying they are enriching him, but through his ego madness that thought never comes to mind. Then the philosophers will say that is man.

But such a remark is bullshit, only protecting the beast in himself or herself as they protect him, our unkind and selfish millionaire.

While a human (as is the case with all other living things), holds life being intertwined with singularity there is a distinction we all know of between animal life and human life. Animals are lesser in life than are humans because they cannot sense a cosmic position other than eat and mate, and in that sense we are born with a natural same view animals hold as just merely being senseless objects in space-time, without having the ability to distinguish between what is permanent and what is of a passing nature and our position as cosmic objects occupying space-time. We maintain our position as if we do not realise we truly are not positioned in the centre of the Universe as we seem to be, because animals have not the sense to distinguish such realisation. Everything else in the cosmos is not there for our satisfaction including other humans. Having the ability to distinguish makes us intelligent, and moreover, extellegent as humans might be. As long as the atheist, the criminal, the ego-centric maniac and all self-appointed gods connected to singularity and space-time consider them to be the star with all other objects in the Universe as heat in space-time, there for one purpose and that is to be used, he will be an animal. Our purpose is to move away from the animal, because through a lack of better knowledge, that is the attitude of the animal. The animal holds the outlook that all objects are there for its pleasure or displeasure pivoting around the animal's view of eats or be eaten. The Universe focus around the animal as he sees the universe, and has no ability to see other forms of life with equal importance.

As human life we have to see that there are billions of singularity that included with ours, always encircle a more important singularity as lesser singularity revolves around us, and in the epitome, there is the ultimate singularity connecting all. We are all planets, only planets, with some being bigger and some smaller, but we maintain a queue from the planets and objects in our solar system, not counting the number or size in the personal solar system. As successful humans we must see our position in relevance to the Universe we occupy and our goal to serve by improving such position with evolutionary time laps. When we get the urge to grab what we do not own, we must realise the animal in us urging us to return to animal form. The fortune of man is his ability to recognise the stars above, but much more: to recognise time in space and his role in that. Animals have not received the ability to recognise stars, and therefore has no ability recognising the possibility of singularity and the ultimate Creator of singularity being God as such. Without that they become space-time for man to use and dismiss. In that man and mans superiority becomes the complex issue we do not understand. Above al, when we recognise singularity, our aim should be distinguishing life from singularity and parting their different purpose to the dimensions controlling creation.

Even when a planet finds itself to be just another planet without finding other planets to guide, and holds no relation to other planets as is the case with Mercury; its orbital pattern is very obscure. Since we humans are part of the cosmos with life intertwined with singularity (I delve into this aspect in another book by the name of **A Cosmic Birth...Dismissing Nothing** I. S. B. N. 0 - 620 – 31609 – 8) we must find similarities in life and the cosmos, because the cosmos takes after life as life would take after the Creator. When the Cosmos is gone to singularity, returned to the dot once again, life as energy would still be, because the Universe is heat, and heat goes in three forms, but life is another energy, apart from heat in the cosmos very indestructible. By recognising the lay out of the cosmos, we can determine with an intellectual eye, what our purpose should be and we should place our objective to the future accordingly. What is the use of money-mania and the accumulating thereof, if one would die in any case and lose control over it afterwards? I could never understand that part of the gaining game.

Through your relevancy with God you become the God that makes you your own God as much as you're own devil and your own forgiver as much as your own accuser. You make the relevancy and the relevancy proves you. Being positive is loathing and being negative is damning and being neutral is obstinate and in everything you confirm the applying of norms be they right or wrong, good or bad as it is the free choice you make proving your distance you have as a factor claiming life and location as a being from the animal. In the case of the animal the ride is free for the animal knows no better but man does and is liable through conscience placing relevancy to deeds. On the physical aspect in the being named man, the human body is an accumulation of singularity positioning divinity as one comprising of three identities without ever separating the trinity. The one holds the straight line the other forms the triangle and the third is the dome, the inclusive sphere, and the container having seven sides to the entire outside world. You cannot be in two sides of the Universe simultaneously but you may observe both sides by being centre.

CLAIMING A POSITION IN SINGULARITY'S INFINITY IS LIFE'S DIVINITY.
YOU LEARN TO LIVE AS MUCH AS YOU LIVE TO LEARN

Singularity has three in parts as does divinity have three in places. On the other side of the divide is the divine and as things are in the one side of the divide in total relation to the other side because all relevancies align as much as match in equilibrium and in as much as it is the cosmos forming one half of the divide it has to be the divine duplicating to establish equilibrium.

The purpose of man in life is to know your other entities, to learn to live in recognising the other entities, to take from them strength they can give but also to detach from their weaknesses, and above everything else recognise yourself in what you do as pure or spoilt. Gain knowledge the knowledge you have in the good you can acquire and the knowledge in the bad you have to detest. Our life task is to come to terms and come to control the entities we have a lifetime to struggle with and which we must for one life term fight with. Then only you may serve a life of gain. Did I prove anything, well you are my judge! As I am the one promoting all things connect, so I am the man believing all things connect. Without a zero there cannot be death. Without a nothing there cannot be no God. Accept one and you cannot except one.

If it is that simple then why is it complicated.

BEST WISHES,

PETRUS. (PEET) S. J. SCHUTTE

FOR other related information, PLEASE VISIT THE WEB SITE www.singulartyveracity.com , FOR YOUR CONVENIENCE

To find more about other books please visit... www.singulartyveracity.com

The Books published in e-book format:
Book 1 The Absolute Relevancy of Singularity in terms of Applying Physics
Book 2 The Absolute Relevancy of Singularity in terms of The Sound Barrier
Book 3 The Absolute Relevancy of Singularity in Relation to the Four Phenomena
Book 4 The Absolute Relevancy of Singularity in terms of The Four Cosmic Phenomena
The Books that could be ordered from the Author :
Book 5 The Veracity Of Gravity
Book 6 An open letter On Gravity Part 1 Volume 1 + 2
Book 7 An open letter On Gravity Part 2 Volume 1 + 2
Book 8 Newton's Mythology
Book 9 Sir Isaac Newton: A Conspiracy to Defraud Science
Book 10 An open letter Announcing Gravity's Recipe
Book 11 An open letter Addressing Gravity's Formula
Book 12 An open letter About Gravity's Prescription
Book 13 An open letter Explaining Gravity's Rules
Book 14 An open letter To Selected Academics
Book 15 A Cosmic Birth Dismissing Nothing
Book 16 An open letter About Investigating Kepler
Book 17 The Dissertation about the Absolute Basis of Physics forming Gravity

Po Box 1093,
Ellisras
0555
R. South Africa.

P.S. J. SCHUTTE (PEET SCHUTTE)

WRITTEN BY PEET (P.S.J.) SCHUTTE

mailto:orders@sirnewtonsfraud.com
or mailto:info@questioneblescience.net
naturescosmicconcept / ANaturesCosmiConcept /
NaturesCosmiConcept-E-Z
/ NaturesCosmiConcept-E-Z-R
http://www.titius-bode-law-explain.co.za/index.html /
www.sirnewtonsfraud.com / www.questioneblescience.net